Y0-ACE-854

AN INTRODUCTION TO THE FINITE ELEMENT METHOD WITH APPLICATIONS TO NONLINEAR PROBLEMS

AN INTRODUCTION TO THE FINITE ELEMENT METHOD WITH APPLICATIONS TO NONLINEAR PROBLEMS

R. E. WHITE
North Carolina State University
Raleigh, North Carolina

A Wiley-Interscience Publication
JOHN WILEY & SONS
New York • Chichester • Brisbane • Toronto • Singapore

Library of Congress Cataloging in Publication Data:

White, R. E. (Robert E.)
An introduction to the finite element method with applications to nonlinear problems.

"A Wiley-Interscience publication."
Bibliography: p.
Includes index.
1. Finite element method. 2. Nonlinear theories.
I. Title.

TA347.F5W49 1985 620'.001'515353 85-6440
ISBN 0-471-80909-8

Printed in the United States of America

10 9 8 7 6 5 4 3 2 1

PREFACE

This text has evolved from lecture notes for my 1981–1985 spring semester courses on the finite element method. The students were mostly from the graduate-level engineering programs at North Carolina State University. Consequently, the most important objectives included (1) giving the student the ability to modify existing finite element codes or to create new codes, (2) giving the student some appreciation for the error estimates, and (3) giving a summary and illustration of nonlinear algorithms. The present text has been written so that readers can choose either the methods aspects or the theoretical considerations as their main interests.

At the end of every chapter I have indicated additional readings, and I have pointed to certain exercises that former students have found helpful. These include some programming problems as well as some theory problems. Readers will soon discover that these programming problems can be very time consuming; consequently, I recommend working with a partner. This helps the debugging process and gives students an opportunity to talk about the course.

The programs in this text are not meant to be optimal or elegant, but I hope they will be instructive. There are many optimized codes that one should try to use in "production" work. Any computer center should have relevant manuals, such as J. Rice's *Numerical Methods Software and Analysis: IMSL Reference Edition* (McGraw-Hill, 1983).

I would like to acknowledge the students, especially Maurizio Benassi, who have made many useful remarks on the contents of this book. Many thanks go to my friends who have listened to me concerning the more mundane aspects of writing this text. Finally, let me thank the staff of the mathematics department and, in particular, Nancy Burke, who did the typing of the manuscript.

R. E. WHITE

Raleigh, North Carolina
August 1985

CONTENTS

AN INTRODUCTION TO THE FINITE ELEMENT METHOD WITH APPLICATIONS TO NONLINEAR PROBLEMS

INTRODUCTION

In this text we describe the finite element method with an emphasis on approximating solutions to second-order linear and nonlinear partial differential equations. The advantages of the finite element method over the finite difference method are (1) usually a more "accurate" approximation is obtained and (2) irregular shaped domains may be considered in the context of one program. The following examples illustrate the latter point.

Example 1. Ideal fluid flow around a pipe (see Figure 1). By the expected symmetry and the fact that the stream lines change most near the pipe, we may be interested in the nodes as distributed in

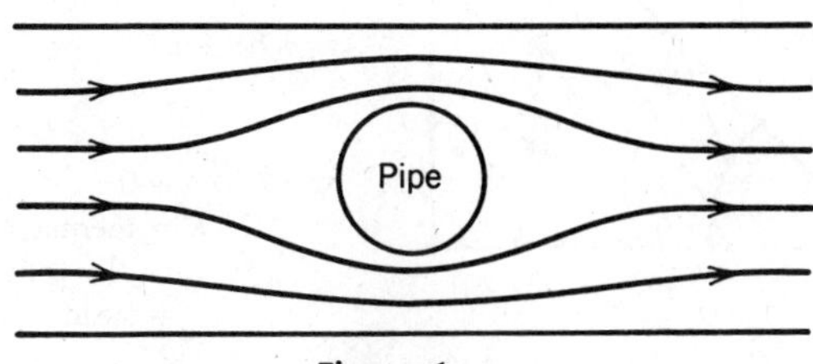

Figure 1

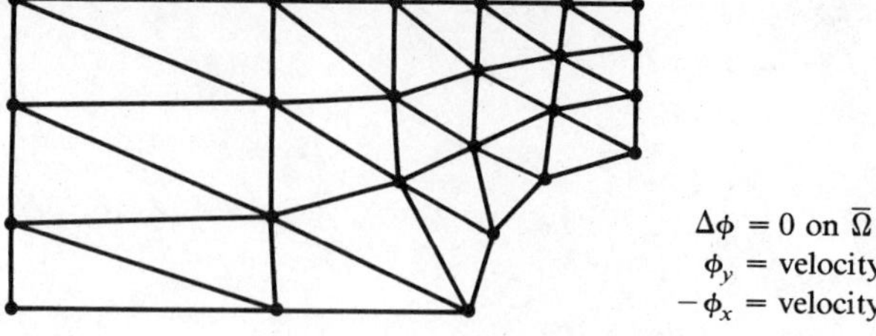

Figure 2

Figure 2. Note $\bar{\Omega}$ is the union of the triangles and approximates Ω, the upper left region of fluid flow, more accurately than a union of rectangles with a similar number of nodes. The triangular regions are called elements for $\bar{\Omega}$.

Example 2. Steady-state heat flow. For example, consider an insulated steam pipe as illustrated in Figure 3. By using the symmetry we may reduce the number of nodes by $\frac{1}{8}$ (see Figure 4). The same

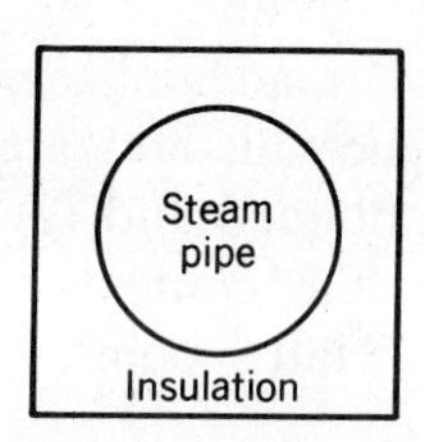

Figure 3

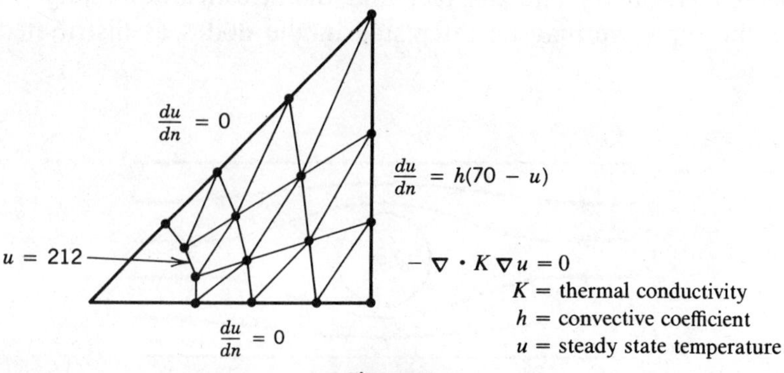

Figure 4

finite element program can be used to approximate the solution to both examples. One must input different data for both examples. Also, time-dependent problems may be considered.

Two main objectives of this text are (1) to present enough material so that readers can write their own finite element programs or alter existing codes and (2) to present some techniques for solving nonlinear problems. In the latter case we shall consider incompressible viscous fluid flow problems and nonlinear heat transfer problems such as the Stefan problem.

1

THE ENERGY AND WEAK FORMULATIONS

In order to introduce the finite element method (FEM), we consider a one-variable model problem. In the first three sections we illustrate the three equivalent formulations of this model problem. These are the classical, energy, and weak formulations. In Section 1.4 we discuss how they are related to one another. Sections 1.5 and 1.6 contain a description of two methods of assembling the system matrix, namely, assembly by nodes and assembly by elements. Section 1.7 contains a general outline for the finite element method.

1.1 THE CLASSICAL FORMULATION AND THE FINITE DIFFERENCE METHOD

The model problem that we shall use in this chapter is a mass subject to gravitational force and another force that is proportional to the displacement and whose positions at time $t = 0$ and $t = L$ are given. The classical formulation uses Newton's law and has the form

$$-m\ddot{y}(t) = mg - ky(t), \tag{1.1.1}$$

$$y(0) = a, \tag{1.1.2}$$

$$y(L) = b, \tag{1.1.3}$$

where m is the mass, k the proportionality constant, and g the acceleration due to gravity. $ky(t) - mg$ represents the external force.

There are other physical problems that have the same form as (1.1.1)–(1.1.3). For example, the steady-state deflection of an ideal string has the form

$$-(Tu_x(x))_x = f,$$

where T is the tension, u the displacement, and f the loading pressure. In this case the independent variable is a space variable x. Another example is steady-state heat conduction. A linearized version of the problem in exercise 1-25 has the form of (1.1.1).

Definition. We shall say that $y(t)$ is a *classical solution of the continuum problem* (1.1.1)–(1.1.3) if and only if $y \in C^2[0, L]$ and equations (1.1.1)–(1.1.3) are satisfied. ($C^2[0, L]$ is the set of functions on $[0, L]$ that have two continuous derivatives.)

Of course, if $m, k > 0$ are constants, then the classical solution of (1.1.1)–(1.1.3) is easy to find. If m, k are dependent on t, y or if equation (1.1.1) is more complicated, then one may not be able to find an explicit formula for the classical solution. One way to handle the more complicated problems is to approximate the continuum problem by a discrete model. The following is one such model called the finite difference method (FDM). Make the following approximations of y and $\ddot{y}$:

$$y(t) \to y_i \quad \text{where } y_i \simeq y(i\,\Delta t),\, \Delta t = L/N = h$$

and

$$\ddot{y}(t) \to \left(\frac{y_{i+1} - y_i}{h} - \frac{y_i - y_{i-1}}{h}\right)\frac{1}{h}.$$

Then equations (1.1.1)–(1.1.3) are approximated by

$$-m\frac{y_{i+1} - 2y_i + y_{i-1}}{h^2} = mg - ky_i, \qquad 1 \le i \le N-1 \tag{1.1.4}$$

$$y_0 = a, \tag{1.1.5}$$

$$y_N = b. \tag{1.1.6}$$

For $m = k = 1$, $g = 32$, equation (1.1.4) may be written in the form

$$-\frac{1}{h}y_{i-1} + \left(\frac{2}{h} + h\right)y_i - \frac{1}{h}y_{i+1} = 32h. \tag{1.1.7}$$

Consequently, we have $N - 1$ unknowns and $N - 1$ equations. For $N = 4$, these may be written in matrix form

$$\begin{pmatrix} 1 & 0 & 0 & 0 & 0 \\ \frac{-1}{h} & \frac{2}{h} + h & \frac{-1}{h} & 0 & 0 \\ 0 & \frac{-1}{h} & \frac{2}{h} + h & \frac{-1}{h} & 0 \\ 0 & 0 & \frac{-1}{h} & \frac{2}{h} + h & \frac{-1}{h} \\ 0 & 0 & 0 & 0 & 1 \end{pmatrix} \begin{pmatrix} y_0 \\ y_1 \\ y_2 \\ y_3 \\ y_4 \end{pmatrix} = \begin{pmatrix} a \\ 32h \\ 32h \\ 32h \\ b \end{pmatrix}. \tag{1.1.8}$$

Definition. The discrete formulation of (1.1.1)–(1.1.3) given by (1.1.4)–(1.1.6) or in matrix form (for $N = 4$) by (1.1.8) is called the *finite difference model* of the classical formulation. The matrix in (1.1.8) is often called the *system matrix*.

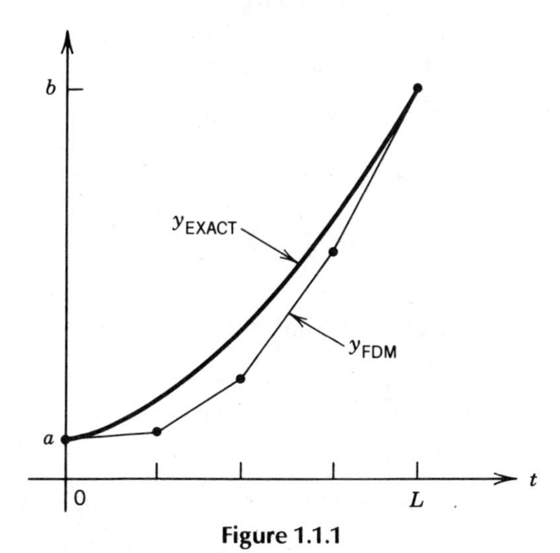

Figure 1.1.1

One can associate a continuous function with the solution of (1.1.8). It is hoped that as N gets large this function will converge to the solution of the continuum problem (1.1.1)–(1.1.3). Figure 1.1.1 illustrates how one can find such a continuous function, y_{FDM}. y_{EXACT} is the solution of (1.1.1)–(1.1.3).

1.2 THE ENERGY FORMULATION AND VARIATIONAL FINITE ELEMENT METHOD

The principle of least action is the governing physical law for the energy formulation. *Principle of Least Action*: The motion $y(t)$ of a given mass ought to be such that the total difference in kinetic energy and potential energy is a minimum over a given time interval. Several examples of this principle are discussed in the text by Sagan [20, Chapter 1].

We restrict our discussion to the model problem of this chapter. Since there is only one mass, m, we have

$$\mathrm{KE}(t) = \text{the kinetic energy} = \tfrac{1}{2}m\dot{y}^2. \tag{1.2.1}$$

Since the external force is $ky(t) - mg$,

$$\begin{aligned}\mathrm{PE}(t) &= \text{the potential energy}\\ &= -\int_0^{y(t)} (k\bar{y} - mg)\, d\bar{y} = -\tfrac{1}{2}ky^2 + mgy.\end{aligned} \tag{1.2.2}$$

The action, or total energy change, is

$$\text{Total energy} = \int_0^L \mathrm{KE}(t) - \mathrm{PE}(t)\, dt. \tag{1.2.3}$$

Thus, (1.2.1)–(1.2.3) yield the total energy change

$$X(y(t)) \equiv \frac{1}{2}\int_0^L (m\dot{y}^2 + ky^2 - 2mgy)\, dt. \tag{1.2.4}$$

The principle of least action prompts the following definition.

Definition. $\hat{y}(t)$ will be called a *solution of the energy formulation* of (1.1.1)–(1.1.3) if and only if

$$X(\hat{y}) = \min_{y \in S} X(y) \quad \text{where } \hat{y} \in S$$

and

$$S \equiv \{y: [0, L] \to \mathbb{R} \,|\, y(0) = a,\ y(L) = b,$$
$$y \text{ is "suitable" such that } X(y) < \infty\}.$$

Remark. If $y \in C^1[0, L]$, then $X(y) < \infty$. In Chapter 5 we shall discuss in more detail the definition of the set S, which is often called an *admissible* set of functions. Until Chapter 5 we shall assume y has a piecewise-continuous derivative, in which case $X(y) < \infty$.

A natural question is, "How does one find a continuous solution of the energy formulation?" This is, in general, difficult, and usually one only attempts to approximate the energy solution. One method is the Rayleigh–Ritz method.

Rayleigh – Ritz Method. Let $\hat{y}(t)$ be approximated by $\sum_{j=1}^{M} a_j u_j(t)$, where $a_j \in \mathbb{R}$ are unknowns and $u_j(t)$ are from a given suitable class of functions, for example, the first M terms of the Fourier series.

Step I: Define $H(a_1, \ldots, a_M) \equiv X(\sum_{j=1}^{M} a_j u_j(t))$.
Step II: Choose $a_1, \ldots, a_M$ so that H is a minimum. Consequently, $\partial H/\partial a_1 = \cdots = \partial H/\partial a_M = 0$.

Remarks

1. Step II gives an algebraic system with M unknowns, a_j, and M equations, $\partial H/\partial a_j = 0$.
2. The Rayleigh–Ritz method gives only approximate solutions because it finds a function

$$\hat{u} = \sum_{j=1}^{M} \hat{a}_j u_j(t) \quad \text{with } \frac{\partial H}{\partial a_i}(\hat{a}_1, \ldots, \hat{a}_M) = 0$$

so that

$$X(\hat{u}) = \min_{u \in \hat{S}} X(u)$$

$$\text{where } \hat{S} = \left\{ u = \sum_{j=1}^{M} a_j u_j(t) | a_1, \ldots, a_M \in \mathbb{R}, \right.$$

$$\left. u_j(t) \text{ are suitable functions} \right\}.$$

Since $\hat{S}$ is a proper subset of S, $X(\hat{y}) \leq X(\hat{u})$. It is hoped that, as M gets large, $\hat{u}$ will converge to $\hat{y}$.

The energy formulation may be discretized by using the Rayleigh–Ritz method for a particular choice of $u_i(t)$. One choice of $u_i(t)$ is defined by the following shape functions.

Definitions. Ψ_i, $N_1^{e_i}$, and $N_2^{e_i}$ are defined by Figures 1.2.1, 1.2.2, and 1.2.3, respectively. The Ψ_i are called *test functions* at the nodes t_i and may be written

$$\Psi_i(t) = N_2^{e_{i-1}} + N_1^{e_i}.$$

The intervals $[t_i, t_{i+1}]$ are called *elements of* $[0, L]$ and are referred to as e_i. $N_1^{e_i}$ and $N_2^{e_i}$ are *linear shape functions* of the elements e_i.

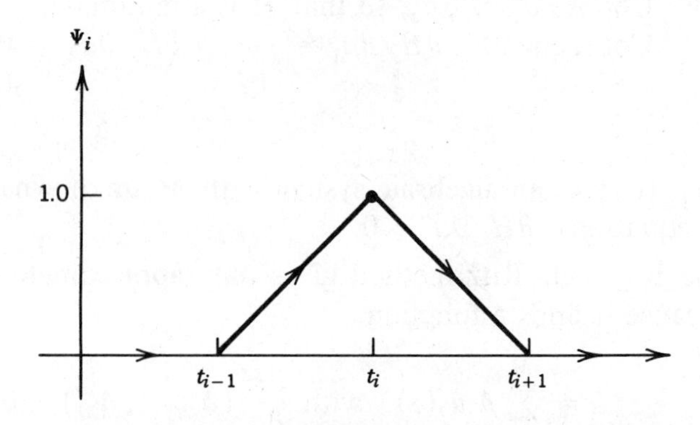

Figure 1.2.1

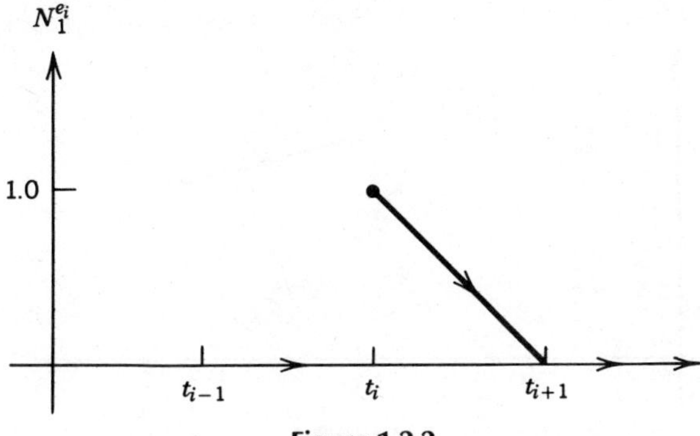

Figure 1.2.2

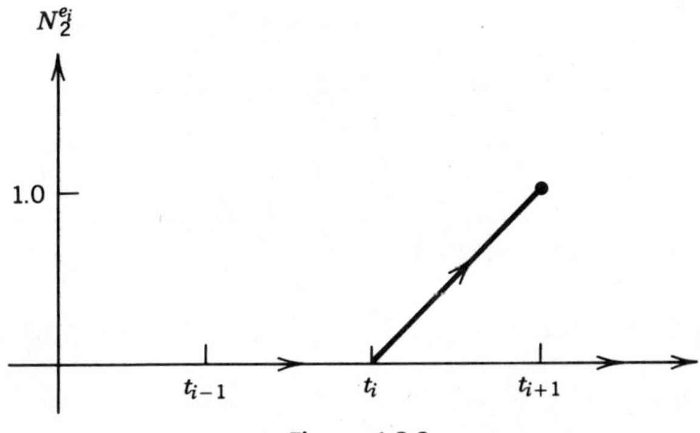

Figure 1.2.3

Let y_i approximate $y(t)$ at $t = t_i$. Then as in the FDM we may approximate $y(t)$ by $\hat{y}(t)$ where $\hat{y} = y_{\text{FDM}}$ is given in Figure 1.1.1. The functions Ψ_i, $N_1^{e_i}$, and $N_2^{e_i}$ can be used to describe $\hat{y}$.

$$\hat{y}(t) = \sum_{j=0}^{N} \hat{y}_j \Psi_j \quad \text{(sum with respect to the nodes)}, \tag{1.2.5}$$

$$= \sum_{j=0}^{N-1} y^{e_j} \quad \text{(sum with respect to the elements)}, \tag{1.2.6}$$

where $y^{e_i} \equiv \hat{y}_i N_1^{e_i} + \hat{y}_{i+1} N_2^{e_i}$ and whose graph is given in Figure 1.2.4. In the Rayleigh–Ritz method we may use $u_i(t) = \Psi_i(t)$.

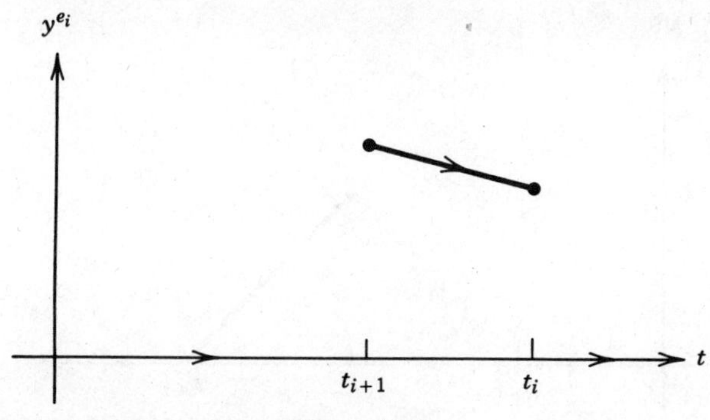

Figure 1.2.4

Definition. Let $H(y_1, \ldots, y_{N-1}) \equiv X(\Sigma_{j=0}^{N} y_j \Psi_j)$. Let $\hat{y}_i$ satisfy

$$\frac{\partial H}{\partial y_i}(y_1, \ldots, y_{N-1}) = 0$$

for $i = 1, \ldots, N-1$. Then $\hat{y} = \Sigma_{j=0}^{N} \hat{y}_j \Psi_j$ is called a *variational finite element solution of* (1.1.1)–(1.1.3).

Example. Consider (1.1.1)–(1.1.3) with $m = k = 1$ and $g = 32$.

$$H(y_1, \ldots, y_{N-1}) = X\left(\sum_{j=0}^{N-1} y^{e_j}\right)$$

$$= \sum_{j=0}^{N-1} \int_{t_j}^{t_{j+1}} \tfrac{1}{2}(\dot{y}^{e_j})^2 + \tfrac{1}{2}(y^{e_j})^2 - 32 y^{e_j} \quad (1.2.7)$$

As the only terms in (1.2.7) that have y_i are y^{e_i} and $y^{e_{i-1}}$, $\partial H/\partial y_i = 0$ simplifies as

$$\frac{\partial H}{\partial y_i}(y_1, \ldots, y_{N-1}) = 0 = \frac{\partial}{\partial y_i}\int_{t_{i-1}}^{t_i} \frac{1}{2}(\dot{y}^{e_{i-1}})^2 + \frac{1}{2}(y^{e_{i-1}})^2 - 32 y^{e_{i-1}}$$

$$+ \frac{\partial}{\partial y_i}\int_{t_i}^{t_{i+1}} \frac{1}{2}(\dot{y}^{e_i})^2 + \frac{1}{2}(y^{e_i})^2 - 32 y^{e_i}$$

$$(1.2.8)$$

The first term on the right-hand side of (1.2.8) is

$$\int_{t_{i-1}}^{t_i} \left[\left(y_{i-1}\dot{N}_1^{e_{i-1}} + y_i \dot{N}_2^{e_{i-1}} \right) \dot{N}_2^{e_{i-1}} \right.$$

$$\left. + \left(y_{i-1} N_1^{e_{i-1}} + y_i N_2^{e_{i-1}} \right) N_2^{e_{i-1}} - 32 N_2^{e_{i-1}} \right]$$

$$= y_{i-1} \int_{t_{i-1}}^{t_i} \left(\dot{N}_1^{e_{i-1}} \dot{N}_2^{e_{i-1}} + N_1^{e_{i-1}} N_2^{e_{i-1}} \right)$$

$$+ y_i \int_{t_{i-1}}^{t_i} \left(\dot{N}_2^{e_{i-1}} \dot{N}_2^{e_{i-1}} + N_2^{e_{i-1}} N_2^{e_{i-1}} \right) - 32 \int_{t_{i-1}}^{t_i} N_2^{e_{i-1}}.$$

Since $N_1^{e_{i-1}}$ and $N_2^{e_{i-1}}$ are linear, the integrals are easily computed (see exercise 1-2) by the following formulas.

Integration Formulas. Let m and k be nonnegative integers.

$$\int_e N_1^m N_2^n = \frac{m!n!}{(m+n+1)!} h.$$

Thus, the first term on the right-hand side of (1.2.8) is

$$y_{i-1}\left(-\frac{1}{h} + \frac{h}{6} \right) + y_i \left(\frac{1}{h} + \frac{h}{3} \right) - 32\frac{h}{2}. \tag{1.2.9}$$

In a similar manner the second term on the right-hand side of (1.2.8) becomes

$$y_i \left(\frac{1}{h} + \frac{h}{3} \right) + y_{i+1} \left(\frac{-1}{h} + \frac{h}{6} \right) - 32\frac{h}{2}. \tag{1.2.10}$$

Thus, by placing (1.2.9) and (1.2.10) into (1.2.8) we have for *each node* t_i the equation

$$\frac{\partial H}{\partial y_i}(y_1, \ldots, y_{N-1}) = 0 = y_{i-1}\left(\frac{-1}{h} + \frac{h}{6} \right)$$

$$+ y_i \left(\frac{2}{h} + 2\frac{h}{3} \right) + y_{i+1}\left(\frac{-1}{h} + \frac{h}{6} \right) - 32h. \tag{1.2.11}$$

By incorporating the conditions $y_0 = a$ and $y_N = b$ into (1.2.11) and letting $N = 4$ we may write the resulting system of algebraic equations as

$$\begin{pmatrix} 1 & 0 & 0 & 0 & 0 \\ \frac{-1}{h} + \frac{h}{6} & \frac{2}{h} + \frac{2h}{3} & \frac{-1}{h} + \frac{h}{6} & 0 & 0 \\ 0 & \frac{-1}{h} + \frac{h}{6} & \frac{2}{h} + \frac{2h}{3} & \frac{-1}{h} + \frac{h}{6} & 0 \\ 0 & 0 & \frac{-1}{h} + \frac{h}{6} & \frac{2}{h} + \frac{2h}{3} & \frac{-1}{h} + \frac{h}{6} \\ 0 & 0 & 0 & 0 & 1 \end{pmatrix} \begin{pmatrix} y_0 \\ y_1 \\ y_2 \\ y_3 \\ y_4 \end{pmatrix}$$

$$= \begin{pmatrix} a \\ 32h \\ 32h \\ 32h \\ b \end{pmatrix}. \tag{1.2.12}$$

The equations of (1.2.12) could be obtained by making the following substitutions in (1.1.1).

$$\ddot{y} \to \frac{y_{i-1} - 2y_i + y_{i+1}}{h^2}$$

$$y \to \frac{1}{2h}\frac{h}{3}(y_{i-1} + 4y_i + y_{i+1})$$

$$= \text{average of } y \text{ over } [t_{i-1}, t_{i+1}] \text{ (by Simpson's rule)}$$

$$mg \to \left(\int_{t_i}^{t_{i+1}} mg(t) N_1^{e_i} + \int_{t_{i-1}}^{t_i} mg(t) N_2^{e_{i-1}}\right)$$

$$= \text{an average (not } mg(t) \to mg(t_i)).$$

TABLE 1.2.1

t	EXACT	FDM	FEM
0.5	9.1903	9.0703	9.3171
1.0	12.5584	12.4082	12.7169
1.5	10.9639	10.8481	11.0863

Table 1.2.1 compares the solutions of (1.1.1)–(1.1.3), (1.1.8), and (1.2.12) when $g = 32$, $a = 0$, $b = 4$, $m = k = 1$, $L = 2$, and $h = 0.5$.

One can show for both FEM and FDM that as the mesh h goes to zero, the numerical approximation will converge to the exact solution. In Chapter 5 this is proved for the FEM. The reader may find it interesting to do the above calculations for smaller mesh, say $h = 0.25$ and $h = 0.125$.

1.3 THE WEAK FORMULATION AND GALERKIN FINITE ELEMENT METHOD

The weak formulation of a boundary-value problem with prescribed values at the boundaries may be obtained from the equations by the following two step procedure.

Step I: Multiply equation (1.1.1) by $\Psi \in C^2[0, L]$ where $\Psi(0) = 0 = \Psi(L)$. (Such Ψ are called *test functions*.)

Step II: Integrate the resulting equation.

For our model problem, these steps yield

$$\int_0^L - m\ddot{y}\Psi + \int_0^L ky\Psi = \int_0^L mg\Psi,$$

$$- m\dot{y}\Psi|_0^L + \int_0^L m\dot{y}\dot{\Psi} + \int_0^L ky\Psi = \int_0^L mg\Psi,$$

$$\int_0^L (m\dot{y}\dot{\Psi} + ky\Psi) = \int_0^L mg\Psi. \tag{1.3.1}$$

Note that in order for (1.3.1) to have finite integrals, it is sufficient for y, $\Psi \in C^1[0, L]$. In Chapter 5 we shall relax this smoothness constraint on y and Ψ, and presently, y and Ψ need only have piecewise-continuous derivatives.

Definition. $y \in C[0, L]$ is a *weak solution of* (1.1.1)–(1.1.3) if and only if

1. y has piecewise-continuous derivative.
2. The *weak equation*, line (1.3.1), holds for all Ψ that have piecewise-continuous derivative and $\Psi(0) = 0 = \Psi(L)$.
3. $y(0) = a$ and $y(L) = b$.

The derivation of line (1.3.1) given above shows that any classical solution is also a weak solution. The adjective weak is used because the solution does not require two continuous derivatives. Not all differential equations have classical solutions. A simple example is steady-state heat conduction in which the thermal conductivity is not continuous (see Section 5.1). In this example the temperature is continuous, but its derivative has a jump discontinuity. Many physical problems, such as those in Chapter 9 on variational inequalities, are more easily studied in the context minimizing an energy integral. So, each formulation has its own advantages and disadvantages.

Line (1.3.1) will yield a discrete problem when

$$y \to \sum_{j=0}^{N} y_j \Psi_j \tag{1.3.2}$$

and

$$\Psi \to \Psi_i, \qquad 1 \le i \le N - 1. \tag{1.3.3}$$

By placing (1.3.2) and (1.3.3) into (1.3.1) we obtain

$$\int_0^L \left(m\left(\sum_{j=0}^{N} y_j \dot{\Psi}_j \right) \dot{\Psi}_i + k\left(\sum_{j=0}^{N} y_j \Psi_j \right) \Psi_i \right) = \int_0^L mg\Psi_i,$$

$$\sum_{j=0}^{N} \left(\int_0^L \left(m\dot{\Psi}_i \dot{\Psi}_j + k\Psi_i\Psi_j \right) \right) y_j = \int_0^L mg\Psi_i. \tag{1.3.4}$$

Let $A = (a_{ij})$ be a $(N-1) \times (N-1)$ system matrix with $a_{ij} \equiv \int_0^L m\dot{\Psi}_i\dot{\Psi}_j + k\Psi_i\Psi_j$. Since y_0 and y_N are given, $a_{1,0}y_0$ and $a_{N-1,N}y_N$ may be incorporated into the right-hand side of (1.3.4). Let

$$f = (f_i) = \begin{cases} \int_0^L mg\Psi_1 - a_{1,0}y_0, & i = 1 \\ \int_0^L mg\Psi_i, & 1 < i < N-1 \\ \int_0^L mg\Psi_{N-1} - a_{N-1,N}y_N, & i = N-1. \end{cases}$$

Line (1.3.4) may be written in matrix form

$$A\hat{y} = f \quad \text{where } \hat{y} = (\hat{y}_i) \tag{1.3.5}$$

This system is analogous to the systems (1.1.8) and (1.2.12). In fact, when Ψ_i are the linear shape functions of the previous section, (1.2.12) and (1.3.5) are equivalent.

Definition. The *Galerkin finite element solution* of (1.1.1)–(1.1.3) is $y = \sum_{j=0}^N \hat{y}_j\Psi_j$ where $\hat{y}$ satisfies (1.3.5).

Remarks

1. The Galerkin finite element method derives the algebraic system from the weak equation (1.3.1) and the substitutions (1.3.2) and (1.3.3). These steps may always be done regardless of the original equation.
2. The variational finite element method requires the existence of an appropriate energy integral, $X(y)$. It is not always obvious what form it should have, or whether it even exists. Does the equation $-\ddot{y} + \dot{y} + y = 0$ have an energy integral?

In Section 1.4 we discuss equivalence of the three formulations and a test to see whether or not $X(y)$ is an appropriate energy integral for a given boundary-value problem.

1.4 COMPARISON OF THE THREE FORMULATIONS

From the derivation of the weak formulation (1.3.1), we have seen that a classical solution of (1.1.1)–(1.1.3) is a weak solution. In fact, as one hopes for the model problem, the classical formulation, the energy formulation, and the weak formulation give the same unique solution of (1.1.1)–(1.1.3). We shall discuss this in detail in Chapter 5. For the moment, we prove the following theorem, whose proof will also help us define "suitable" energy integral and "suitable" test function.

Theorem 1.4.1. Any energy solution of (1.1.1)–(1.1.3) is also a weak solution. Moreover, there is only one weak solution. Consequently, the energy solution and the classical solution are equal.

Proof. Let $\hat{y}$ be an energy solution, that is, $\hat{y} \in S$ and $X(\hat{y}) = \min_{y \in S} X(y)$ where $S = \{y\colon [0, L] \to \mathbb{R} \mid y(0) = a, y(L) = b, X(y) < \infty\}$. We must show for all $\Psi \in C[0, L]$ which have piecewise-continuous derivatives and $\Psi(0) = 0 = \Psi(L)$,

$$\int_0^L (m\dot{\hat{y}}\dot{\Psi} + k\hat{y}\Psi) = \int_0^L mg\Psi. \tag{1.4.1}$$

Define $F(\lambda) = X(\hat{y} + \lambda\Psi)$, where $\lambda \in (-1, 1)$. Note, as $\Psi(0) = 0 = \Psi(L)$, $\hat{y} + \lambda\Psi \in S$. In particular,

$$F(0) = X(\hat{y}) = \min_{y \in S} X(y) = \min_{\lambda \in (-1,1)} X(\hat{y} + \lambda\Psi). \tag{1.4.2}$$

Thus, $F'(0) = 0$. Now, $F'(\lambda)$ is easily computed as

$$F'(\lambda) = \int_0^L (m(\dot{\hat{y}} + \lambda\dot{\Psi})\dot{\Psi} + k(\hat{y} + \lambda\Psi)\Psi - mg\Psi). \tag{1.4.3}$$

So (1.4.3) and $F'(0) = 0$ imply (1.4.1).

In order to show uniqueness of a weak solution, let (1.4.1) hold for $\hat{y}$ and $\hat{\hat{y}} \in S$. By subtracting the two equations, we obtain for all test functions Ψ

$$\int_0^L \left(m\dot{\hat{y}}\dot{\Psi} - m\dot{\hat{\hat{y}}}\dot{\Psi} + k\hat{y}\Psi - k\hat{\hat{y}}\Psi\right) = \int_0^L mg\Psi - \int_0^L mg\Psi$$

$$\int_0^L m\left(\dot{\hat{y}} - \dot{\hat{\hat{y}}}\right)\dot{\Psi} + k\left(\hat{y} - \hat{\hat{y}}\right)\Psi = 0. \tag{1.4.4}$$

Since $(\hat{y} - \hat{\hat{y}})(0) = a - a = 0$ and $(\hat{y} - \hat{\hat{y}})(L) = b - b = 0$, and, therefore, $\hat{y} - \hat{\hat{y}}$ satisfies the conditions of the test function Ψ. Let $\Psi = \hat{y} - \hat{\hat{y}}$ and, therefore, (1.4.4) yields

$$\int_0^L m\left(\dot{\hat{y}} - \dot{\hat{\hat{y}}}\right)^2 + k\left(\hat{y} - \hat{\hat{y}}\right)^2 = 0. \tag{1.4.5}$$

Since $m, k > 0$, the integrand is nonnegative. Thus, $\hat{y} - \hat{\hat{y}} = 0$ and the theorem is proved.

The proof of this first part of the theorem suggests a test for the correctness of the choice for an energy integral.

Test for the Energy Integral. Consider any boundary-value problem, and let equation W represent the equation that describes the weak formulation. $X(y)$ is *admissible* for the given boundary-value problem if and only if $F'(0) = 0$ yields equation W, where $F(\lambda) \equiv X(\hat{y} + \lambda\Psi)$, $\hat{y}$ is an energy solution, $\lambda \in (-1, 1)$, and Ψ is a "suitable" test function.

Example

$$-\ddot{y} + y = f(t), \tag{1.4.6}$$

$$y(0) = 1, \tag{1.4.7}$$

$$\dot{y}(2) = s\left(y_s - y(2)\right), \qquad s, y_s = \text{const.} \tag{1.4.8}$$

In order to find the equation W, which is analogous to (1.3.1), we duplicate the two steps given at the beginning of Section 1.3. First, multiply (1.4.6) by a test function $\Psi \in C^2[0, 2]$, where $\Psi(0) = 0$. Note, Ψ is zero only on that portion of the boundary where y is given. Second, perform integration by parts:

$$\int_0^2 (-\ddot{y} + y)\Psi = \int_0^2 f\Psi,$$

$$-\dot{y}\Psi|_0^2 + \int_0^2 (\dot{y}\dot{\Psi} + y\Psi) = \int_0^2 f\Psi,$$

$$-\dot{y}(2)\Psi(2) + \int_0^2 (\dot{y}\dot{\Psi} + y\Psi) = \int_0^2 f\Psi,$$

$$-s(y_s - y(2))\Psi(2) + \int_0^2 (\dot{y}\dot{\Psi} + y\Psi) = \int_0^2 f\Psi.$$

Or,

$$\int_0^2 (\dot{y}\dot{\Psi} + y\Psi) + sy(2)\Psi(2) = \int_0^2 f\Psi + sy_s\Psi(2). \quad (1.4.9)$$

Equation (1.4.9) is the desired weak equation, W.

The choice of $X(y)$ will be such that $F'(0) = 0$ implies (1.4.9). After some trial and error we obtain

$$X(y) \equiv \frac{1}{2}\int_0^2 (\dot{y}^2 + y^2 - 2yf) + \tfrac{1}{2}sy(2)^2 - sy_s y(2). \quad (1.4.10)$$

This $X(y)$ is admissible for (1.4.6)–(1.4.8) when the following set is used:

$$\bar{S} = \{y: [0, L] \to \mathbb{R} \mid y(0) = a,\ X(y) < \infty\}.$$

For problem (1.1.1)–(1.1.3), the test functions Ψ were required to be zero at both boundaries, and in problem (1.4.6)–(1.4.8) we only required the test functions to be zero at the one boundary with prescribed value of the unknown function. Also, note in the problem (1.4.6)–(1.4.8) with $s \equiv 0$ the energy integral is the same as the energy integral for problem (1.1.1)–(1.1.3); however, for problem (1.4.6)–(1.4.8), we have a larger class of test functions and the energy integral is minimized over a larger set $\bar{S}$ which properly contains S. When $s = 0$, that is, $y'(2) = 0$, then this is called a *natural* boundary condition.

In order to define what we mean by a "suitable" class of test function, the reader should consider the proof of the second part of Theorem 1.4.1. Note how $\Psi(0) = 0 = \Psi(L)$ are used to show uniqueness of the weak solution. Similar arguments may be made for boundary conditions such as in line (1.4.8), see exercise 1-14 and exercise 1-15. Consequently, as "suitable" class of test functions should imply, if possible, that a weak solution is unique. Another property is that all integrals involving the test functions should be finite. For the moment, we use these two properties as a guide for choosing "suitable" test functions. This topic will be carefully discussed in Chapter 5.

1.5 ASSEMBLY BY NODES

In the variational finite element method the system matrix of (1.2.12) was assembled one row at a time. Each row of the system matrix corresponds to $\partial H/\partial y_i$, where y_i is the unknown approximate value at each node. The process given by lines (1.2.7)–(1.2.12) is called *assembly by nodes*. This technique for finding the system matrix may be outlined as follows.

Assembly by Nodes

1. Find all nodes surrounding the ith node.
2. Find $\partial H/\partial y_i = 0$, where $H(y_1, \ldots, y_{N-1}) \equiv X(\Sigma_{j=0}^{N-1} y^{e_j})$.
3. Determine one entire row of the system matrix.
4. Adjust the matrix for any boundary conditions.

In Section 1.6 we discuss another technique, called *assembly by elements*, for finding the system matrix. After more examples are given in Chapter 2, we shall discuss the merits of each method.

1.6 ASSEMBLY BY ELEMENTS

Recall that the elements of (1.1.1)–(1.1.3) correspond to the intervals $[t_i, t_{i+1}]$. Hence, there are N elements and $N - 1$ unspecified nodes. In order to establish a consistent notation, we define the following for (1.2.12):

Definitions

System matrix = SM, the 5×5 matrix

System nodes = $\{t_0, t_1, t_2, t_3, t_4\}$

Element nodes (identifiers) for element $e_i = \{t_i, t_{i+1}\}$

Element matrix for element e_i = SME, a 2×2 matrix given by

$$\frac{\partial X(y^{e_i})}{\partial y_i} = k_{11}y_i + k_{12}y_{i+1} - R_1, \tag{1.6.1}$$

$$\frac{\partial X(y^{e}i)}{\partial y_{i+1}} = k_{21}y_i + k_{22}y_{i+1} - R_2, \tag{1.6.2}$$

$$\text{SME} \equiv \begin{pmatrix} k_{11} & k_{12} \\ k_{21} & k_{22} \end{pmatrix}.$$

Note that the element e_i is fixed and y_i and y_{i+1} are the two surrounding nodal values to the element e_i. t_i and t_{i+1} are identified with the *first and second element node numbers*, 1 and 2. An array NOD(ILT, I) where ILT = $i + 1$ for element e^i and $I = 1, 2$ will identify each element node number with a system node number as follows:

ILT	NOD(ILT, 1)	NOD(ILT, 2)
1	1	2
2	2	3
3	3	4
4	4	5

$$\begin{array}{ccccccccc} 1 & & 2 & & 3 & & 4 & & 5 \\ \vdash & e_0 & + & e_1 & + & e_2 & + & e_3 & \dashv \end{array}$$

The system node numbers $\{1,\ldots,5\}$ correspond to the system nodes $\{t_0,\ldots,t_4\}$, respectively.

Example. Let $X(y^{e_i})$ be from the model problem (see line (1.2.7))

$$X(y^{e_i}) = \int_{t_i}^{t_{i+1}} \left(\tfrac{1}{2}(\dot{y}^{e_i})^2 + \tfrac{1}{2}(y^{e_i})^2 - 32y^{e_i}\right). \tag{1.6.3}$$

As in the computation in lines (1.29) and (1.2.10), lines (1.6.1) and (1.6.2) are

$$\frac{\partial X(y^{e_i})}{\partial y_i} = \left(\frac{1}{h} + \frac{h}{3}\right)y_i + \left(\frac{-1}{h} + \frac{h}{6}\right)y_{i+1} - \left(32\frac{h}{2}\right), \tag{1.6.4}$$

$$\frac{\partial X(y^{e_i})}{\partial y_{i+1}} = \left(\frac{-1}{h} + \frac{h}{6}\right)y_i + \left(\frac{1}{h} + \frac{h}{3}\right)y_{i+1} - \left(32\frac{h}{2}\right). \tag{1.6.5}$$

Thus

$$k_{11} = k_{22} = \frac{1}{h} + \frac{h}{3}, \tag{1.6.6}$$

$$k_{12} = k_{21} = \frac{-1}{h} + \frac{h}{6}, \tag{1.6.7}$$

$$R_1 = R_2 = 32\frac{h}{2}. \tag{1.6.8}$$

Note $h = t_{i+1} - t_i$ may change from one element to the next.

The next step is to combine each of these element matrices in such a manner that one obtains the system matrix. By carefully examining the assembly by elements procedure as illustrated by lines (1.2.7) to (1.2.11), we may conclude that the following program segment is the proper way to "add" the element matrices to obtain the system matrix. Henceforth, the expression "program segment" is just a task and not a complete program or subroutine.

Program by Elements

```
      NE = 4
      DO 30 ILT = 1,NE
         H = ABS(T(ILT+1) - T(ILT))
         SME(1,1) = 1/H + H/3
         SME(2,2) = SME(1,1)
         SME(1,2) = -1/H + H/6
         SME(2,1) = SME(1,2)
         DO 20 I = 1,2
            DO 10 J = 1,2
               L = NOD(ILT,I)
               M = NOD(ILT,J)
               SM(L,M) = SM(L,M) + SME(I,J)
10          CONTINUE
20       CONTINUE
30    CONTINUE
```

After going through the loops the proposed system matrix will have the form

$$\begin{pmatrix} < & < & 0 & 0 & 0 \\ < & \substack{<\\ \times} & \times & 0 & 0 \\ 0 & \times & \substack{\times \\ -} & - & 0 \\ 0 & 0 & - & \mp & + \\ 0 & 0 & 0 & + & + \end{pmatrix}. \tag{1.6.9}$$

The matrix $\begin{pmatrix} < & < \\ < & < \end{pmatrix}$ is the first element matrix, and the matrix $\begin{pmatrix} + & + \\ + & + \end{pmatrix}$ is the fourth element matrix. Any row that corresponds to a node with prescribed values should be set equal to zero and then a 1 placed in the diagonal position. The matrix (1.6.9) becomes

$$\begin{pmatrix} 1 & 0 & 0 & 0 & 0 \\ < & \substack{<\\ \times} & \times & 0 & 0 \\ 0 & \times & \substack{\times \\ -} & - & 0 \\ 0 & 0 & - & \mp & + \\ 0 & 0 & 0 & 0 & 1 \end{pmatrix}. \tag{1.6.10}$$

The right-hand side of (1.2.12) may also be formed by using NOD(ILT, I). Let RHS(L) be the Lth system node number of the right-hand side of (1.2.12). Then in loop 20, and not in loop 10, $L =$ NOD(ILT, I) and

$$\text{RHS}(L) = R_I + \text{RHS}(L).$$

In order to insert the boundary condition in rows 1 and 5, let RHS(1) $= a$ and RHS(5) $= b$.

The procedure described above is called *assembly by elements* and may be outlined as follows.

Assembly by Elements

1. Define the array NOD(ILT, I).
2. Find the element matrices.
3. Compute the preliminary version of the system matrix by "adding" the element matrices one element at a time.
4. Insert the boundary conditions.

Remarks

1. Note the differences in step 3 of the assembly-by-elements outline and the assembly-by-nodes outline. Hence, their names are natural.
2. Prescribed boundary conditions are inserted in the same way for both assembly methods. Boundary conditions such as (1.4.8) are inserted into the energy integral. Note, in (1.4.8) if $s = 0$, $X(y)$ in (1.4.10) is unchanged. If the Galerkin formulation is used on problems with boundary conditions of the form (1.4.8), then there must be an additional test function(s), ψ_N, in line (1.3.3).

1.7 GENERAL OUTLINE OF FEM

This section contains an outline for the FEM. The model problem has illustrated some of the alternative steps in this outline. The examples in the following chapters will illustrate other alternatives.

Outline of FEM

I. Input data, for example, coordinates of nodes, NOD matrix.

II. Choose either energy (variational) or weak (Galerkin) methods and the shape functions, for example, linear, quadratic.

III. Assemble the system matrix. Use either assembly by nodes or assembly by elements.

IV. Insert the boundary condition.

V. Solve the linear system. Use either a direct method, for example, Gaussian elimination, or an iterative method, for example, Gauss–Seidel method.

VI. Output the computed data, for example, table or graphical form or stream functions.

Step I for our present model problem is not too complicated. In Chapter 2 we shall see an increasing amount of input data. Usually one tries to input some of this information by established subroutines.

In our model problem we have, in Step II, used only the linear shape function $y^{e_i} = y_i N_1^{e_i} + y_{i+1} N_2^{e_i}$. By inserting half nodes in the element e^i, we may replace the linear shape function by a quadratic function

$$y^{e_i}(t) = \alpha_1 + \alpha_2 t + \alpha_3 t^2. \qquad (1.7.1)$$

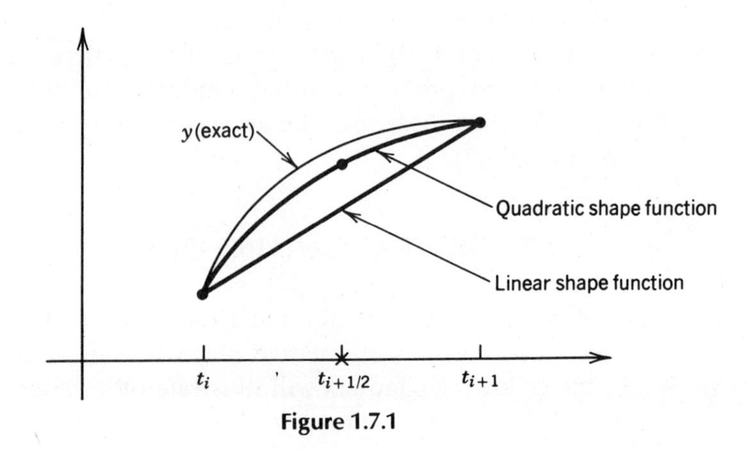

Figure 1.7.1

If $y^{e_i}(t)$ is given at all three nodes $t_i, t_{i+1/2}, t_{i+1}$, then $\alpha_1, \alpha_2, \alpha_3$ can be uniquely determined. One objective of using the quadratic shape function is that one hopes to obtain more accuracy by approximating the exact solution by a curve as compared to the straight line (see Figure 1.7.1). This will be discussed in Chapter 4.

Step V can be done either directly or indirectly. The most common direct method is Gaussian elimination, which we illustrate by a 3×3 system:

$$\begin{pmatrix} 2 & -1 & 0 \\ -1 & 2 & -1 \\ 1 & 1 & -2 \end{pmatrix} \begin{pmatrix} y_1 \\ y_2 \\ y_3 \end{pmatrix} = \begin{pmatrix} 1 \\ 1 \\ 1 \end{pmatrix}. \tag{1.7.2}$$

The system matrix is 3×3, SM. It is common to incorporate the right-hand side of (1.7.2) into an augmented 3×4 system matrix, also denoted by SM, and given in (1.7.3).

$$\left(\begin{array}{ccc|c} 2 & -1 & 0 & 1 \\ -1 & 2 & -1 & 1 \\ 1 & 1 & -2 & 1 \end{array}\right). \tag{1.7.3}$$

Gaussian elimination has two parts: first, use row operations on the augmented matrix so that the original matrix becomes identically zero in the lower left matrix, and, second, use back substitution to find the unknowns. These parts are illustrated for (1.7.3)

Part I. Put (1.7.3) into the form (1.7.4)

$$\left(\begin{array}{ccc|c} \cdot & \cdot & \cdot & \cdot \\ 0 & \cdot & \cdot & \cdot \\ 0 & 0 & \cdot & \cdot \end{array}\right) \tag{1.7.4}$$

(Row 1)/2:

$$\left(\begin{array}{ccc|c} 1 & -\frac{1}{2} & 0 & \frac{1}{2} \\ -1 & 2 & -1 & 1 \\ 1 & 1 & -2 & 1 \end{array}\right)$$

$$\begin{array}{l} \text{(Row 2)} + \text{(Row 1)}: \\ \text{(Row 3)} - \text{(Row 1)}: \\ \text{(Row 3)} - \text{(Row 2)}: \end{array} \left(\begin{array}{ccc|c} 1 & -\frac{1}{2} & 0 & \frac{1}{2} \\ 0 & \frac{3}{2} & -1 & \frac{3}{2} \\ 0 & 0 & -1 & -1 \end{array}\right) \tag{1.7.5}$$

Part II. Use back substitution to find the unknowns. Line (1.7.5) is a shorthand way of writing

$$y_1 - \tfrac{1}{2}y_2 + 0y_3 = \tfrac{1}{2}, \tag{1.7.6}$$

$$0y_1 + \tfrac{3}{2}y_2 - y_3 = \tfrac{3}{2}, \tag{1.7.7}$$

$$0y_1 + 0y_2 - y_3 = -1. \tag{1.7.8}$$

(1.7.8) gives $y_3 = 1$. Then (1.7.7) gives $y_2 = \frac{5}{3}$. Then (1.7.6) gives $y_1 = \frac{4}{3}$.

The Gaussian elimination method and related methods may be programmed. In Chapter 2 the finite element program FEMI has a Gaussian elimination with partial pivoting subroutine called IMAT.

A common indirect method is the Gauss–Seidel iterative method. Iterative methods are often used on large systems. For example, if Ω is a cube with 30 nodes in each direction, then the total number of nodes is $30^3 = 27 \cdot 10^3$. The augmented system matrix must have $27 \cdot 10^3 \cdot (27 \cdot 10^3 + 1)$ components. Thus the Gaussian elimination method is not practical. In large systems arising from partial differential equations the system matrix often has many zeros for components. The Gauss–Seidel method attempts to exploit this fact. We very briefly state the Gauss–Seidel algorithm and the successive overrelaxation (SOR) technique.

Gauss–Seidel Algorithm. Let SM $= A = (a_{ij})$ be an $N \times N$ matrix and consider the problem, for $u,\ R \in \mathbb{R}^N$,

$$Au = R, \tag{1.7.9}$$

$$u_i^{k+1/2} = \frac{-\sum\limits_{j<i} a_{ij}u_j^{k+1} - \sum\limits_{j>i}^{N} a_{ij}u_j^k + R_i}{a_{ii}}, \tag{1.7.10}$$

$$u_i^{k+1} = (1-w)u_i^k + wu_i^{k+1/2}, \qquad 1 \le w < 2. \tag{1.7.11}$$

If $w = 1$, then (1.7.10), (1.7.11) is called the *Gauss–Seidel algorithm*. If $1 < w < 2$, then (1.7.10), (1.7.11) is called the *Gauss–Seidel algorithm with SOR*.

Remarks

1. In practice the summations in (1.7.10) are done only for those j with $a_{ij} \neq 0$.
2. The proper choice of the SOR parameter w can accelerate the convergence of this algorithm by a factor of 10.
3. One can show for certain matrices that as $k \to \infty$ $u_i^{k+1} \to u_i$ for $i = 1, \ldots, N$. If the system matrix is an M-matrix (see [15, pp. 108 and 120]), then it convergences to a unique solution. An example of an M matrix is one that is strictly diagonally dominant, that is, $|a_{ii}| \geq \Sigma_{j \neq i}|a_{ij}| + \delta$ with $\delta > 0$ for all i, $a_{ii} > 0$, and $a_{ij} \leq 0$ $i \neq j$. Another condition that will imply convergence is that A be symmetric ($A^{\mathrm{T}} = A$) and positive definite ($x^T A x > 0$ for all real column vectors and A real and symmetric) (see [15, p. 127]). Another reference is Berman and Plemmons [4].
4. In practice one stops the iteration when some criterion is satisfied. One commonly used criterion is relative error. We say that $u_i^{k+1} \to u_i$ *with respect to the relative error*, ER, when for all $i = 1, \ldots, N$ (1.7.12) holds

$$|u_i^{k+1} - u_i^k|/|u_i^{k+1}| \leq \mathrm{ER}. \tag{1.7.12}$$

1.8 OBSERVATIONS AND REFERENCES

The Galerkin formulation of the finite element method appears to be easy to describe and holds for a large class of problems. The reasons for discussing the energy formulation evolve around theoretical considerations and a number of physical problems. In Chapter 5 we shall use the energy formulation to establish the existence of a solution and certain error estimates. In a number of plate or beam problems, the nonlinearities are so severe that the partial differential equation is more difficult to solve than the energy formulation. Moreover, in porous media problems, such as illustrated in Chapter 9, variational inequalities, a type of energy formulation, can be used to model the physical problem.

Several textbooks that the reader may find useful are by Bathe [2], Norrie and deVires [3], Segerlind [23], Strang and Fix [28] and the series by Becker, Carey and Oden [3, 5, and 6]. The text by Strang and Fix tends to be more theoretical than the first three. The series by Becker, Carey, and Oden is quite extensive.

Of the following exercises students have found 1-1, 1-2, 1-5, 1-9, 1-10, 1-15, 1-16, and 1-18 to be particularly helpful in learning this material. Problems that are more theoretical include 1-11, 1-13, and 1-14. Problems 1-12, 1-25 and 1-26 lead to topics in subsequent chapters.

EXERCISES

1-1 Consider problem (1.1.1)–(1.1.3) with $m = 1$, $k = 1$, and g replaced by $t^2 + 1$. Find the exact solution and the finite difference solution with $\Delta t = 2/4$, $L = 2$, $N = 4$.

1-2 Let $N_1(t)$ and $N_2(t)$ be as in Figures 1.2.2 and 1.2.3. Assume $h = t_{i+1} - t_i = t_i - t_{i-1}$. Show the following:

$$\int_{t_i}^{t_{i+1}} \dot{N}_m \dot{N}_n = \begin{cases} -1/h, & m \neq n, \\ 1/h, & m = n, \end{cases} \qquad m, n = 1 \text{ or } 2,$$

$$\int_{t_i}^{t_{i+1}} N_m N_n = \begin{cases} h/6, & m \neq n, \\ h/3, & m = n, \end{cases} \qquad m, n = 1 \text{ or } 2,$$

$$\int_{t_i}^{t_{i+1}} N_m = h/2, \qquad m = 1 \text{ or } 2.$$

1-3 Let m, n be nonnegative integers. Establish the integration formulas

$$\int N_1^m N_2^n = \frac{m!n!}{(m+n+1)!} h.$$

1-4 Verify line (1.2.10) and then line (1.2.11).

1-5 Let $\Psi_i(t)$ be as in Figure 1.2.1. Use the results of 1-2 to show the following:

$$\int \dot{\Psi}_m \dot{\Psi}_n = \begin{cases} 0, & |m-n| > 1 \\ -1/h, & m = n \pm 1 \\ 2/h, & m = n, \end{cases}$$

$$\int \Psi_m \Psi_n = \begin{cases} 0, & |m-n| > 1 \\ h/6, & m = n \pm 1 \\ 2h/3, & m = n. \end{cases}$$

1-6 Verify that the algebraic systems (1.2.12) and (1.3.5) are equivalent.

1-7 In 1-2 and 1-3 what are the integrals when $h_i \neq h_{i-1}$ where

$$h_i \equiv t_{i+1} - t_i \text{ and } h_{i-1} \equiv t_i - t_{i-1}?$$

1-8 Let $f: [a, b] \to R$ be continuous. The *linear interpolation* of $f(t)$ with respect to $t_0 = a,\ t_1, \ldots, t_N = b$ is defined to be the function

$$f_I: [a, b] \to \mathbb{R},$$

$$f_I(t) = \sum_{j=0}^{N} f(t_j)\Psi_j(t),$$

where

$$\Psi_0(t) = N_1^{e_0}(t),$$

$$\Psi_i(t) = N_2^{e_{i-1}}(t) + N_1^{e_i}(t), \qquad 0 < i < N,$$

$$\Psi_N(t) = N_2^{e_{N-1}}(t).$$

The following graphs compare $f(t)$ and $f_I(t)$ when $N = 4$.

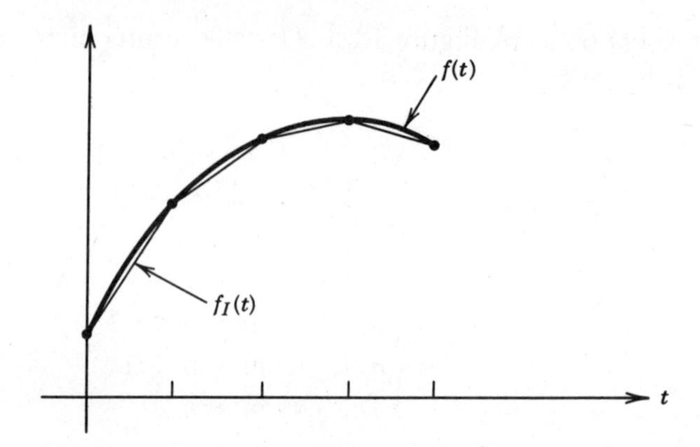

Compute $\int f_I(t) N_1^{e_i}(t)$, $\int f_I(t) N_2^{e_{i-1}}(t)$, and $\int f_I(t) \Psi_i(t)$.

1-9 Consider the problem in 1-1. Define an "appropriate" energy integral for this problem. Let $N = 4$ with $t_1 = 0.5$, $t_2 = 1.0$, and $t_3 = 1.5$. Use the energy formulation of the FEM and form an algebraic system that is analogous to the one in (1.2.12). If $g(t) = f(t) = t^2 + 1$ is approximated $f_I(t)$, find the algebraic system that is analogous to (1.2.12). Compute the solutions to both systems and compare these results with the computations in 1-1.

1-10 Consider the problem in 1-9. State the definition of a weak solution. Use the Galerkin formulation of the FEM to determine the algebraic systems for $f(t) = t^2 + 1$ and for $f_I(t)$.

1-11 Consider $-\ddot{y} + \dot{y} + y = f(t)$, $y(0) = 0$, and $y(1) = 0$. State a definition of a weak solution. Use the Galerkin formulation of the FEM to form an algebraic system. How does this system compare with the algebraic system that follows from FDM? Is it possible to prove uniqueness of a weak solution? (See Theorem 1.4.1.)

1-12 Consider $-u_{xx} - u_{yy} + u = f(x, y)$, $u = u(x, y)$, with $(x, y) \in \{(x, y) | x^2 + y^2 < 1\} \equiv \Omega$ and $u(x, y) = 0$ on $\partial\Omega$. Find a suitable energy integral and show it is admissible. (You must use Green's theorem to state an "appropriate" definition of a weak solution.)

1-13 Consider the problem in 1-12. State and prove a theorem that will be analogous to Theorem 1.4.1.

1-14 Show $X(y)$ in (1.4.10) is admissible for the problem (1.4.6)–(1.4.8). Prove a theorem that will be analogous to Theorem 1.4.1; assume $s \geq 0$.

1-15 Consider the problem (1.4.6)–(1.4.8). Change (1.4.7) to $-\dot{y}(0) = T(y_T - y(0))$, T, $y_T = \text{const}$. State a definition of a weak solution to this problem. Find an energy integral and show it is admissible. Prove a theorem that will be analogous to Theorem 1.4.1; assume $T \geq 0$.

1-16 Consider problem (1.4.6)–(1.4.8) with $f(t) = t^2 + 1$, $s = 2$, and $y_s = 10$. Approximate $f(t)$ by $f_I(t)$ with respect to $t_0 = 0.0$, $t_1 = 0.5$, $t_2 = 1.0$, $t_3 = 1.5$, $t_4 = 2.0$. Use the Galerkin formulation of the FEM to form an algebraic system. Solve this system and compare its solution with the exact solution.

1-17 Verify lines (1.6.4) and (1.6.5).

1-18 Consider the problem in line (1.6.3). Use the *Program by Elements* to assemble the algebraic system when two different sets of nodes are used

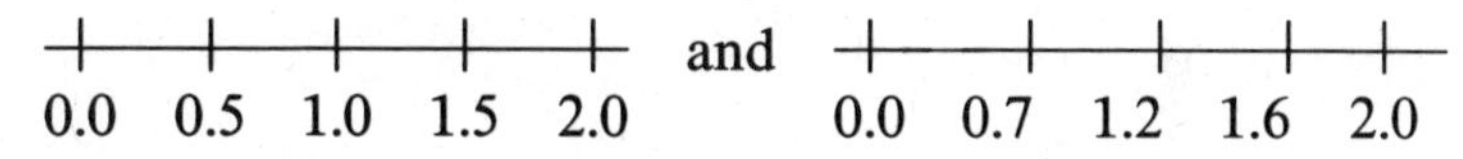

1-19 Consider the algebraic problem

$$\begin{pmatrix} 2 & -1 & 0 \\ -1 & 2 & -1 \\ 0 & -1 & 2 \end{pmatrix} \begin{pmatrix} y_1 \\ y_2 \\ y_3 \end{pmatrix} = \begin{pmatrix} 0 \\ 1 \\ 2 \end{pmatrix}.$$

Use Gaussian elimination to solve for y_1, y_2, and y_3.

1-20 Consider the problem in 1-19. Use the Gauss–Seidel-SOR algorithm to approximate its solution. Experiment with w and ER.

1-21 Consider the problem in 1-15 with $f(t) = 32.0$. Write a program to approximate its solution by the energy formulation of the FEM. Write it so that any nodes may be input. Use assembly by elements, and note that the two boundary conditions must be inserted after the *Program by Elements* is used.

1-22 Consider the problem 1-21. How is the program modified if the right-hand side of (1.4.6) is replaced by $f(t)$?

1-23 Consider the problem 1-21. How is the problem modified if in (1.4.6) we consider the differential equation $-(m\dot{y})\dot{} + ky = f$ and k is replaced by a function of t, $k(t)$? What happens if m is replaced by a function of t, $m(t)$?

1-24 Consider (1.1.1)–(1.1.3) with g replaced by $g(y)$. Show that $X(y) \equiv \frac{1}{2}\int_0^L[\dot{y}^2 + y^2 - 2\int_0^y g(\bar{y})\,d\bar{y}]$ is admissible. Approximate $g(y)$ by $g(y)_I \equiv \Sigma_{j=0}^N g(y_j)\psi_j$. Use the FEM (either energy or Galerkin formulations) to form an algebraic system for this problem. Why is it easier to approximate $g(y)$ by $g(y)_I$ instead of $g(y_I) \equiv g(\Sigma_{j=0}^N y_j\psi_j)$?

1-25 Consider the steady-state cooling problem where the forcing term $g(u)$ is given by the heat loss from radiation.

$$-(Ku_x)_x = \pi d \epsilon\, \sigma_{SB}\left(u_s^4 - u^4\right),$$

$$u(0) = u_0 > u_s,$$

$$u(L) = u_s,$$

where K = thermal conductivity,
d = diameter of the thin rod or wire,
ϵ = emissivity,
σ_{SB} = Stefan–Boltzmann constant,
u_s = surrounding temperature,
u = temperature in a thin rod,

Do 1-24 for this $g(u)$. Also, linearize $g(u)$ and approximate the above problem by one of the form (1.1.1)–(1.1.3).

1-26 Consider the steady-state heat conduction problem $\Omega = \{(x, y)|x^2 + y^2 < 1\}$ (or a steady-state membrane problem). When axial symmetry is assumed, the equation for $u = u(r)$ = temperature (or deformation) and with $r = \sqrt{x^2 + y^2}$ has the form

$$-\Delta u = -\frac{1}{r}(ru_{,r})_{,r} = f(r) \quad \text{on } \Omega$$

$$u_{,r}(0) = 0$$

$$u(1) = \text{given.}$$

Multiply by r and a "suitable" test function and find the weak formulation of this problem. Prove an analog of Theorem 1.4.1. Use the Galerkin formulation of FEM with linear shape functions and find the corresponding algebraic problem. You must compute the integrals that will be analogous to those in 1-5.

1-27 Repeat 1-26 with $u(1) =$ given replaced by $u_{,r}(1) = s(u_s - u(1))$ where $s, u_s =$ given.

2

THE FINITE ELEMENT METHOD FOR TWO-SPACE-VARIABLE PROBLEMS

In this chapter we consider the finite element method for second-order elliptic boundary-value problems with two space variables and with several types of boundary conditions. Section 2.1 contains a discussion of triangular elements and linear shape functions. The variational or energy formulation is given for a simple problem in Section 2.2. The construction of element matrices is done in Section 2.3. In Section 2.4 the program FEMI is discussed. Finally, Section 2.5 contains a generalization to a more sophisticated problem.

2.1 TRIANGULAR ELEMENTS AND LINEAR SHAPE FUNCTIONS

We must develop a function in two variables, $u^e(x, y)$, which is analogous to y^{e_i} of Chapter 1. If $u^e(x, y)$ is to be linear, then it must have the form

$$u^e(x, y) \equiv \begin{cases} \alpha_1 + \alpha_2 x + \alpha_3 y, & (x, y) \in e \\ 0, & (x, y) \notin e, \end{cases} \tag{2.1.1}$$

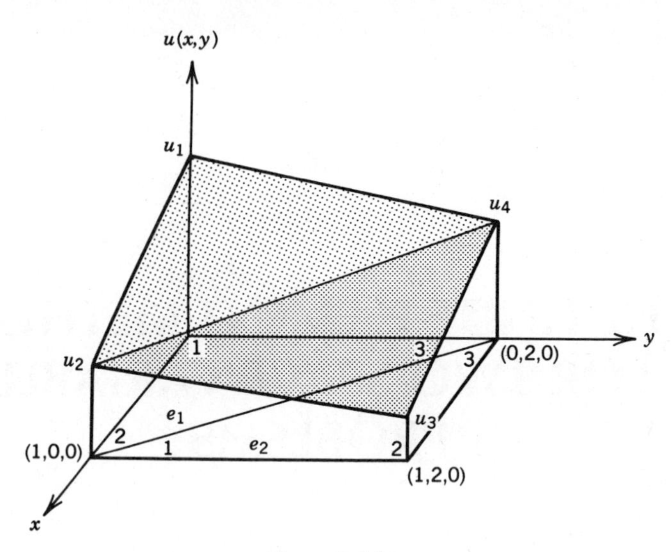

Figure 2.1.1

where $\alpha_1, \alpha_2, \alpha_3$ are to be found. Thus, $u^e(x, y)$ must be given at three nodes $(x_1, y_1), (x_2, y_2), (x_3, y_3)$. Hence, it seems reasonable to let the elements of Ω be triangles. Figure 2.1.1 indicates a simple triangulation of $\Omega = (0, 1) \times (0, 2)$, $u^e(x, y)$ for two elements e_1 and e_2, and $u(x, y) = u^{e_1}(x, y) + u^{e_2}(x, y)$.

It will be useful to have functions $N_1^e(x, y)$, $N_2^e(x, y)$, and $N_3^e(x, y)$ that are analogous to $N_1^{e_i}(t)$ and $N_2^{e_i}(t)$. As we shall see in Proposition 2.1.1, the linear functions given by Figures 2.1.2, 2.1.3, and 2.1.4 will be the proper choice.

Notation. (x_1, y_1), (x_2, y_2), (x_3, y_3) are the (x, y) components of the *element node numbers* 1, 2, 3, respectively. They will change according to each element. For each element e and element node number I, there is a *system node number* $\equiv$ NOD(e, I). For example, in Figure 2.1.1 let the system node numbers 1, 2, 3, 4 correspond to the nodes (0, 0, 0), (1, 0, 0), (1, 2, 0), and (0, 2, 0), respectively. Then the array NOD has values as given in Table 2.1.1. It is a common notation to let the letters i, j, m correspond to the element node numbers 1, 2, 3.

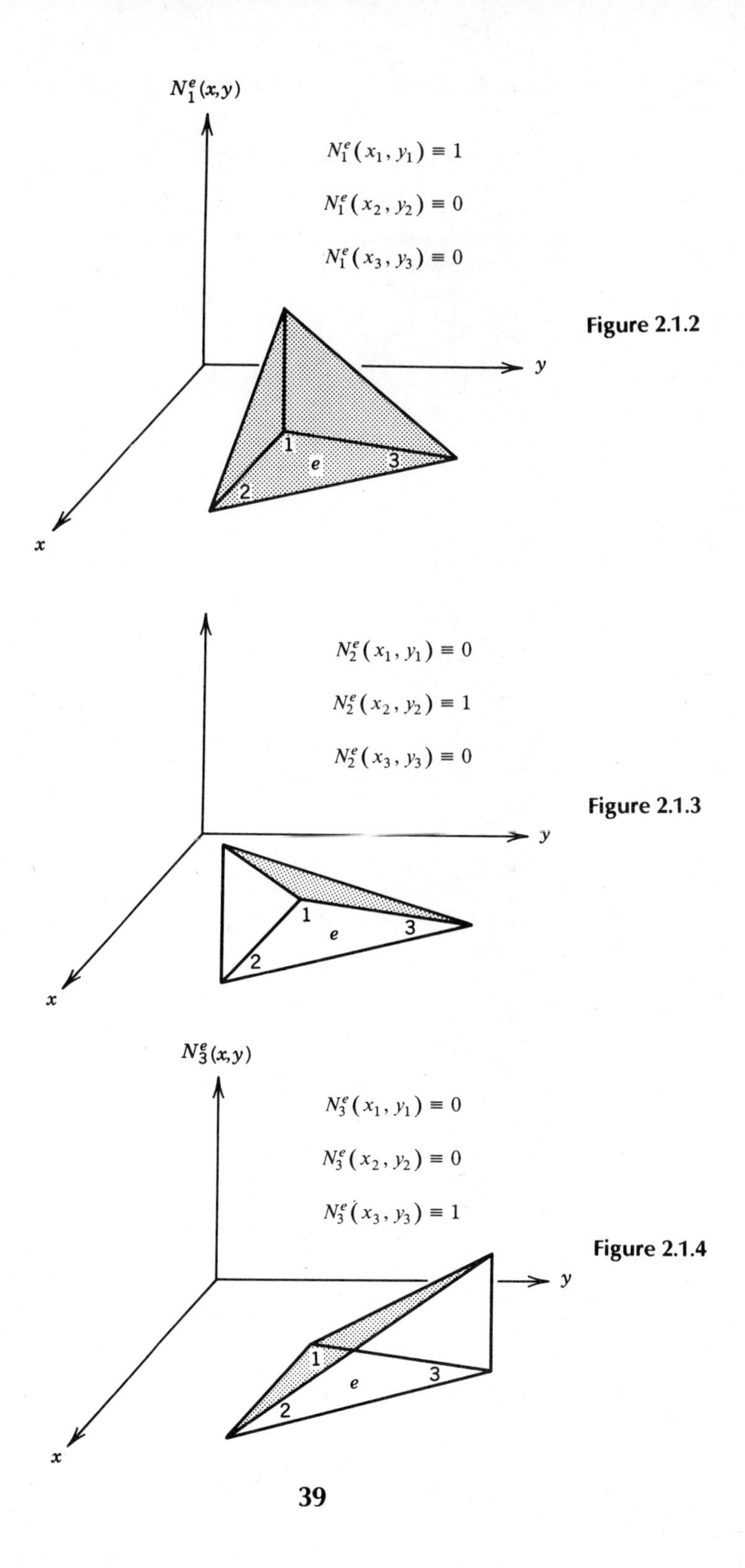

Figure 2.1.2

Figure 2.1.3

Figure 2.1.4

TABLE 2.1.1

e	I	NOD(e, I)
$e^1 = 1$	1	1
$e^1 = 1$	2	2
$e^1 = 1$	3	4
$e^2 = 2$	1	2
$e^2 = 2$	2	3
$e^2 = 2$	3	4

Proposition 2.1.1. Let u_1^e, u_2^e, u_3^e be the values of u^e at the nodes $(x_1, y_1), (x_2, y_2), (x_3, y_3)$. If $u^e(x, y) = u_1^e N_1^e(x, y) + u_2^e N_2^e(x, y) + u_3^e N_3^e(x, y)$, then

$$N_i^e(x, y) = \frac{a_i + b_i x + c_i y}{2\Delta} \tag{2.1.2}$$

where

$$a_i = x_j y_m - x_m y_j,$$

$$b_i = y_j - y_m,$$

$$c_i = x_m - x_j,$$

$$\Delta = \tfrac{1}{2}\det\begin{pmatrix} 1 & x_1 & y_1 \\ 1 & x_2 & y_2 \\ 1 & x_3 & y_3 \end{pmatrix}$$

$$= \pm \text{area of element } e.$$

We use the cycle notation between i, j, and m:

when $i = 1$, then $j = 2$ and $m = 3$;
when $i = 2$, then $j = 3$ and $m = 1$;
when $i = 3$, then $j = 1$ and $m = 2$.

Proof. The proof is by straightforward substitution:

$$u_1^e = \alpha_1 + \alpha_2 x_1 + \alpha_3 y_1, \tag{2.1.3}$$

$$u_2^e = \alpha_1 + \alpha_2 x_2 + \alpha_3 y_2, \tag{2.1.4}$$

$$u_3^e = \alpha_1 + \alpha_2 x_3 + \alpha_3 y_3. \tag{2.1.5}$$

Lines (2.1.3), (2.1.4) and (2.1.5) may be written in matrix form

$$\begin{pmatrix} u_1^e \\ u_2^e \\ u_3^e \end{pmatrix} = \begin{pmatrix} 1 & x_1 & y_1 \\ 1 & x_2 & y_2 \\ 1 & x_3 & y_3 \end{pmatrix} \begin{pmatrix} \alpha_1 \\ \alpha_2 \\ \alpha_3 \end{pmatrix} \tag{2.1.6}$$

Provided the determinant of the 3×3 matrix in (2.1.6) is nonzero, one can solve by Cramer's rule for α_1, α_2, and α_3. By doing this, placing $\alpha_1, \alpha_2, \alpha_3$ into (2.1.1), and collecting all the terms that are coefficients of u_i^e, we obtain $N_i^e(x, y)$ as given in (2.1.2). If $i = 1$, then this manipulation is as follows:

$$\alpha_1 = \det \begin{pmatrix} u_1^e & x_1 & y_1 \\ u_2^e & x_2 & y_2 \\ u_3^e & x_3 & y_3 \end{pmatrix} \Bigg/ 2\Delta$$

$$= \left[u_1^e \det \begin{pmatrix} x_2 & y_2 \\ x_3 & y_3 \end{pmatrix} - u_2^e(\cdot) + u_3^e(\cdot\,\cdot) \right] \Big/ 2\Delta,$$

$$\alpha_2 = \det \begin{pmatrix} 1 & u_1^e & y_1 \\ 1 & u_2^e & y_2 \\ 1 & u_3^e & y_3 \end{pmatrix} \Bigg/ 2\Delta$$

$$= \left[-u_1^e \det \begin{pmatrix} 1 & y_2 \\ 1 & y_3 \end{pmatrix} + u_2^e(-) - u_3^e(-\,-) \right] \Big/ 2\Delta,$$

$$\alpha_3 = \det\begin{pmatrix} 1 & x_1 & u_1^e \\ 1 & x_2 & u_2^e \\ 1 & x_3 & u_3^e \end{pmatrix} \Big/ 2\Delta$$

$$= \left[u_1^e \det\begin{pmatrix} 1 & x_2 \\ 1 & x_3 \end{pmatrix} - u_2^e(\sim) + u_3^e(\sim\sim) \right] \Big/ 2\Delta,$$

$$u^e = \alpha_1 + \alpha_2 x + \alpha_3 y$$

$$= u_1^e\left(\frac{x_2 y_3 - y_2 x_3}{2\Delta} + \frac{y_2 - y_3}{2\Delta}x + \frac{x_3 - x_2}{2\Delta}y \right) + u_2^e(\bigstar) + u_3^e(\bigstar\bigstar).$$

So, $a_1 = x_2 y_3 - y_2 x_3$, $b_1 = y_2 - y_3$, and $c_1 = x_3 - x_2$. The proofs for N_2^e and N_3^e are similar.

The remaining item is to show that

$$\left| \det\begin{pmatrix} 1 & x_1 & y_1 \\ 1 & x_2 & y_2 \\ 1 & x_3 & y_3 \end{pmatrix} \right| = 2(\text{area of } e) \neq 0.$$

By Figure 2.4.2 and the fact that the area of a triangle equals $\frac{1}{2}|\mathbf{A} \times \mathbf{B}|$, where $\mathbf{A}$ and $\mathbf{B}$ are vectors on the sides of the triangle, we have

$$\text{area of } e = \tfrac{1}{2}\left| \det\begin{pmatrix} \mathbf{i} & \mathbf{j} & \mathbf{k} \\ x_j - x_i & y_j - y_i & 0 \\ x_m - x_i & y_m - y_i & 0 \end{pmatrix} \right|$$

$$= \tfrac{1}{2}\left| \left[(x_j - x_i)(y_m - y_i) - (x_m - x_i)(y_j - y_i) \right] \right|$$

$$= \tfrac{1}{2}\left| \left[x_j y_m - x_j y_i - x_i y_m + x_i y_i \right.\right.$$
$$\left.\left. - x_m y_j + x_m y_i + x_i y_j - x_i y_i \right] \right|$$

$$= \tfrac{1}{2}\left| \left[(x_j y_m - x_m y_j) - (x_i y_m - x_m y_i) + (x_i y_j - x_j y_i) \right] \right|$$

$$= \tfrac{1}{2}\left| \det\begin{pmatrix} 1 & x_i & y_i \\ 1 & x_j & y_j \\ 1 & x_m & y_m \end{pmatrix} \right|.$$

This proves Proposition 2.1.1.

2.2 THE ENERGY INTEGRAL FOR A SIMPLE PROBLEM

We consider the elliptic boundary-value problem

$$-\Delta u = f(x, y) \qquad \text{on } \Omega, \tag{2.2.1}$$

$$u = g_1(x, y) \qquad \text{on } \partial\Omega_1, \tag{2.2.2}$$

$$\frac{du}{dn} = 0 \qquad \text{on } \partial\Omega_3, \tag{2.2.3}$$

where

$$\frac{du}{dn} \equiv \nabla u \cdot \mathbf{n} \text{ is the } \textit{normal derivative},$$

$$\mathbf{n} = \text{outward unit normal to } \partial\Omega_3,$$

$$\partial\Omega = \partial\Omega_1 \cup \partial\Omega_3, \quad \partial\Omega_1 \cap \partial\Omega_3 = \varnothing.$$

A particular choice of Ω, $\partial\Omega_1$, and $\partial\Omega_3$ is given in Figure 2.2.1. Another possible choice of Ω is given in Appendix A.1. The problem given by Figure 2.2.1 could be viewed as a steady-state heat-conduction problem for a thin plate with perfect insulation for $x = 0$ and $x = 2$, and fixed temperatures at $y = 0$ and $y = 2$.

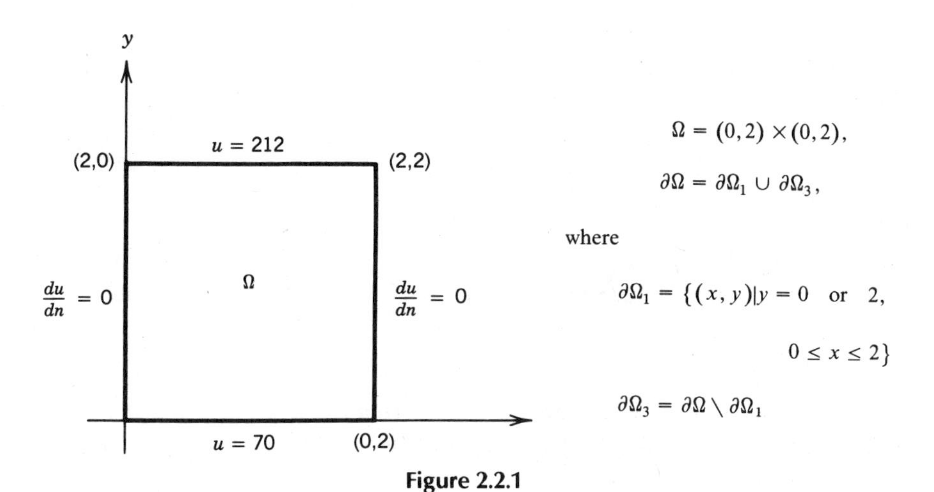

Figure 2.2.1

By considering the problem $-m\ddot{y} + ky = +mg$ for $k = 0$ and using the energy integral $X(y) = \frac{1}{2}\int_0^2 (m\dot{y}^2 + ky^2 - (2mgy))$ as a guide, we expect that

$$X(u) = \frac{1}{2} \iint_{\Omega} \left(u_x^2 + u_y^2 - 2fu\right) \tag{2.2.4}$$

is the proper choice of an energy integral for problem (2.2.1)–(2.2.3). Following the example given at the end of Section 1.4, the boundary condition (2.2.3) is natural or implicit in $X(u)$. By "proper choice" of $X(u)$ we mean $X(u)$ should be admissible with respect to the boundary-value problem (2.2.1)–(2.2.3) (see the test for the energy integral, which is stated before the example in (1.4.6)–(1.4.8)).

In order to use this test, we must find the weak formulation of (2.2.1)–(2.2.3). In two dimensions the analog of integration by parts is given by Green's theorem and Green's identity. In the following we have not been precise about the assumptions on P, Q, Ω, K, u, and Ψ; the reader should consult Taylor and Mann [29].

Green's Theorem. For suitable $P(x, y)$, $Q(x, y)$, and Ω,

$$\int_{\partial\Omega} P(x, y)\,dx + Q(x, y)\,dy = \iint_{\Omega} \left(Q_x - P_y\right) dx\,dy. \tag{2.2.5}$$

Green's Identity. For suitable $K(x, y)$, $u(x, y)$, $\Psi(x, y)$, and Ω,

$$\int_{\partial\Omega} \frac{du}{d\nu} \Psi\,d\sigma = \iint_{\Omega} \Psi \nabla \cdot K \nabla u + \iint_{\Omega} K \nabla u \cdot \nabla \Psi, \tag{2.2.6}$$

where $du/d\nu \equiv K \nabla u \cdot \mathbf{n} =$ *conormal derivative of* u, $\mathbf{n}$ = unit outward normal to $\partial\Omega$ (see Figure 2.2.2), and $d\sigma$ = infinitesimal arclength.

Derivation of Green's Identity. We shall assume (2.2.5) is true and derive (2.2.6) from (2.2.5). Simply let $P = -Ku_y\psi$ and $Q = Ku_x\Psi$.

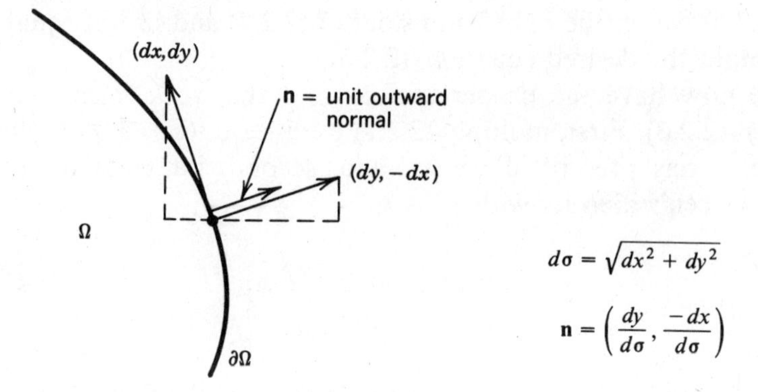

Figure 2.2.2

The left-hand side of (2.2.5) becomes

$$\begin{aligned}\int_{\partial\Omega}(P\,dx + Q\,dy) &= \int_{\partial\Omega}(-Ku_y\Psi\,dx + Ku_x\Psi\,dy)\\ &= \int_{\partial\Omega}\left(Ku_y\Psi\left(-\frac{dx}{d\sigma}\right)d\sigma + Ku_x\Psi\frac{dy}{d\sigma}\,d\sigma\right)\\ &= \int_{\partial\Omega}(Ku_x, Ku_y)\cdot\mathbf{n}\Psi\,d\sigma\\ &= \int_{\partial\Omega}\frac{du}{d\nu}\Psi\,d\sigma. \end{aligned} \tag{2.2.7}$$

The right-hand side of (2.2.5) is

$$\begin{aligned}\iint_{\Omega}(Q_x - P_y)\,dx\,dy &= \iint_{\Omega}(Ku_x\Psi)_x - (-Ku_y\Psi)_y\,dx\,dy\\ &= \iint_{\Omega}\Big[(Ku_x)_x\Psi + Ku_x\Psi_x\\ &\qquad + (Ku_y)_y\Psi + Ku_y\Psi_y\Big]\,dx\,dy\\ &= \iint_{\Omega}\Psi\nabla\cdot K\nabla u + \iint_{\Omega}K\nabla u\cdot\nabla\Psi. \end{aligned} \tag{2.2.8}$$

By (2.2.5) we set the right-hand sides of (2.2.7) and (2.2.8) equal, and we obtain the desired equation, (2.2.6).

We now have the proper tool to find the weak formulation of (2.2.1)–(2.2.3). First, multiply (2.2.1) by $\Psi \in C^2(\overline{\Omega})$, $\Psi \equiv 0$ on $\partial\Omega_1$, where u has prescribed values, and, second, integrate both sides using Green's identity with $K \equiv 1$.

$$\iint_\Omega -\Delta u \Psi = \iint_\Omega f\Psi \tag{2.2.9}$$

The left-hand side of (2.2.9) is the negative of the first term on the right-hand side of (2.2.6) and so (2.2.9) becomes

$$-\int_{\partial\Omega} \frac{du}{dn}\Psi + \iint_\Omega \nabla u \cdot \nabla \Psi = \iint_\Omega f\Psi. \tag{2.2.10}$$

Since $\Psi \equiv 0$ on $\partial\Omega_1$ and (2.2.3) gives $du/dn = 0$ on $\partial\Omega_3 = \partial\Omega \setminus \partial\Omega_1$,

$$\iint_\Omega \nabla u \cdot \nabla \Psi = \iint_\Omega f\Psi. \tag{2.2.11}$$

Definition. $u \in C(\overline{\Omega})$ with piecewise-continuous first-order partial derivatives is a *weak solution of* (2.2.1)–(2.2.3) if and only if (2.2.2) holds and (2.2.11) holds for all $\Psi \in C(\overline{\Omega})$ with piecewise-continuous first-order partial derivatives and $\Psi \equiv 0$ on $\partial\Omega_1$.

u and Ψ can belong to larger classes of functions, which we will not discuss. The important restriction on u and Ψ is that the integrals in (2.2.11) are finite. Also, as in Chapter 1 we have set $\Psi = 0$ on the portion of the boundary where u has prescribed values. This has the attribute that there will only be one weak solution (see exercise 2-16).

Proposition 2.2.1. The energy integral (2.2.4) is admissible with respect to the problem (2.2.1)–(2.2.3).

Proof. Let $F(\lambda) \equiv X(u + \lambda\Psi)$, where Ψ is from the definition of a weak solution. We must show that $F'(0) = 0$ implies (2.2.11).

$$F(\lambda) = \frac{1}{2} \iint_{\Omega} \left((u + \lambda\Psi)_x^2 + (u + \lambda\Psi)_y^2 - 2(u + \lambda\Psi)f \right),$$

$$F'(\lambda) = \frac{1}{2} \iint_{\Omega} \left(2(u_x + \lambda\Psi_x)\Psi_x + 2(u_y + \lambda\Psi_y)\Psi_y - 2\Psi f \right),$$

$$F'(0) = \iint_{\Omega} \left(u_x\Psi_x + u_y\Psi_y - \Psi f \right).$$

Therefore, $F'(0) = 0$ implies the weak equation (2.2.11) and by definition the energy integral is admissible.

2.3 THE CONSTRUCTION OF THE ELEMENT MATRICES

In order to define an element matrix for a triangular element, we must recall the definition, in lines (1.6.1) and (1.6.2), for the element matrix of an interval element. In the interval case there are two surrounding nodes and the element matrix is determined by computing the partial derivatives of $X(y^{e_i})$ with respect to the approximated values of y, y_i and y_{i+1}. With this as motivation we state the next definition.

Definition. Let u^e be defined by (2.1.1) and $X(u)$ for problem (2.2.1)–(2.2.3) be defined by (2.2.4). The *element matrix, SME or k^e*, is a 3×3 matrix formed by $k^e = (k_{ij}^e)$, where

$$\frac{\partial X(u^e)}{\partial u_1^e} = k_{11}^e u_1^e + k_{12}^e u_2^e + k_{13}^e u_3^e - R_1^e, \tag{2.3.1}$$

$$\frac{\partial X(u^e)}{\partial u_2^e} = k_{21}^e u_1^e + k_{22}^e u_2^e + k_{23}^e u_3^e - R_2^e, \tag{2.3.2}$$

$$\frac{\partial X(u^e)}{\partial u_3^e} = k_{31}^e u_1^e + k_{32}^e u_2^e + k_{33}^e u_3^e - R_3^e, \tag{2.3.3}$$

where u_1^e, u_2^e, u_3^e are the approximated values of u at the three surrounding element nodes corresponding to the element node numbers 1, 2, 3.

The k_{ij}^e and R_i^e are computed as in lines (1.6.3)–(1.6.8) for the one-variable case. One simply computes the left-hand sides of (2.3.1)–(2.3.3).

Consider (2.3.1)

$$\frac{\partial X(u^e)}{\partial u_1^e} = \frac{\partial}{\partial u_1^e}\frac{1}{2}\iint_\Omega \left(u_x^{e^2} + u_y^{e^2} - 2u^e f\right)$$

$$= \iint_\Omega \left(u_x^e \frac{\partial u_x^e}{\partial u_1^e} + u_y^e \frac{\partial u_y^e}{\partial u_1^e} - \frac{\partial u^e}{\partial u_1^e} f \right). \qquad (2.3.4)$$

But $u^e = \alpha_1 + \alpha_2 x + \alpha_3 y$, and by Proposition 2.1.1

$$u^e = u_1^e N_1^e(x, y) + u_2^e N_2^e(x, y) + u_3^e N_3^e(x, y). \qquad (2.3.5)$$

By placing (2.3.5) into (2.3.4) we have

$$\frac{\partial X(u^e)}{\partial u_1^e} = \iint_\Omega \left(u_1^e N_{1x}^e + u_2^e N_{2x}^e + u_3^e N_{3x}^e\right) N_{1x}^e$$

$$+ \iint_\Omega \left(u_1^e N_{1y}^e + u_2^e N_{2y}^e + u_3^e N_{3y}^e\right) N_{1y}^e$$

$$+ \iint_\Omega - N_1^e f$$

$$= u_1^e \left(\iint_\Omega N_{1x}^e N_{1x}^e + N_{1y}^e N_{1y}^e \right)$$

$$+ u_2^e \left(\iint_\Omega N_{2x}^e N_{1x}^e + N_{2y}^e N_{1y}^e \right)$$

$$+ u_3^e \left(\iint_\Omega N_{3x}^e N_{1x}^e + N_{3y}^e N_{1y}^e \right)$$

$$- \iint_\Omega N_1^e f \qquad (2.3.6)$$

Thus

$$k_{1j}^e = \iint_\Omega (N_{1x}^e N_{jx}^e + N_{1y}^e N_{jy}^e)$$

and

$$R_1^e = \iint_\Omega N_1^e f.$$

These integrals can be explicitly computed by using (2.1.2) of Proposition 2.1.1. For example,

$$\iint N_{1x}^e N_{jx}^e = \iint_e \frac{b_1}{2\Delta} \frac{b_j}{2\Delta} = \frac{b_1 b_j}{4\Delta^2} \cdot \Delta = \frac{b_1 b_j}{4\Delta}$$

and

$$\iint_\Omega N_{1y}^e N_{jy}^e = \iint_e \frac{c_1}{2\Delta} \frac{c_j}{2\Delta} = \frac{c_i c_j}{4\Delta^2} \Delta = \frac{c_1 c_j}{4\Delta}.$$

Thus $k_{1j}^e = (b_1 b_j + c_1 c_j)/4\Delta$.

The other two lines (2.3.2) and (2.3.3) may be examined in a similar manner. These results are summarized in the next proposition.

Proposition 2.3.1. The element matrix given by (2.3.1)–(2.3.3) has components

$$k_{ij}^e = \frac{b_i b_j + c_i c_j}{4\Delta}. \tag{2.3.7}$$

Moreover,

$$R_i^e = \iint_\Omega N_i^e f.$$

Remarks

1. Δ depends on the choice of e.
2. The integrals

$$R_i^e = \iint_{\Omega} N_i^e f(x, y)$$

may be difficult to compute if $f(x, y)$ is complicated. In this case the integration is usually done numerically as is illustrated in Bathe [2, Chapter 5]. One method involves approximating $f(x, y)$ by its linear interpolation, $f_I(x, y)$.

Definition. $f_I(x, y) \equiv f_1^e N_1^e + f_2^e N_2^e + f_3^e N_3^e$ on e where $f_i^e = f(x_i, y_i)$, $i = 1, 2, 3$, are the element node numbers and (x_i, y_i) are the (x, y) coordinates of the element nodes. See Figure 2.3.1 where the plane is the *linear interpolation of f on e*, $f_I(x, y)$.

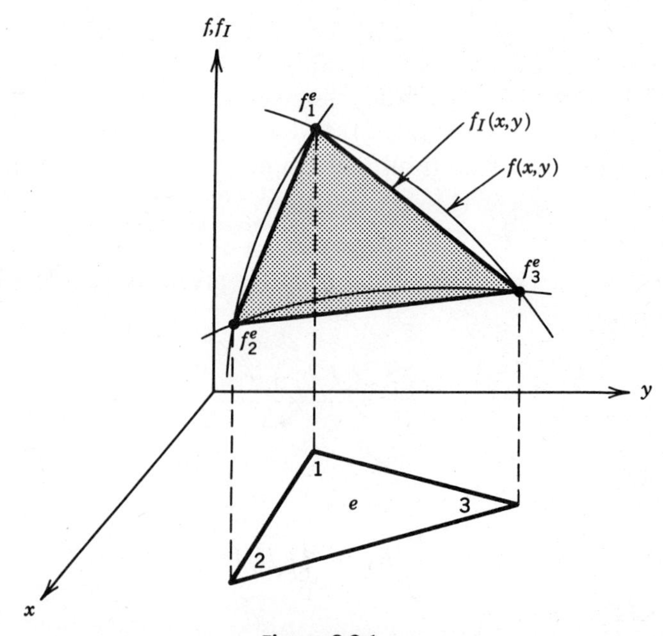

Figure 2.3.1

Approximation of $\iint_\Omega N_i^e f = R_i^e$

$$
\begin{aligned}
R_i^e &= \iint_\Omega N_i^e f \\
&\simeq \iint_\Omega N_i^e f_I \equiv \bar{R}_i^e \\
&= \iint_\Omega N_i^e (f_1^e N_1^e + f_2^e N_2^e + f_3^e N_3^e) \\
&= f_1^e \iint_\Omega N_1^e N_i^e + f_2^e \iint_\Omega N_2^e N_i^e + f_3^e \iint_\Omega N_3^e N_i^e .
\end{aligned} \tag{2.3.8}
$$

The integrals in (2.3.8) may be computed by using the integration formula, whose derivation is discussed in Norrie and de Vries [13, Chapter 9].

Integration Formulas. Let e be a triangular element and m, n, l are nonnegative integers.

$$
\iint_e (N_1^e)^m (N_2^e)^n (N_3^e)^l = \frac{m!n!l!}{(m+n+l+2)!} 2\Delta . \tag{2.3.9}
$$

Proposition 2.3.2. Let $\bar{R}_i^e \equiv \iint_\Omega N_i^e f_I$. Then

$$
\bar{R}_1^e = f_1^e \frac{\Delta}{6} + f_2^e \frac{\Delta}{12} + f_3^e \frac{\Delta}{12}, \tag{2.3.10}
$$

$$
\bar{R}_2^e = f_1^e \frac{\Delta}{12} + f_2^e \frac{\Delta}{6} + f_3^e \frac{\Delta}{12}, \tag{2.3.11}
$$

$$
\bar{R}_3^e = f_1^e \frac{\Delta}{12} + f_2^e \frac{\Delta}{12} + f_3^e \frac{\Delta}{6}. \tag{2.3.12}
$$

Proof. Simply use the integrals in (2.3.9) and put them into (2.3.8).

The formula (2.3.7) for the element matrix may be derived directly from the weak formulation (2.2.11) by using the Galerkin approach. In (2.2.11) simply replace u by $\sum_j u_j \Psi_j$ and Ψ by Ψ_i, where i corresponds to the nodes without prescribed values. The Ψ_j and Ψ_i

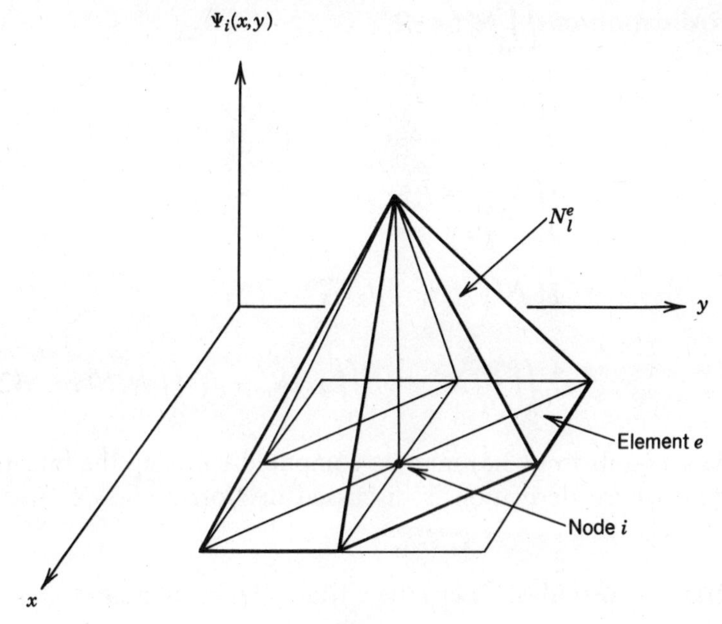

Figure 2.3.2

are continuous and piecewise-linear functions. They are formed by a sum of N_l^e, $l = 1, 2, 3$, where the sum is with respect to e, the elements surrounding the given node i (see Figure 2.3.2). In particular, (2.2.11) yields when $f \simeq f_I \equiv \Sigma_j f_j \Psi_j$

$$\iint_\Omega \sum_j u_j \nabla \Psi_j \nabla \Psi_i = \iint_\Omega \sum_j f_j \Psi_j \Psi_i,$$

$$\sum_j \left(\iint_\Omega \nabla \Psi_j \cdot \nabla \Psi_i \right) u_j = \sum_j \left(\iint_\Omega \Psi_j \Psi_i \right) f_j. \qquad (2.3.13)$$

If the integrals in (2.3.13) are done by elements, we obtain

$$\sum_e \sum_j \left(\iint_e \nabla N_l^e \cdot \nabla N_k^e \right) u_j = \sum_e \sum_j \left(\iint_e N_l^e N_k^e \right) f_j, \quad (2.3.14)$$

where l, k are the element node numbers for element e. The only nonzero terms in Σ_j are for those j that are system node numbers corresponding to the element node numbers of element e. Thus, $k_{l,k}^e = \iint_e \nabla N_l^e \cdot \nabla N_k^e$ for $l, k = 1, 2, 3$ is the same as in line (2.3.7).

2.4 THE FINITE ELEMENT PROGRAM FEMI

FEMI is a program of the finite element method with triangular elements and linear shape functions for the problem (2.2.1)–(2.2.3) when $f(x, y) \equiv 0$. FEMI is written in FORTRAN and uses only the basic components of this language. The present input data ((x, y) components of the nodes, nodes with prescribed boundary data, number of nodes, number of elements, NOD) correspond to that indicated in Figure 2.4.1. Appendix A.1 has FEMI written in Pascal, with the input data given by subroutines and with a more interesting Ω.

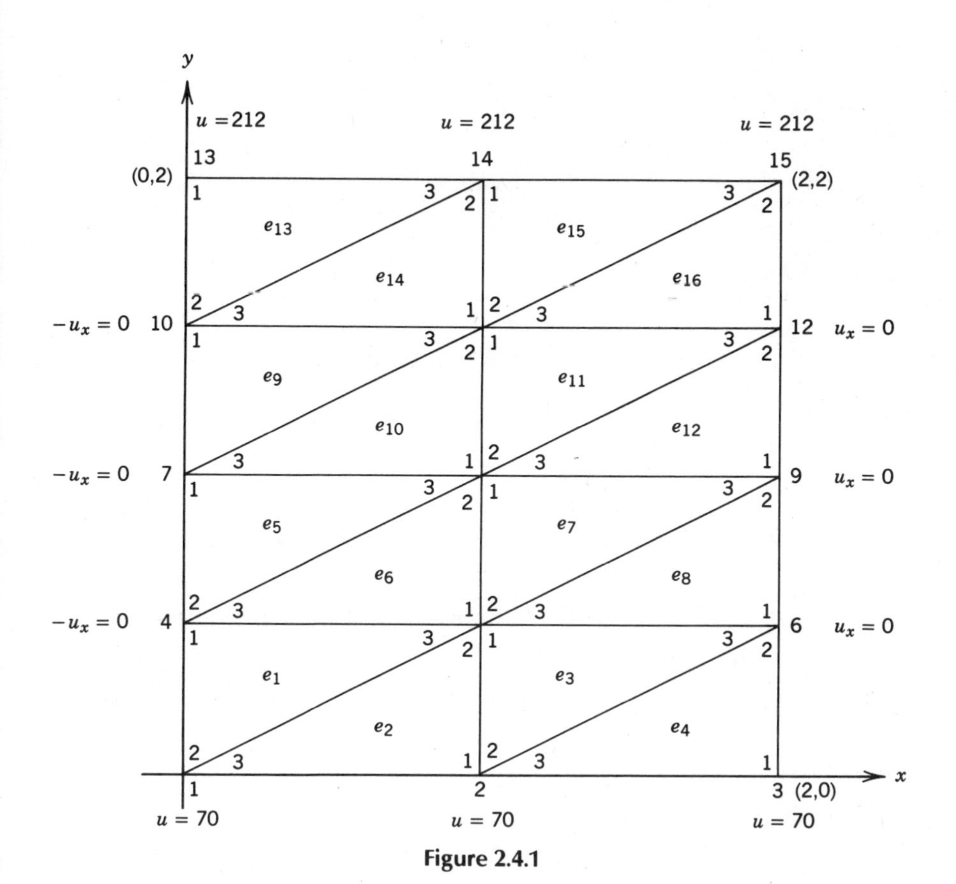

Figure 2.4.1

NOMENCLATURE FOR FEMI

N	the number of nodes
SM	system matrix, and SM($I, N+1$), for $1 \le I \le N$, is the right-hand side of the algebraic problem
NOD	node matrix
X,Y	x and y coordinates for each node
XX,YY	x and y coordinates in a given element
A,B,C	numbers used to compute shape functions
NPT	gives system nodes where the solution is prescribed
G1	gives values at the prescribed nodes
SOL	gives values at nodes as computed by subroutine IMAT
NE	number of elements
NPRES	number of prescribed values

Steps of FEM

I. Input data is given through loop 30.

II. Energy method is used.

III. System matrix SM(I, J), $1 \le I,\ J \le N$, is assembled by elements in loop 60.

- **(a)** For each element A, B, C, Delta $= \Delta =$ area are computed in loop 45.
- **(b)** For each element the element matrix is computed and assembled into SM in loops 50 and 55.

IV. The boundary conditions are directly inserted into SM in loop 70.

V. The algebraic problem is solved by the subroutine. This is a Gaussian elimination method (a direct method).

VI. The output is given in loop 75.

Program FEMI

```
      PROGRAM FEMI
C  THIS PROGRAM SOLVES THE PDE BY THE FINITE ELEMENT METHOD
C            (U)xx + (U)yy = 0.0      ON OMEGA
C                        U = G1       ON BOUNDARY OF OMEGA(PART 1)
C                     (U)n = 0.0      ON THE REMAINING BOUNDARY
      DIMENSION SM(33,34),NOD(44,3),X(33),Y(33),XX(3),YY(3)
      DIMENSION A(3),B(3),C(3),NPT(10),G1(10),SOL(33)
      COMMON N,SOL
      OPEN(5,FILE='*:INPUT.TEXT')
      OPEN(6,FILE='*:OUTPUT.TEXT')
C
C  STEP I:      INPUT THE DATA
C
      READ(5,2) N,NE,NPRES
    2 FORMAT(1X,3I4)
      WRITE(6,4) N,NE,NPRES
    4 FORMAT(1X,'N=',I4,5X,'NE=',I4,4X,'NPRES=',I4)
C  READ IN THE (X,Y) COMPONENTS
      DO 7 K=1,6
        III=K*5
        II=III-4
        READ(5,6) (X(I),Y(I),I=II,III)
    6   FORMAT(1X,10F6.2)
    7 CONTINUE
      DO 10 I=1,N
        WRITE(6,8) I,X(I),Y(I)
    8   FORMAT(1X,'NODE',I4,'='4X,'X=',F6.2,4X,'Y=',F6.2)
   10 CONTINUE
C  READ IN THE ELEMENT NODE NUMBERS IN THE ARRAY NOD
      DO 13 K=1,6
        III=K*7
        II=III-6
        READ(5,12) (NOD(I,1),NOD(I,2),NOD(I,3),I=II,III)
   12   FORMAT(1X,21I3)
   13 CONTINUE
      DO 16 I=1,NE
        WRITE(6,14) I,NOD(I,1),NOD(I,2),NOD(I,3)
   14   FORMAT(1X,'ELT',I3,4X,'NOD 1=',I3,4X,'NOD 2=',I3,4X,
     +  'NOD 3=',I3)
   16 CONTINUE
C  READ IN THE NODES WHERE THE SOLUTION HAS PRESCRIBED VALUES AND
C  READ IN THESE VALUES ( THE VALUES OF G1 )
      DO 25 K=1,2
        III=K*5
        II=III-4
        READ(5,24) (NPT(I),G1(I),I=II,III)
   24   FORMAT(1X,5(I3,F6.2))
   25 CONTINUE
      DO 28 I=1,NPRES
        WRITE(6,26) I,NPT(I),G1(I)
   26   FORMAT(1X,'I=',I3,4X,'NPT=',I3,4X,'G1=',F6.2)
   28 CONTINUE
      NP1=N+1
      DO 30 I=1,N
```

```
          DO 30 J=1,NP1
          SM(I,J) = 0.0
   30   CONTINUE
C
C  STEP  II:   USE THE ENERGY FORMULATION OF THE FEM
C
C
C  STEP  III:  USE ASSEMBLY BY ELEMENTS AND USE LINEAR SHAPE FUNCTIONS.
C              THE ELEMENT MATRICES ARE NOT STORED SINCE WE ONLY NEED THE
C              VALUES OF DELTA TO COMPUTE THEIR COMPONENTS.
C
        DO 60 ILT=1,NE
          DO 40 J=1,3
            JJ=NOD(ILT,J)
            XX(J)=X(JJ)
            YY(J)=Y(JJ)
   40     CONTINUE
          DO 45 I=1,3
            J=MOD(I+1,3)
            IF (J.EQ.0) J=3
            M=MOD(I+2,3)
            IF (M.EQ.0) M=3
            A(I)=XX(J)*YY(M)-XX(M)*YY(J)
            B(I)=YY(J)-YY(M)
            C(I)=XX(M)-XX(J)
   45     CONTINUE
          DELTA=(C(3)*B(2)-C(2)*B(3))/2.0
C  THE ASSEMBLY IS DONE
          DO 55 IR=1,3
            DO 50 IC=1,3
              AK=(B(IR)*B(IC)+C(IR)*C(IC))/(4*DELTA)
              II=NOD(ILT,IR)
              JJ=NOD(ILT,IC)
              SM(II,JJ)=SM(II,JJ)+AK
   50       CONTINUE
   55     CONTINUE
   60   CONTINUE
C
C  STEP  IV:   THE BOUNDARY CONDITIONS ARE INSERTED
C
        DO 70 I=1,NPRES
          NODE=NPT(I)
          DO 65 K=1,N
            SM(NODE,K)=0.0
   65     CONTINUE
          SM(NODE,NODE)=1.0
          SM(NODE,NP1)=G1(I)
   70   CONTINUE
C
C  STEP  V:   THE ALGEBRAIC SYSTEM IS SOLVED BY GAUSSIAN ELIMINATION
C
        CALL IMAT(SM)
C
C  STEP  VI:  THE OUTPUT IS GIVEN IN TABLE FORM
```

```
C
      DO 75 I=1,N
        WRITE(6,72) I,SOL(I)
   72   FORMAT(1X,'NODE',I3,'=',F6.2)
   75 CONTINUE
      STOP
      END
C
      SUBROUTINE IMAT(A)
      DIMENSION A(33,34),SOL(33)
      COMMON N,SOL
      NP1=N+1
      NM1=N-1
      DO 13 K=1,NM1
        JJ=K
        BIG=ABS(A(K,K))
        KP1=K+1
        DO 7 I=KP1,N
          AB=ABS(A(I,K))
          IF (BIG-AB) 6,7,7
    6     BIG=AB
          JJ=I
    7   CONTINUE
        IF (JJ-K) 8,10,8
    8   DO 9 J=K,NP1
          TEMP=A(JJ,J)
          A(JJ,J)=A(K,J)
          A(K,J)=TEMP
    9   CONTINUE
   10   DO 11 I=KP1,N
          QUOT=A(I,K)/A(K,K)
          DO 11 J=KP1,NP1
            A(I,J)=A(I,J)-QUOT*A(K,J)
   11   CONTINUE
        DO 12 I=KP1,N
          A(I,K)=0.0
   12   CONTINUE
   13 CONTINUE
      SOL(N)=A(N,NP1)/A(N,N)
      DO 15 NN=1,NM1
        SUM=0.0
        I=N-NN
        IP1=I+1
        DO 14 J=IP1,N
          SUM=SUM+A(I,J)*SOL(J)
   14   CONTINUE
        SOL(I)=(A(I,NP1)-SUM)/A(I,I)
   15 CONTINUE
      RETURN
      END
 C
 C        OUTPUT
 C
 N=  15      NE=  16    NPRES=   6
```

```
NODE   1=    X=   .00    Y=   .00
NODE   2=    X=  1.00    Y=   .00
NODE   3=    X=  2.00    Y=   .00
NODE   4=    X=   .00    Y=   .50
NODE   5=    X=  1.00    Y=   .50
NODE   6=    X=  2.00    Y=   .50
NODE   7=    X=   .00    Y=  1.00
NODE   8=    X=  1.00    Y=  1.00
NODE   9=    X=  2.00    Y=  1.00
NODE  10=    X=   .00    Y=  1.50
NODE  11=    X=  1.00    Y=  1.50
NODE  12=    X=  2.00    Y=  1.50
NODE  13=    X=   .00    Y=  2.00
NODE  14=    X=  1.00    Y=  2.00
NODE  15=    X=  2.00    Y=  2.00
ELT  1     NOD 1=  4     NOD 2=  1     NOD 3=  5
ELT  2     NOD 1=  2     NOD 2=  5     NOD 3=  1
ELT  3     NOD 1=  5     NOD 2=  2     NOD 3=  6
ELT  4     NOD 1=  3     NOD 2=  6     NOD 3=  2
ELT  5     NOD 1=  7     NOD 2=  4     NOD 3=  8
ELT  6     NOD 1=  5     NOD 2=  8     NOD 3=  4
ELT  7     NOD 1=  8     NOD 2=  5     NOD 3=  9
ELT  8     NOD 1=  6     NOD 2=  9     NOD 3=  5
ELT  9     NOD 1= 10     NOD 2=  7     NOD 3= 11
ELT 10     NOD 1=  8     NOD 2= 11     NOD 3=  7
ELT 11     NOD 1= 11     NOD 2=  8     NOD 3= 12
ELT 12     NOD 1=  9     NOD 2= 12     NOD 3=  8
ELT 13     NOD 1= 13     NOD 2= 10     NOD 3= 14
ELT 14     NOD 1= 11     NOD 2= 14     NOD 3= 10
ELT 15     NOD 1= 14     NOD 2= 11     NOD 3= 15
ELT 16     NOD 1= 12     NOD 2= 15     NOD 3= 11
I=  1     NPT=  1     G1= 70.00
I=  2     NPT=  2     G1= 70.00
I=  3     NPT=  3     G1= 70.00
I=  4     NPT= 13     G1=212.00
I=  5     NPT= 14     G1=212.00
I=  6     NPT= 15     G1=212.00
NODE  1= 70.00
NODE  2= 70.00
NODE  3= 70.00
NODE  4=105.50
NODE  5=105.50
NODE  6=105.50
NODE  7=141.00
NODE  8=141.00
NODE  9=141.00
NODE 10=176.50
NODE 11=176.50
NODE 12=176.50
NODE 13=212.00
NODE 14=212.00
NODE 15=212.00
```

Except for the prescribed boundary conditions, the input data should be self explanatory. In the prescribed boundary data we have the symbols NPRES, NPT, and G1. NPT is used to reorder the system nodes numbers so that the nodes with prescribed values may be addressed sequentially in a loop. Table 2.4.1 lists these data. Loops 25 and 28 illustrate the implementation of this notation.

Since the energy method is used, the results of Proposition 2.3.1 and Proposition 2.1.1 can be used to compute the element matrices. For $I = 1, 2, 3$, the a_i, b_i, c_i, and Δ are computed for each element, ILT. The following discussion explains how the cyclic rotation of $i, j, m(1, 2, 3)$ is programmed. This is done in loop 45 where $I = 1, 2, 3$.

$I=1$: $J = 2$, $M = 3$

$$A(1) = a_i = x_j y_m - x_m y_j = XX(2) * YY(3) - XX(3) * YY(2)$$

$$B(1) = b_i = y_j - y_m = YY(2) - YY(3)$$

$$C(1) = c_i = x_m - x_j = XX(3) - XX(2)$$

$I=2$: $J = 3$, $M = 1$

$$A(2) = a_j = x_m y_i - x_i y_m = XX(3) * YY(1) - XX(1) * YY(3)$$

$$B(2) = b_j = y_m - y_i = YY(3) - YY(1)$$

$$C(2) = c_j = x_i - x_m = XX(1) - XX(3)$$

$I=3$: $J = 1$, $M = 2$

$$A(3) = a_m = x_i y_j - x_j y_i = XX(1) * YY(2) - XX(2) * YY(1)$$

$$B(3) = b_m = y_i - y_j = YY(1) - YY(2)$$

$$C(3) = c_m = x_j - x_i = XX(2) - XX(1)$$

The Δ, the area of e, is then computed by $\frac{1}{2}|\mathbf{A} \times \mathbf{B}|$ where $\mathbf{A}, \mathbf{B}$ are the vectors on the sides of the triangular element. Figure 2.4.2

TABLE 2.4.1

I	NPT(I)	G1(I)
1	1	70
2	2	70
3	3	70
4	13	212
5	14	212
NPRES = 6	15	212

indicates the proper choice of **A** and **B** so that $\Delta = (C(3) * B(2) - C(2) * B(3))/2$. What happens if the element nodes are numbered in a clockwise direction?

The element matrices are assembled directly into the augmented system matrix, SM, and are not stored in the memory. This is done in loop 60 with respect to ILT, IR, and IC. $AK = k^e_{ij}$, where $i =$ IR and $j =$ IC. Note how NOD is used to identify the element node numbers, IR and IC, with the system node numbers, II and JJ. This procedure is analogous to the *Program by Elements* in Section 1.6.

In loop 70 the prescribed boundary conditions are inserted directly into the augmented system matrix. Note how the array NPT is used to identify the rows, that is, the nodes with prescribed boundary data, which must be changed.

The subroutine IMAT is a Gaussian elimination program that uses an augmented matrix. Note that the right-hand side of the algebraic problem is stored in the last column of SM. You may wish to experiment with other subroutines at your disposal (see Rice [17]).

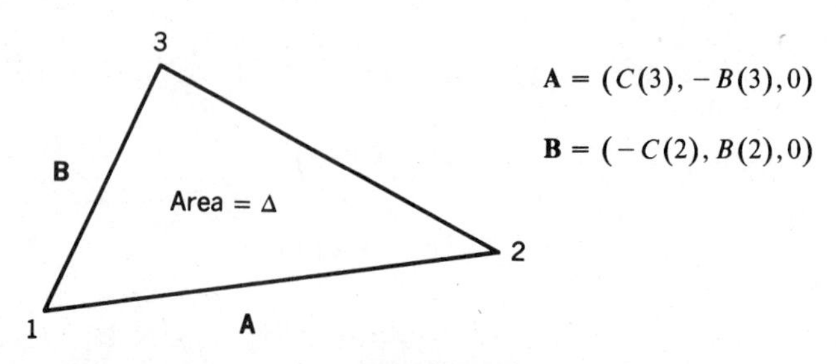

Figure 2.4.2

The output is simply listed by nodes. Usually one needs to do some further manipulations of these numbers so that they become meaningful.

2.5 A MORE GENERAL PROBLEM

In this section we discuss the energy integral, the weak formulation, and the construction of the element matrices for a more sophisticated version of (2.2.1)–(2.2.3).

$$-\nabla \cdot K \nabla u + Cu = f(x, y) \quad \text{on } \Omega, \tag{2.5.1}$$

$$u = g_1(x, y) \quad \text{on } \partial\Omega_1, \tag{2.5.2}$$

$$\frac{du}{d\nu} + su = g_2 \quad \text{on } \partial\Omega_2, \tag{2.5.3}$$

$$\frac{du}{d\nu} = 0 \quad \text{on } \partial\Omega_3, \tag{2.5.4}$$

where C, s, and g_2 are constants, $du/d\nu$ is the conormal derivative on u, and $\partial\Omega = \partial\Omega_1 \cup \partial\Omega_2 \cup \partial\Omega_3$ with $\partial\Omega_1$, $\partial\Omega_2$, $\partial\Omega_3$ disjoint.

By using examples (1.4.6)–(1.4.10) and (2.2.1)–(2.2.4) as a guide, we expect that

$$X(u) = \frac{1}{2} \iint_{\Omega} \left(Ku_x^2 + Ku_y^2 + Cu^2 - 2uf\right) + \frac{1}{2} \int_{\partial\Omega_2} \left(su^2 - 2g_2 u\right) \tag{2.5.5}$$

is an admissible energy integral with respect to (2.5.1)–(2.5.4).

In order to check this, we must first determine the weak formulation of (2.5.1)–(2.5.4). We continue just as in lines (2.2.9)–(2.2.11). Multiply (2.5.1) by $\Psi \in C^2(\overline{\Omega})$, where $\Psi \equiv 0$ on $\partial\Omega_1$. Then integrate by using Green's identity (2.2.6):

$$\iint_\Omega (-\nabla \cdot K \nabla u)\Psi + C \iint_\Omega u\Psi = \iint_\Omega f\Psi$$

$$-\int_{\partial\Omega} \frac{du}{d\nu}\Psi \, d\sigma + \iint_\Omega K \nabla u \cdot \nabla \Psi + C \iint_\Omega u\Psi = \iint f\Psi. \tag{2.5.6}$$

Since $\partial\Omega$ is the disjoint union of $\partial\Omega_1$, $\partial\Omega_2$, $\partial\Omega_3$ and (2.5.3) holds,

$$\begin{aligned}
\int_{\partial\Omega} \frac{du}{d\nu}\Psi \, d\sigma &= \int_{\partial\Omega_1} \frac{du}{d\nu}\Psi \, d\sigma + \int_{\partial\Omega_2} \frac{du}{d\nu}\Psi \, d\sigma + \int_{\partial\Omega_3} \frac{du}{d\nu}\Psi \, d\sigma \\
&= \int_{\partial\Omega_1} \frac{du}{d\nu}\Psi \, d\sigma + \int_{\partial\Omega_2} (g_2 - su)\Psi \, d\sigma + \int_{\partial\Omega_3} \frac{du}{d\nu}\Psi \, d\sigma \\
&= \int_{\partial\Omega_2} (g_2 - su) \, d\sigma
\end{aligned} \tag{2.5.7}$$

by $\Psi \equiv 0$ on $\partial\Omega_1$ and by (2.5.4). Thus, by placing (2.5.7) into (2.5.6), we have for all $\Psi \in C^2(\overline{\Omega})$ with $\Psi \equiv 0$ on $\partial\Omega_1$,

$$\iint_\Omega (K \nabla u \cdot \nabla \Psi + Cu\Psi) + \int_{\partial\Omega_2} su\Psi = \iint_\Omega f\Psi + \int_{\partial\Omega_2} g_2\Psi. \tag{2.5.8}$$

Definition. $u \in C(\overline{\Omega})$ with first-order piecewise-continuous partial derivatives is *a weak solution on* (2.5.1)–(2.5.4) if and only if (2.5.2) holds and the weak equation (2.5.8) holds for all $\Psi \in C(\overline{\Omega})$ with first-order piecewise-continuous partial derivatives and $\Psi \equiv 0$ on $\partial\Omega_1$.

The proof of admissibility of $X(u)$ is similar to that of Proposition 2.2.1.

Proposition 2.5.1. The energy integral (2.5.5) is admissible with respect to the problem (2.5.1)–(2.5.4).

In order to examine the element matrices associated with (2.5.5), let us decompose $X(u)$ into the following terms:

$$X(u) = X_{\Delta}(u) + X_C(u) + X_f(u) + X_2(u), \qquad (2.5.9)$$

where

$$X_{\Delta}(u) = \frac{1}{2} \iint_{\Omega} K u_x^2 + K u_y^2, \qquad (2.5.10)$$

$$X_C(u) = \frac{1}{2} \iint C u^2, \qquad (2.5.11)$$

$$X_f(u) = -\iint uf, \qquad (2.5.12)$$

$$X_2(u) = \frac{1}{2} \int_{\partial\Omega_2} (s u^2 - 2 g_2 u). \qquad (2.5.13)$$

As in (2.3.1)–(2.3.3), we may define the element matrices by computing the partial derivatives of $X(u^e)$ with respect to u_i^e.

Definition. The *element matrix associated with* (2.5.9) is (k_{ij}^e), where k_{ij}^e are the coefficients of u_j^e in

$$\frac{\partial X(u^e)}{\partial u_i^e} = \sum_{j=1}^{3} k_{ij}^e u_j^e - R_i^e, \qquad i, j = 1, 2, 3. \qquad (2.5.14)$$

Since

$$\frac{\partial X(u^e)}{\partial u_i^e} = \frac{\partial X_{\Delta}(u^e)}{\partial u_i^e} + \frac{\partial X_C(u^e)}{\partial u_i^e} + \frac{\partial X_f(u)}{\partial u_i^e} + \frac{\partial X_2(u)}{\partial u_i^e},$$

(k_{ij}^e) can be decomposed as

$$k_{ij}^e = k_{ij}^{\Delta} + k_{ij}^C + k_{ij}^2, \qquad (2.5.15)$$

where k_{ij}^{Δ}, k_{ij}^{C}, k_{ij}^{2} are the element matrices with respect to X_{Δ}, X_C, X_2. (We have suppressed the element, e, dependence of these matrices.) Note X_f depends only linearly on u_i^e, $i = 1, 2, 3$ and so $\partial X_f(u^e)/\partial u_i^e$ are independent of u_i^e. Similarly R_i^e can be decomposed as

$$R_i^e = R_i^f + R_i^2, \tag{2.5.16}$$

where R_i^f and R_i^2 are from $X_f(u^e)$ and $X_2(u^e)$.

From Proposition 2.3.1 we know

$$k_{ij}^{\Delta} = K\frac{b_i b_j + c_i c_j}{4\Delta} \quad \text{and} \quad R_i^f = \iint_{\Omega} N_i^e f. \tag{2.5.17}$$

In order to find k_{ij}^C, we calculate $\partial X_C(u^e)/\partial u_i^e$:

$$\begin{aligned}
\frac{\partial}{\partial u_i^e}\left(\frac{1}{2}\iint_{\Omega} C\left(\sum_{j=1}^{3} u_j^e N_j^e\right)^2\right) &= C\iint_{\Omega}\left(\sum_{j=1}^{3} u_j^e N_j^e\right) N_i^e \\
&= C\sum_{j=1}^{3}\iint_{\Omega} N_j^e N_i^e u_j^e \\
&= \sum_{j=1}^{3}\left(C\iint_{\Omega} N_i^e N_j^e\right) u_j^e.
\end{aligned} \tag{2.5.18}$$

Thus, by (2.5.14) applied to $X_C(u^e)$, we have

$$k_{ij}^C = C\iint_{\Omega} N_i^e N_j^e. \tag{2.5.19}$$

By applying the integral formulas (2.3.9) we have

$$k^C = \frac{C\Delta}{12}\begin{pmatrix} 2 & 1 & 1 \\ 1 & 2 & 1 \\ 1 & 1 & 2 \end{pmatrix}. \tag{2.5.20}$$

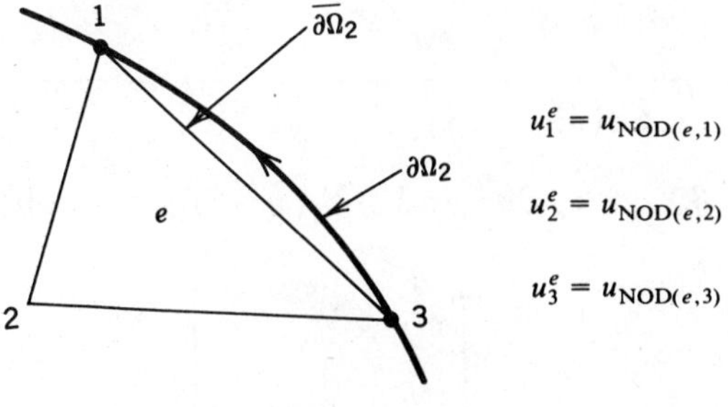

Figure 2.5.1

In order to find k_{ij}^2 and R_i^2, we consider an example element given in Figure 2.5.1 in which the side from element node 3 to element node 1 approximates a segment of $\partial\Omega_2$. If different sides of e approximate $\partial\Omega_2$, then the matrix (k_{ij}^2) and the column vector (R_i^2) will have their nonzero components located in a different position.

An important observation is that $N_1^e|_{\overline{\partial\Omega_2}}$ and $N_3^e|_{\overline{\partial\Omega_2}}$ look like $N_2^{e_i}(t)$ and $N_1^{e_i}(t)$, respectively, when a counterclockwise integration is assumed. Note $N_2^e|_{\overline{\partial\Omega_2}} \equiv 0$. We shall simply write N_2 and N_1. Note h is now the length between the element nodes 1 and 3.

$$\begin{aligned}
\frac{\partial X_2(u^e)}{\partial u_1^e} &\cong \frac{\partial}{\partial u_1^e}\frac{1}{2}\int_{\overline{\partial\Omega_2}}\left(s u^{e^2} - 2g_2 u^e\right) \\
&= \int_{\overline{\partial\Omega_2}} s u^e \frac{\partial u^e}{\partial u_1^e} - g_2 \int_{\overline{\partial\Omega_2}} \frac{\partial u^e}{\partial u_1^e} \\
&= \int_{\overline{\partial\Omega_2}} s\left(\sum_{j=1}^{3} u_j^e N_j^e\right) N_1^e - g_2 \int_{\overline{\partial\Omega_2}} N_1^e \\
&= s\left(\int_{\overline{\partial\Omega_2}} N_1^e|_{\overline{\partial\Omega_2}} N_1^e|_{\overline{\partial\Omega_2}}\right) u_1^e + s\left(\int_{\overline{\partial\Omega_2}} N_2^e|_{\overline{\partial\Omega_2}} N_1^e|_{\overline{\partial\Omega_2}}\right) u_2^e \\
&\quad + s\left(\int_{\overline{\partial\Omega_2}} N_3^e|_{\overline{\partial\Omega_2}} N_{1\,\overline{\partial\Omega_2}}^e\right) u_3^e - g_2 \int_{\overline{\partial\Omega_2}} N_1^e|_{\overline{\partial\Omega_1}} \\
&= s\frac{h}{3} u_1^e + 0 u_2^e + s\frac{h}{6} u_3^e - g_2 \frac{h}{2}
\end{aligned} \tag{2.5.21}$$

where

$$h = \sqrt{(XX(3) - XX(1))^2 + (YY(3) - YY(1))^2}.$$

By computing $\partial X_2(u^e)/\partial u_2^e$ and $\partial X_2(u^e)/\partial u_3^e$, we obtain

$$k^2 = \begin{pmatrix} s\frac{h}{3} & 0 & s\frac{h}{6} \\ 0 & 0 & 0 \\ s\frac{h}{6} & 0 & s\frac{h}{3} \end{pmatrix}, \tag{2.5.22}$$

$$R^2 = \begin{pmatrix} g_2\frac{h}{2} \\ 0 \\ g_2\frac{h}{2} \end{pmatrix}, \tag{2.5.23}$$

when e is an element whose side 1-3 approximates $\partial\Omega_2$.

We summarize these results in Proposition 2.5.2.

Proposition 2.5.2. Consider the problem given by (2.5.1)–(2.5.4). Let e be an element as indicated in Figure 2.5.1. Then (k_{ij}^e) and (R_i^e) of (2.5.14) are given by (2.5.15) and (2.5.16), where k_{ij}^{Δ} is defined in (2.5.17), k_{ij}^C is defined in (2.5.20), (k_{ij}^2) is defined in (2.5.22), R_i^f is defined in (2.5.17), and R_i^2 is defined in (2.5.23).

Remarks

1. If f is replaced by f_I, the linear interpolation of f, then R_i^f must be replaced by R_i^e of lines (2.3.10)–(2.3.12).
2. The results of Proposition 2.5.2 may also be derived from the weak formulation, (2.5.8), and the Galerkin approach.

2.6 OBSERVATIONS AND REFERENCES

There are other ways to program the problems in this chapter. Some program listings are given in Bathe [2], Segerlind [23], and Norrie and deVires [13]. The programs in this chapter and the entire text are not intended to be optimal, but they are mainly for instructional purposes. Most large computing facilities have at least some reasonable finite element code. For routine applications of the finite element method, the reader probably would be wise to use the existing code.

Exercises 2-1, 2-7, 2-9, 2-10, 2-11, and 2-14 have been particularly helpful to students. Some exercises that are more computationally oriented include 2-18, 2-19, 2-20, and 2-21. Some theoretical problems are 2-3, 2-15, 2-16, and 2-17.

EXERCISES

2-1 Let Ω be given in the figure. Assign element numbers, element node numbers, and fill out, as in Table 2.1.1, the array NOD(e, I).

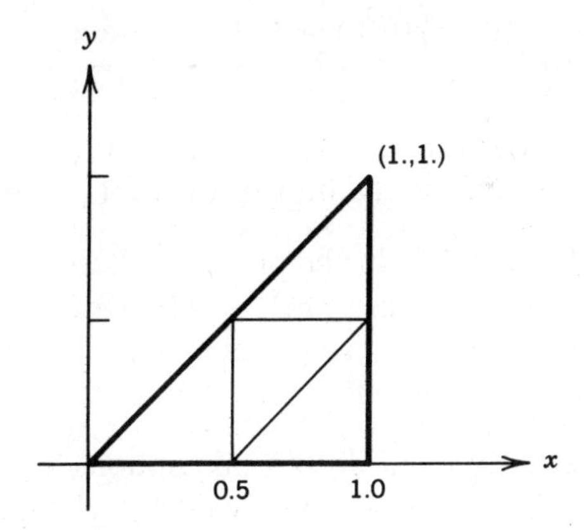

2-2 Consider the proof of Proposition 2.1.1. Do the computations for α_1, α_2, α_3. Derive the formulas for $N_2^e(x, y)$ and $N_3^e(x, y)$.

2-3 Derive the formula (2.3.9) for $a = b = 1$ and $c = 0$. Use the element e_0 given in the figure. Why does this computation hold for the general element e? (Use a change of variable due to translation and rotation.)

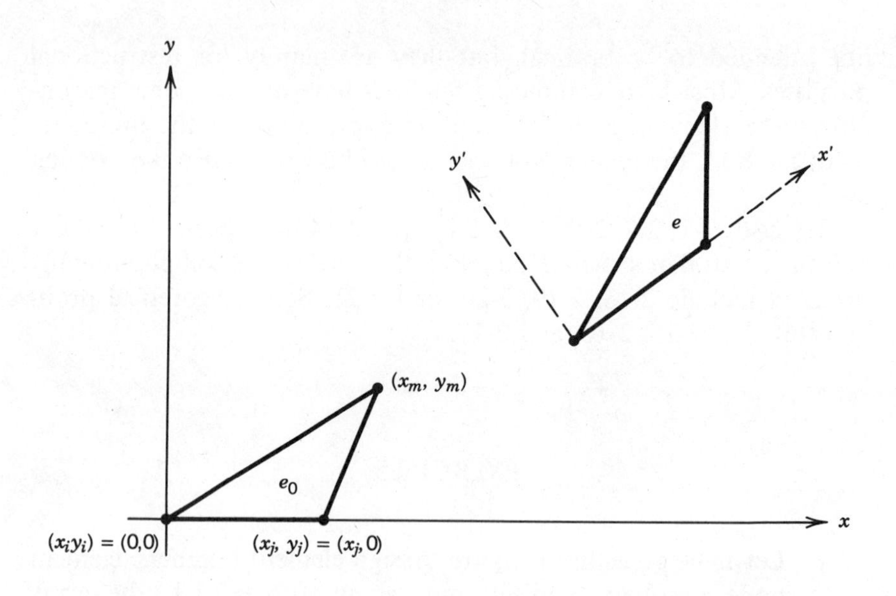

2-4 Let $f(x, y) = e^{xy}$ and the element $e = \{(x, y) | 0 \leq x \leq 1, 0 \leq y \leq x\}$. Let $N_1^e(0,0) = 1$. Compute and compare $\int\int_e f N_1^e$ and $\int\int_e f_I N_1^e$.

2-5 For the element e and $f(x, y)$ in 2-4, find (R_i^e) and $(\overline{R}_i^e)$ for $i = 1, 2, 3$. Assume $N_2^e(1,0) = 1$ and $N_3^e(1,1) = 1$.

2-6 Consider Ω as in 2-1. In the problem (2.2.1)–(2.2.3) let $\partial\Omega_1 = \{(1, y) | 0 \leq y \leq 1\}$ and $\partial\Omega_3 = \partial\Omega \setminus \partial\Omega_1$ and $g_1(1, y) = \cos(\pi y/2)$. Label the system nodes and use NOD(e, I) as in 2-1. Label the arrays NPT, $G1$, X, and Y.

2-7 Consider 2-6. By hand do the calculations in loops 60 and 70 in FEMI. Note that DELTA does not change for these elements. When would it change?

2-8 In 2-7 solve the algebraic problem. Compare this numerical solution with exact solution (consider 2-6 on $(-1, 1) \times (-1, 1)$!).

2-9 Print the value of the system matrix in FEMI. Do this at each step of the assembly by elements, that is, after loop 55 and in loop 60 for each ILT, and after loop 70.

2-10 Select an interesting problem that has the form (2.2.1)–(2.2.3) and use FEMI to approximate its solution.

2-11 Let Ω be as in 2-1. Suppose one wants to triangulate Ω in the same manner but with more elements. Write a subroutine to do this if just given the number of nodes in the x and y direction, NX and NY. Identify the element numbers, the components of the nodes, and the array NOD(e, I). You may want to modify the following subroutine that will generate NOD in FEMI when $NX = 3$, $NY = 5$.

```
      SUBROUTINE GENNOD (NX,NY)
      DIMENSION NOD(44,3)
      COMMON NOD
      L = -1
      JJ = -1
      K = NX-1
      NYM = NY-1
      NXM = NX-1
      DO 20 I = 1,NYM
          K = K+1
          L = L+1
          DO 10 J = 1,NXM
              K = K+1
              L = L+1
              JJ = JJ+2
              NOD(JJ,1) = K
              NOD(JJ,2) = L
              NOD(JJ,3) = K+1
              NOD(JJ+1,1) = L+1
              NOD(JJ+1,2) = K+1
              NOD(JJ+1,3) = L
10        CONTINUE
20    CONTINUE
      RETURN
      END
```

2-12 Do 2-11 for another Ω, say Ω from 2-10.

2-13 In 2-10 experiment with different elements and different numbers of elements. What happens to the algebraic solution as the "mesh" goes to zero?

2-14 Modify FEMI so that the three new complications in (2.5.1)–(2.5.4) are considered. In order to test your program, do this problem:

$$f(x, y) = 1.0, \qquad u = 50.0 \quad \text{for} \quad y = 0.0,$$

$$\frac{du}{dn} + u = 1.0 \quad \text{for} \quad y = 2.0, \qquad C = 1.0,$$

and $du/dn = 0.0$ for $x = 0.0$ and $x = 2.0$. The exact solution is $49e^{-y} + 1$, and the following numbers should appear:

50.00,	50.00,	50.00,	30.99,	30.57,	30.14
19.32,	18.85,	18.37,	12.23,	11.78,	11.33
8.02,	7.52,	7.05.			

Use the modified FEMI on a problem that is interesting to you.

2-15 Consider 2-14. Suppose $C = C(x, y)$ is not constant. Modify the program to account for this case. How do you modify the program if K, s, and g_2 depend on (x, y)?

2-16 Prove Proposition 2.5.1. Prove an analog of Theorem 1.4.1; assume $s \geq 0$.

2-17 Consider the steady-state clamped-plate problem

$$\Delta(K\Delta u) = f \qquad \text{on } \Omega,$$

$$u = 0 \quad \text{and} \quad \frac{du}{dn} = 0 \quad \text{on } \partial\Omega.$$

By using Green's theorem show that an "appropriate" definition of a weak solution is

$u \in C^1(\overline{\Omega})$ with piecewise continuous second derivatives is a *weak solution* if and only if

(i) $u = 0$ and $du/dn = 0$ on $\partial\Omega$

(ii) $\int_\Omega K\Delta u\Delta\Psi = \int_\Omega f\Psi$ for all

$$\Psi \in C^1(\bar{\Omega}) \quad \text{with} \quad \Psi = 0, \qquad \frac{d\Psi}{dn} = 0 \text{ on } \partial\Omega,$$

and piecewise-continuous second derivative. Show $X(u) \equiv \frac{1}{2}\int_\Omega(K(\Delta u)^2 - 2uf)$ is admissible. In order to use the Galerkin formulation of FEM, is it appropriate to use the linear shape functions N_i^e? Why?

2-18 Consider a problem in three dimensions with axial symmetry. Then $-\Delta u = f$ becomes $-(ru_r)_r - (ru_z)_z = rf$. Assume $u = g_1$ on $\partial\Omega_1$ and $du/dn = 0$ on $\partial\Omega \setminus \partial\Omega_1$. Find an admissible energy integral. For triangular elements and linear shape functions, find the element matrices and modify FEMI for this problem.

2-19 Consider the problem in 2-18. Modify it so that it has the form given in (2.5.1)–(2.5.4). Repeat the steps in 2-18 for this problem.

2-20 Use FEMI to analyze the following heat-conduction problem. Vary the number of nodes and compare the solutions. Indicate the isothermal curves. Note that the subroutine GENNOD in 2-11 may be used to generate NOD where the elements have the form as given below.

Use the modified FEMI in 2-14 to consider the same problem but with $u = 70$ replaced by $du/dn = s(70 - u)$.

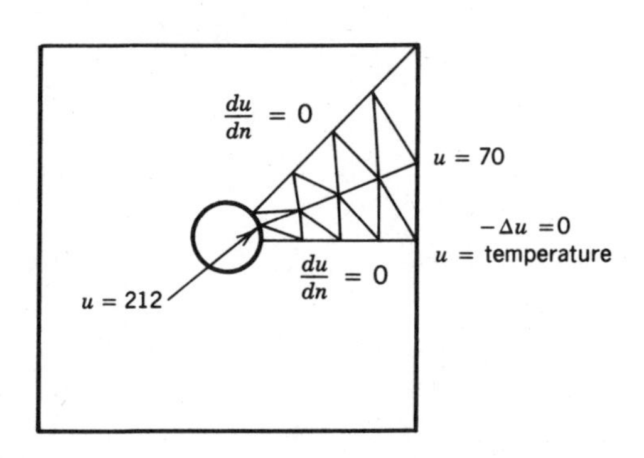

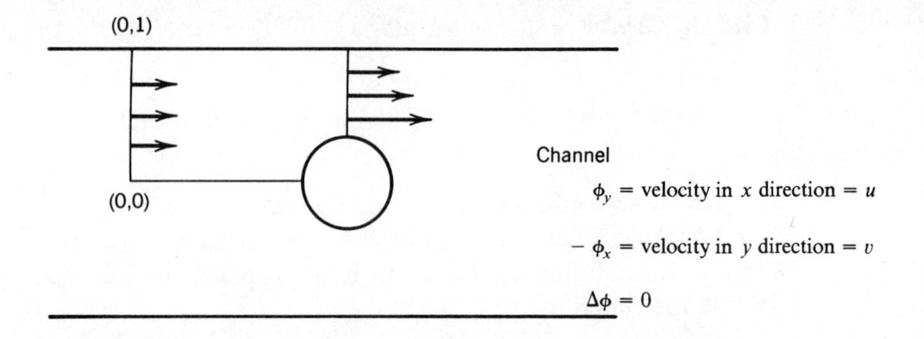

2-21 Use FEMI to analyze the ideal fluid flow in a channel about an obstacle. Indicate the stream lines.

Use the following boundary conditions:

$x = 0$: $(u, v) = (u_0, 0)$, $\phi(0, y) = u_0 y$,
$y = 1$: $\phi(x, 1) = u_0$,
lower boundary: $\phi = 0$,
right side boundary: $v = 0$, $-\phi_x = \dfrac{-d\phi}{dn} = 0$.

3

ASSEMBLY BY NODES AND A REDUCED SYSTEM MATRIX

The primary objectives in this chapter are to illustrate assembly by nodes and to indicate how a reduced system matrix can be used in an iterative scheme to solve the algebraic system. Section 3.1 contains some programming considerations, and in Section 3.2 we discuss some comparisons of assembly by nodes and assembly by elements. The last two sections contain a description of a reduced system matrix and a description of how it may be used in the Gauss–Seidel iterative algorithm. The reduced system matrix, which stores only the nonzero components of a system matrix, is useful because most of the algebraic problems from elliptic and parabolic partial differential equations are sparse and can be solved iteratively.

3.1 PROGRAMMING ASSEMBLY BY NODES

In the one-variable case we described the assembly-by-nodes method (see Section 1.5) by four steps. The first step is to list for each node all surrounding nodes. This can be done implicitly by listing for each node all surrounding elements. For example, in Figure 2.4.1 system node 1 has two surrounding elements, 1 and 2, and system node 8 has six surrounding elements, 5, 6, 7, 10, 11, 12. It is an important observation that the ith row of the system matrix will have nonzero components corresponding to only the element node numbers in the surrounding elements of the ith system node. The following array is introduced to record the surrounding elements.

Notation. NESUR(NODE, J)

NESUR(NODE, 1) = number of elements surrounding system node number NODE

NESUR(NODE, J) = elements surrounding NODE
$2 \leq J \leq \text{NESUR(NODE, 1)} + 1$

Example. Consider nodes NODE = 1 and 8 in Figure 2.4.1.

$$\text{NESUR}(1,1) = 2, \quad \text{NESUR}(1,2) = 1, \quad \text{NESUR}(1,3) = 2,$$
$$\text{NESUR}(8,1) = 6, \quad \text{NESUR}(8,3) = 6, \quad \text{NESUR}(8,7) = 12.$$

The second and third steps of Section 1.5 can be performed once one observes that the terms of $\partial H/\partial y_i = 0$ consist of components in the element matrices of the surrounding elements. This is also true for the two-variable case. The ith row, given by NODE, of the system matrix may be constructed by the segment *Program by Nodes*. We are assuming the node NODE does not have prescribed values and that the element matrices SME(ILT, IR, IC) have been computed where ILT = a surrounding element, IR, IC = element node numbers.

Program by Nodes

```
DO 40 NODE = 1,N
   NES = NESUR(NODE,1) + 1
   DO 30 J = 2,NES
      ILT = NESUR(NODE,J)
      DO 20 IR = 1,3
         IF (NOD(ILT,IR).NE.NODE) GO TO 20
         DO 10 IC = 1,3
            JJ = NOD(ILT,IC)
            SM(NODE,JJ) = SM(NODE,JJ) + SME(ILT,IR,IC)
```

```
10                CONTINUE
20             CONTINUE
30          CONTINUE
40       CONTINUE
```

Program by Nodes is done for each node NODE, which does not have prescribed values. The fourth step is to insert the prescribed boundary condition in each row that corresponds to a node with prescribed values. This last step is similar to the last step in assembly by elements.

3.2 COMPARISON OF ASSEMBLY BY NODES AND BY ELEMENTS

As previously noted assembly by nodes involves constructing the system matrix one row at a time. (There is a loop with respect to system nodes.) On the other hand, assembly by elements requires that each element matrix be "added" to the system matrix. (There is a loop with respect to the elements.)

If the resulting algebraic problem is to be solved directly, the system matrix, as a whole, must be available. If the algebraic problem is to be solved iteratively, then at each step one may need only a single row of the system matrix. Furthermore, one may only need the nonzero components of the particular row. This is clearly seen to be the case in the Gauss–Seidel algorithm with SOR as stated in lines (1.7.10) and (1.7.11). Consequently, for large systems, where one wants to solve the algebraic equations iteratively and to avoid storing the total system matrix, assembly by nodes has some advantages over assembly by elements.

3.3 A REDUCED SYSTEM MATRIX

In Figure 2.4.1 each node is surrounded by at most six other nodes. Therefore, each row of the augmented matrix SM has at most eight nonzero components. The reduced augmented matrix must be an $N \times 8$ matrix, where N is the number of nodes. The following notation will provide a mechanism for going to and from the reduced augmented matrix, RSM, and the augmented matrix, SM.

Notation. RSM, NOSUR, and INOSUR. Let RSM be an $N \times 8$ matrix called the reduced system matrix. The first column of RSM will contain the $N + 1$ column of the $N \times (N + 1)$ augmented system matrix, SM. The other columns will contain the nonzero components of the $N \times N$ part of SM. In order to identify components of RSM with the components of SM and vice versa, we need the arrays NOSUR and its "inverse" INOSUR.

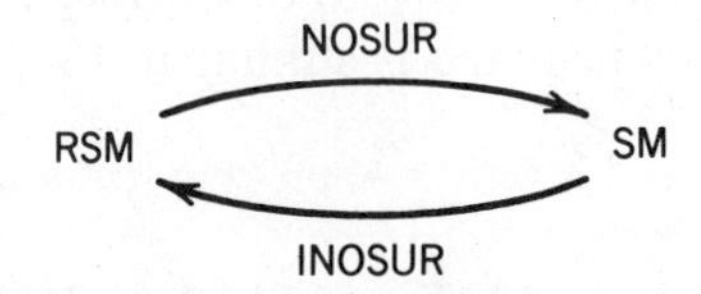

Let NOSUR be an $N \times 8$ matrix with

NOSUR(NODE, 1) = the number of surrounding nodes (includes NODE)

NOSUR(NODE, J) = the system node numbers surrounding node, NODE, listed in increasing order

$$2 \le J \le \text{NOSUR(NODE, 1)} + 1$$

Let INOSUR be an $N \times 1$ matrix defined as an inverse of NOSUR, that is, INOSUR(NOSUR(NODE, J)) = J.

Example. Consider Figure 2.4.1 with NODE = 8, $N = 9$, NOSUR(8, 1) = 7, and

NOSUR(8, 2) = 4,	INOSUR(4) = 2,
NOSUR(8, 3) = 5,	INOSUR(5) = 3,
NOSUR(8, 4) = 7,	INOSUR(7) = 4,
NOSUR(8, 5) = 8,	INOSUR(8) = 5,
NOSUR(8, 6) = 9,	INOSUR(9) = 6,
NOSUR(8, 7) = 11,	INOSUR(11) = 7,
NOSUR(8, 8) = 12,	INOSUR(12) = 8.

The matrix RSM requires only $N \times 8$ storage locations, whereas the matrix SM requires $N \times (N + 1)$. For large N, RSM is much smaller than SM. The segment *Program by Nodes* in Section 3.2 can be altered.

Program by Nodes — Reduced

```
      DO 40 NODE=1,N
         NOS=NOSUR(NODE,1)+1
         DO 5 J=2,NOS
            INOSUR(NOSUR(NODE,J))=J
5        CONTINUE
         NES=NESUR(NODE,1)+1
         DO 30 J=2,NES
            ILT=NESUR(NODE,J)
            DO 20 IR=1,3
               IF (NOD(ILT,IR).NE.NODE) GO TO 20
               DO 10 IC=1,3
                  JJ=INOSUR(NOD(ILT,IC))
                  RSM(NODE,JJ)=RSM(NODE,JJ)+SME(ILT,IR,IC)
10             CONTINUE
20          CONTINUE
30       CONTINUE
40    CONTINUE
```

Remarks

1. The only differences between the two versions are the definitions of JJ and RSM.
2. In loop 5 INOSUR is calculated for each NODE. At most only seven components of INOSUR are needed.
3. $2 \leq \text{JJ} \leq 8$; therefore, the first column of RSM is free to store SM(NODE, $N + 1$).

3.4 SOLUTION OF THE ALGEBRAIC PROBLEM BY ITERATION

In this section we examine the Gauss–Seidel algorithm with SOR, (1.7.10) and (1.7.11), when RSM is used, i = NODE is a fixed row of RSM, and R_i = RSM(NODE, 1). We shall assume RSM has its

i = NODE row computed as in either *Program by Nodes—Reduced* or *Program by Elements—Reduced* (see exercise 3-5). Let U(NOSUR(NODE, J)) represent u_j in lines (1.7.10) and (1.7.11). The segment *Program Gauss–Seidel*—SOR is algorithm (1.7.10), (1.7.11) with RSM.

Program Gauss – Seidel — SOR

```
         W=1.6
         DO 40 NODE=1,N
            NOS=NOSUR(NODE,1)+1
            DO 10 J=2,NOS
               INOSUR(NOSUR(NODE,J))=J
10          CONTINUE
            JNOS=INOSUR(NODE)
            DEN=RSM(NODE,JNOS)
            UU=RSM(NODE,1)/DEN
            DO 20 J=2,JNOS-1
               UU=UU-RSM(NODE,J)*U(NOSUR(NODE,J))/DEN
20          CONTINUE
            DO 30 J=JNOS+1,NOS
               UU=UU-RSM(NODE,J)*U(NOSUR(NODE,J))/DEN
30          CONTINUE
            U(NODE)=(1-W)*U(NODE)+W*UU
40       CONTINUE
```

Example. Consider Figure 2.4.1 with NODE = 8 and NOSUR and INOSUR as listed in Section 3.2. Then in the preceding program segment the following computation in terms of the system matrix SM = (a_{ij}) is

$$u_8^{k+1/2} = R_8/a_{8,8} - \left(a_{8,4}u_4^{k+1} + a_{8,5}u_5^{k+1} + a_{8,7}u_7^{k+1}\right)/a_{8,8}$$
$$- \left(a_{8,9}u_9^k + a_{8,11}u_{11}^k + a_{8,12}u_{12}^k\right)/a_{8,8},$$
$$u_8^{k+1} = (1-w)u_8^k + wu_8^{k+1/2}$$

Remark. In *Program Gauss–Seidel–SOR* it is not necessary that RSM be assembled by nodes. In fact, in *Program by Elements* or in

FEMI it is easy to replace SM by RSM and to use NOSUR and INOSUR to locate the components (see exercise 3-5).

3.5 OBSERVATIONS AND REFERENCES

The use of reduced system matrices is important in problems where there are a large number of unknowns. Also, some problems that are nonlinear and can be solved by the nonlinear Gauss–Seidel algorithm are particularly suited for the use of a reduced system matrix (see Chapter 8). In Appendixes A.3 and A.4 this is illustrated for the Stefan problem and the fluid flow in a porous medium problem, respectively. In Appendix A.3 a reduced system matrix of the form RSM of Section 3.3 is used. In Appendix A.4, where a rectangular grid is used, a rectangular reduced system matrix of the form RRSM(i, j, k) is used. $N = n \cdot m$, where $i = 1, \ldots, n$, $j = 1, \ldots, m$, and $k = 1, \ldots, 5$ (see exercise 3-9).

Exercises 3-1, 3-2, and 3-5 have proved to be instructive.

EXERCISES

3-1 Consider the example given by Figure 2.4.1. For NODE = 4, list NESUR. By hand do *Program by Nodes* for NODE = 4. You may wish to compute the element matrices, SME, by printing them out when FEMI is used. Also, print the system matrix, SM, from FEMI and compare its fourth row with the above computations.

3-2 Consider the example given by Figure 2.4.1. For NODE = 4, list NOSUR and INOSUR. By hand do *Program by Nodes—Reduced* for NODE = 4.

3-3 Consider the example given by Figure 2.4.1. For NODE = 4 state the line given for the Gauss–Seidel algorithm in *Program Gauss–Seidel—SOR*.

3-4 In FEMI replace the subroutine IMAT with a subroutine using the Gauss–Seidel algorithm. Use the same system matrix, SM.

3-5 In FEMI replace the system matrix by a reduced system matrix, RSM. Also, replace the subroutine IMAT with a subroutine using the Gauss–Seidel algorithm and RSM. Continue to use assembly by elements.

3-6 Repeat 3-5 using assembly by nodes.

3-7 In a problem that is of interest to you, use FEMI (or a modification of FEMI) with both IMAT and the subroutine from exercise 3-5. Compare the computing times for different $1 \leq w < 2$. For a "larger" number of unknown nodes and for a good choice of $1 \leq w < 2$, the subroutine from exercise 3-5 should be faster than IMAT.

3-8 The matrix SM will be banded. The band width will depend on the choice of node numbering. Use a packaged routine from your computing facility for solving banded linear systems. Compare the computing time with those in exercise 3-7. Also, use the banded nature of SM to shorten the subroutine in exercise 3-4 and compare the computing times.

3-9 Consider exercises 2-18 and 2-19. Assume the grid is given by rectangles and each rectangle is divided into two triangular elements. In this case the system matrix may be written $\text{SM}(i, j; I, J)$, where i is the row number of the grid and j is the column number of the grid. Compute SM for $i = I$, $j = J$ and for $i = I \pm 1$, $j = J \pm 1$. Show these are the only nonzero components of SM. Use this to give a rectangular reduced system matrix of the form $\text{RRSM}(i, j, k)$, $k = 1, \ldots, 5$.

4

SHAPE FUNCTIONS

In this chapter we consider other shape functions and different shaped elements. Section 4.1 contains the generalization of linear shape functions to three dimensions. In Sections 4.2 and 4.3 we discuss quadratic shape functions, which were mentioned in Section 1.7. Bilinear shape functions on rectangular elements are studied in Section 4.4. The complete cubic shape functions are briefly examined in Section 4.5. Shape functions that are not linear may give more accurate numerical solutions, or their elements may better fit the space variable domain. Also, some problems such as fluid flow problems in Chapter 8, require mixed shape functions for the stability of the numerical scheme.

4.1 LINEAR SHAPE FUNCTIONS ON TETRAHEDRAL ELEMENTS

The construction of a linear shape function for variables in three dimensions follows in the same way as in one and two dimensions. In the three-dimensional case the element is a tetrahedron, as indicated in Figure 4.1.1. As there are four surrounding nodes, any element matrix will be a 4×4 matrix.

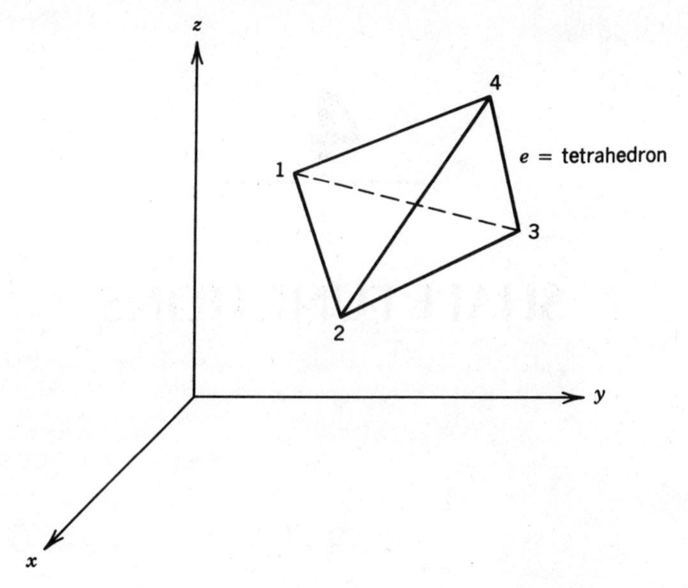

Figure 4.1.1

Any linear shape function u^e defined on e must have the form

$$u^e(x, y, z) = \alpha_1 + \alpha_2 x + \alpha_3 y + \alpha_4 z. \tag{4.1.1}$$

Let (x_i, y_i, z_i), $i = 1, 2, 3, 4$, be the (x, y, z) coordinates of the element nodes $1, 2, 3, 4$, and let u_i^e, $i = 1, 2, 3, 4$, be $u^e(x_i, y_i, z_i)$. Then (4.1.1) when $(x, y, z) = (x_i, y_i, z_i)$ gives four equations that may be written in matrix form:

$$\mathbf{u}^e = A\boldsymbol{\alpha}, \tag{4.1.2}$$

where $\mathbf{u}^e = (u_1^e, u_2^e, u_3^e, u_4^e)^{\mathrm{T}}$, $\boldsymbol{\alpha} = (\alpha_1, \alpha_2, \alpha_3, \alpha_4)^{\mathrm{T}}$, and

$$A \equiv \begin{pmatrix} 1 & x_1 & y_1 & z_1 \\ 1 & x_2 & y_2 & z_2 \\ 1 & x_3 & y_3 & z_3 \\ 1 & x_4 & y_4 & z_4 \end{pmatrix}.$$

We can show that $\det A = 6V \neq 0$, where V is the volume of e.

Therefore, line (4.1.2) implies

$$\boldsymbol{\alpha} = A^{-1}\mathbf{u}^e \tag{4.1.3}$$

By placing α_i from (4.1.3) into (4.1.1) we may write

$$u^e(x, y, z) = \sum_{i=1}^{4} u_i^e N_i^e(x, y, z) \tag{4.1.4}$$

where each $N_i^e(x, y, z)$ is a linear function that is 1 at (x_i, y_i, z_i) and 0 at (x_j, y_j, z_j), $j \neq i$. In fact, the following formula may be derived by simply writing A^{-1} explicitly and doing the substitution described above. The proofs of formulas (4.1.5) and (4.1.6) are analogous to the two-dimensional cases in lines (2.1.2) and (2.3.9), respectively.

Proposition 4.1.1

$$N_i^e(x, y, z) = (a_i + b_i x + c_i y + d_i z)/6V. \tag{4.1.5}$$

The a_i, b_i, c_i, d_i and V are defined by cyclic rotation:

$$a_1 = \det\begin{pmatrix} x_2 & y_2 & z_2 \\ x_3 & y_3 & z_3 \\ x_4 & y_4 & z_4 \end{pmatrix}, \qquad b_1 = -\det\begin{pmatrix} 1 & y_2 & z_2 \\ 1 & y_3 & z_3 \\ 1 & y_4 & z_4 \end{pmatrix},$$

$$c_1 = -\det\begin{pmatrix} x_2 & 1 & z_2 \\ x_3 & 1 & z_3 \\ x_4 & 1 & z_4 \end{pmatrix}, \qquad d_1 = -\det\begin{pmatrix} x_2 & y_2 & 1 \\ x_3 & y_3 & 1 \\ x_4 & y_4 & 1 \end{pmatrix},$$

$$V = \tfrac{1}{6}\det A.$$

The following formula is useful.

Integration Formulas. Let e be a tetrahedral element and m, n, l, and k be integers.

$$\iiint_e N_1^{e^m} N_2^{e^n} N_3^{e^l} N_4^{e^k} = \frac{m!n!l!k!}{(m + n + l + k + 3)!}6V. \tag{4.1.6}$$

Consider constructing the element matrix for the problem

$$-\Delta u = f \quad \text{on } \Omega \subset \mathbb{R}^3, \tag{4.1.7}$$

$$u = 0 \quad \text{on } \partial\Omega. \tag{4.1.8}$$

Then an admissible energy integral is

$$X(u) = \frac{1}{2} \iiint_{\Omega} \left(u_x^2 + u_y^2 + u_z^2 - 2uf\right). \tag{4.1.9}$$

In order to find the element matrix for e, we let u^e be from (4.1.4) and compute for $i = 1, 2, 3, 4$

$$\frac{\partial X(u^e)}{\partial u_i^e}. \tag{4.1.10}$$

As in lines (2.3.4)–(2.3.6), it is easy to show the expression in (4.1.10) is for each $i = 1, 2, 3, 4$ and each element

$$\sum_{j=1}^{4} u_j^e \Big(\iiint \left(N_{ix}^e N_{jx}^e + N_{iy}^e N_{jy}^e + N_{iz}^e N_{jz}^e\right)\Big) - \iiint f N_i^e. \tag{4.1.11}$$

So the element matrix $k^e = (k_{ij}^e)$ and the element portion of the right-hand side $R^e = (R_i^e)$ are defined by

$$k_{ij}^e = \iiint \left(N_{ix}^e N_{jx}^e + N_{iy}^e N_{jy}^e + N_{iz}^e N_{jz}^e\right),$$

$$R_i^e = \iiint f N_i^e.$$

By using the integral formula (4.1.6) we may deduce from (4.1.11) the following proposition.

Proposition 4.1.2. The 4×4 element matrix for (4.1.7)–(4.1.8) is $k^e = (k_{ij})$, where

$$k_{ij} = \frac{b_i b_j + c_i c_j + d_i d_j}{36V}. \tag{4.1.12}$$

Remark. The components R_i^e of the right-hand side of the algebraic problem may be approximated by replacing f by its linear interpolation on e,

$$f_I \equiv \sum_{j=1}^{4} f_j^e N_j^e, \qquad f_j^e \equiv f(x_j, y_j, z_j).$$

4.2 QUADRATIC SHAPE FUNCTIONS ON INTERVAL ELEMENTS

With the hope of obtaining more accuracy, we try to approximate a function by a quadratic shape function

$$y^e(t) = \bar{\alpha}_1 + \bar{\alpha}_2 t + \bar{\alpha}_3 t^2. \tag{4.2.1}$$

In order to find $\bar{\alpha}_i$, $i = 1, 2, 3$, we must have three values of $y^e(t)$ given. Therefore, we introduce half nodes $t_{i+1/2}$ on the element interval $[t_i, t_{i+1}]$.

For ease of calculation, we change over to a general coordinate system,

$$\xi = t - t_i. \tag{4.2.2}$$

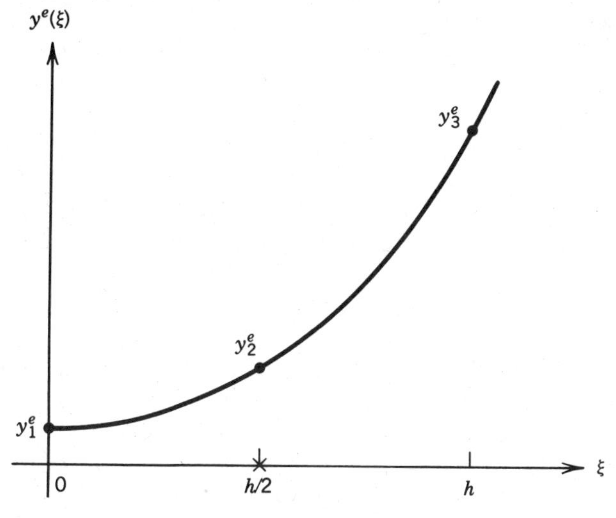

Figure 4.2.1

By the change of variable theorem in calculus,

$$\int_{t_i}^{t_{i+1}} H(t)\,dt = \int_0^h \overline{H}(\xi)\,d\xi, \qquad \overline{H}(\xi) \equiv H(\xi + t_i) \quad (4.2.3)$$

If $H(t) = \frac{1}{2}[(\dot{y}(t))^2 + (y(t))^2] - 32y(t)$, as in the integrand of the energy integral in (4.2.11) for the problem (4.2.8)–(4.2.10), then $X(y^e(t)) = X(y^{\bar{e}}(\xi))$, where $e = [t_i, t_{i+1}]$, $\bar{e} = [0, h]$, $h = t_{i+1} - t_i$, and $y^{\bar{e}}(\xi) = \alpha_1 + \alpha_2\xi + \alpha_3\xi^2$. Consequently, the element matrix derived from the general coordinate system will equal the element matrix.

Figure 4.2.1 illustrates the half node, $\xi = h/2$, the quadratic shape function, and the general coordinate system. If there are n element intervals, then there are $2n + 1$ nodes. As there are three nodes for each element, the element matrices must be 3×3. The derivation of the element matrices is a little different from that for the linear shape functions.

Let $m_1 = 0$, $m_2 = 1$, and $m_3 = 2$, and write (4.2.1) in general coordinate system as

$$y^e(\xi) = \sum_{j=1}^{3} \alpha_j \xi^{m_j}. \quad (4.2.4)$$

Let $y_1^e = y^e(0)$, $y_2^e = y^e(h/2)$, and $y_3^e = y^e(h)$. Then by computing (4.2.4) at $\xi = 0$, $h/2$, and h, we have three equations that may be written in matrix form

$$\mathbf{y}^e = A\boldsymbol{\alpha} \quad (4.2.5)$$

where $\mathbf{y}^e = (y_1^e, y_2^e, y_3^e)^{\mathrm{T}}$, $\boldsymbol{\alpha} = (\alpha_1, \alpha_2, \alpha_3)^{\mathrm{T}}$, and

$$A = \begin{pmatrix} 1 & 0 & 0 \\ 1 & h/2 & h^2/4 \\ 1 & h & h^2 \end{pmatrix}.$$

A^{-1} exists and is easily computed as

$$A^{-1} \equiv B = \begin{pmatrix} 1 & 0 & 0 \\ -3/h & 4/h & -1/h \\ 2/h^2 & -4/h^2 & 2/h^2 \end{pmatrix}. \quad (4.2.6)$$

Consequently,

$$\boldsymbol{\alpha} = A^{-1}\mathbf{y}^e = B\mathbf{y}^e \quad (4.2.7)$$

For a sample problem, let us consider the model problem of Chapter 1:

$$-\ddot{y} + y = 32, \tag{4.2.8}$$

$$y(0) = 0, \tag{4.2.9}$$

$$y(2) = 4. \tag{4.2.10}$$

An admissible energy integral for (4.2.8)–(4.2.10) is

$$X(y) = \frac{1}{2}\int_0^2 (\dot{y}^2 + y^2) - \int_0^2 32y. \tag{4.2.11}$$

By substituting (4.2.4) for y in (4.2.11) and computing

$$\frac{\partial X(y^e)}{\partial y_i^e}, \qquad i = 1, 2, 3, \tag{4.2.12}$$

we can find the 3×3 element matrix.

$$\begin{aligned}
X(y^e) &= \frac{1}{2}\int_0^h \left[\left(\sum_{i=1}^{3} \alpha_i m_i \xi^{m_i - 1}\right)^2 + \left(\sum_{i=1}^{3} \alpha_i \xi^{m_i}\right)^2\right] d\xi \\
&\quad - \int_0^h 32 \sum_{i=1}^{3} \alpha_i \xi^{mi}\, d\xi \\
&= \frac{1}{2}\int_0^h \left[\sum_{i=1}^{3}\sum_{j=1}^{3} \left(\alpha_i m_i \xi^{m_i - 1}\right)\left(\alpha_j m_j \xi^{m_j - 1}\right)\right. \\
&\qquad \left. + \sum_{i=1}^{3}\sum_{j=1}^{3} \left(\alpha_i \xi^{m_i}\right)\left(\alpha_j \xi^{m_j}\right)\right] d\xi \\
&\quad - \int_0^h 32 \sum_{i=1}^{3} \alpha_i \xi^{m_i}\, d\xi \\
&= \frac{1}{2}\sum_{i=1}^{3}\sum_{j=1}^{3} \alpha_i \left(\int_0^h m_i m_j \xi^{m_i + m_j - 2} + \xi^{m_i + m_j}\, d\xi\right)\alpha_j \\
&\quad - \sum_{i=1}^{3} \alpha_i \int_0^h 32 \xi^{m_i}\, d\xi \\
&= \tfrac{1}{2}\boldsymbol{\alpha}^{\mathrm{T}} G \boldsymbol{\alpha} - \boldsymbol{\alpha}^{\mathrm{T}} \mathbf{F}
\end{aligned} \tag{4.2.13}$$

where

$$G = (g_{ij}), \qquad g_{ij} \equiv \int_0^h m_i m_j \xi^{m_i+m_j-2} + \xi^{m_i+m_j}\, d\xi,$$

$$\mathbf{F} = (F_i), \qquad F_i \equiv \int_0^h 32\xi^{m_i}\, d\xi.$$

Lines (4.2.7) and (4.2.13) imply

$$X(y^e) = \tfrac{1}{2}\mathbf{y}^{e\mathrm{T}}B^{\mathrm{T}}GB\mathbf{y}^e - \mathbf{y}^{e\mathrm{T}}B^{\mathrm{T}}\mathbf{F}. \tag{4.2.14}$$

Recall or note

$$\frac{\partial}{\partial \mathbf{y}}\frac{1}{2}\mathbf{y}^{\mathrm{T}}C\mathbf{y} = C\mathbf{y} \tag{4.2.15}$$

where $\mathbf{y} \in \mathbb{R}^N$, C is an $N \times N$ matrix, C is symmetric (i.e., $c_{ij} = c_{ji}$), and

$$\frac{\partial}{\partial \mathbf{y}} \equiv \left(\frac{\partial}{\partial y_1}, \ldots, \frac{\partial}{\partial y_N}\right)^{\mathrm{T}}.$$

Since k^e is given by $\partial X/\partial \mathbf{y}^e$ in (4.2.14) and $C = B^{\mathrm{T}}GB$ is symmetric, line (4.2.15) implies

$$k^e = B^{\mathrm{T}}GB. \tag{4.2.16}$$

Also, R^e is given by $B^{\mathrm{T}}\mathbf{F}$. More precisely, k^e and R^e are explicitly stated in the next proposition.

Proposition 4.2.1. Consider problem (4.2.8)–(4.2.10). The element matrix is

$$k^e = B^{\mathrm{T}}GB$$

$$= \frac{1}{30h}\begin{pmatrix} 4h^2+70 & 2h^2-80 & -h^2+10 \\ 2h^2-80 & 16h^2+160 & 2h^2-80 \\ -h^2+10 & 2h^2-80 & 4h^2+70 \end{pmatrix}, \tag{4.2.17}$$

$$R^e = B^{\mathrm{T}}\mathbf{F} = \frac{h}{3}(16, 64, 16)^{\mathrm{T}}. \tag{4.2.18}$$

Proof. In order to compute B^TGB, we must first find G explicitly from (4.2.13) and then use B from (4.2.6). We claim

$$G = \frac{h}{60}\begin{pmatrix} 60 & 30h & 20h^2 \\ 30h & 20h^2 + 60 & 15h^3 + 60h \\ 20h^2 & 15h^3 + 60h & 12h^4 + 80h^2 \end{pmatrix}. \quad (4.2.19)$$

We content ourselves to compute a few of the components in (4.2.19):

$$g_{11} = \int_0^h m_1 m_1 \xi^{m_1+m_1-2} + \xi^{m_1+m_1}\, d\xi$$

$$= \int_0^h 0 + 1\, d\xi = h,$$

$$g_{12} = \int_0^h m_1 m_2 \xi^{m_1+m_2-2} + \xi^{m_1+m_2}\, d\xi$$

$$= \int_0^h 0 + \xi\, d\xi = \frac{h^2}{2},$$

$$g_{23} = \int_0^h m_2 m_3 \xi^{m_2+m_3-2} + \xi^{m_2+m_3}\, d\xi$$

$$= \int_0^h 1 \cdot 2\xi^1 + \xi^{1+2}\, d\xi = h^2 + \frac{h^4}{4}.$$

The reader should verify the remainder of (4.2.19) and then compute B^TGB to obtain (4.2.17). Also, the computation of (4.2.18) is straightforward.

Example. Consider problem (4.2.8)–(4.2.10) with two elements, and use quadratic shape functions. Suppose each element has equal length, $h = 1$, and hence, by Proposition 4.2.1, both element matrices are equal and (4.2.17) gives

$$k^e = \frac{1}{30}\begin{pmatrix} 74 & -78 & 9 \\ -78 & 176 & -78 \\ 9 & -78 & 74 \end{pmatrix}.$$

Line (4.2.18) gives

$$R^e = \left(\tfrac{16}{3}, \tfrac{64}{3}, \tfrac{16}{3}\right)^T.$$

TABLE 4.2.1

t	EXACT	FDM	FEM (linear)	FEM (quadratic)
.5	9.1903	9.0703	9.3171	9.1972
1.0	12.5584	12.4082	12.7169	12.5476
1.5	10.9639	10.8481	11.0841	10.9699

By using assembly by elements and inserting the boundary conditions we obtain the algebraic problem

$$\begin{pmatrix} 1 & 0 & 0 & 0 & 0 \\ -\frac{78}{30} & \frac{176}{30} & -\frac{78}{30} & 0 & 0 \\ \frac{9}{30} & -\frac{78}{30} & 2\left(\frac{74}{30}\right) & -\frac{78}{30} & \frac{9}{30} \\ 0 & 0 & -\frac{78}{30} & \frac{176}{30} & -\frac{78}{30} \\ 0 & 0 & 0 & 0 & 1 \end{pmatrix} \begin{pmatrix} y_0 \\ y_1 \\ y_2 \\ y_3 \\ y_4 \end{pmatrix} = \begin{pmatrix} 0 \\ \frac{64}{3} \\ 2\left(\frac{16}{3}\right) \\ \frac{64}{3} \\ 4 \end{pmatrix}. \tag{4.2.20}$$

The solution of (4.2.20) is given under the column FEM (quadratic) in Table 4.2.1. The other three columns were given in Table 1.2.1. FEM (linear) refers to FEM with four equal elements and linear shape functions. The reader should find it interesting to compare these computations for this particular problem.

Remark. If the differential equation (4.2.8) was changed, then the matrix G defined by (4.2.13), or $\mathbf{F}$, would change and, hence, k^e or R^e would change.

4.3 QUADRATIC SHAPE FUNCTIONS ON TRIANGULAR ELEMENTS

A quadratic shape function in two variables has the form

$$u^e(x, y) = \bar{\alpha}_1 + \bar{\alpha}_2 x + \bar{\alpha}_3 y + \bar{\alpha}_4 x^2 + \bar{\alpha}_5 xy + \bar{\alpha}_6 y^2. \tag{4.3.1}$$

Since six $\bar{\alpha}_i$, $i = 1, \ldots, 6$, must be determined, six values of $u^e(x, y)$

must be given. When triangular elements are considered, this can be done by placing half nodes on each side of the triangle. Thus, for each element, there will be six element node numbers and, consequently, the element matrix will be a 6×6 matrix.

In order to minimize computations, we shall work in a general coordinate system (ξ, η). (ξ, η) is formed by a translation and rotation with $(\bar{x}, \bar{y})$ equal to the (x, y) components of the new origin, $(\xi, \eta) = (0, 0)$. Figure 4.3.1 illustrates the general coordinates.

$$\begin{pmatrix} x - \bar{x} \\ y - \bar{y} \end{pmatrix} = \begin{pmatrix} \cos\theta & -\sin\theta \\ \sin\theta & \cos\theta \end{pmatrix} \begin{pmatrix} \xi \\ \eta \end{pmatrix}. \tag{4.3.2}$$

The constants a, b, and c may be computed by using Figures 4.3.2, 4.3.3, and 4.3.4:

$$r = \left((x_3 - x_1)^2 + (y_3 - y_1)^2\right)^{1/2}, \tag{4.3.3}$$

$$\cos\theta = (x_3 - x_1)/r,$$

$$\sin\theta = (y_3 - y_1)/r,$$

$$\begin{aligned} a &= (x_3 - x_5)\cos\theta - (y_5 - y_3)\sin\theta \\ &= [(x_3 - x_5)(x_3 - x_1) - (y_5 - y_3)(y_3 - y_1)]/r, \end{aligned} \tag{4.3.4}$$

$$\begin{aligned} b &= (x_5 - x_1)\cos\theta + (y_5 - y_1)\sin\theta \\ &= [(x_5 - x_1)(x_3 - x_1) + (y_5 - y_1)(y_3 - y_1)]/r, \end{aligned} \tag{4.3.5}$$

$$\begin{aligned} c &= (y_5 - y_3)\cos\theta + (x_3 - x_5)\sin\theta \\ &= [(y_5 - y_3)(x_3 - x_1) + (x_3 - x_5)(y_3 - y_1)]/r. \end{aligned} \tag{4.3.6}$$

$u^e(x, y)$ may be written in terms of (ξ, η)

$$u^e(\xi, \eta) = \sum_{i=1}^{6} \alpha_i \xi^{m_i} \eta^{n_i}, \tag{4.3.7}$$

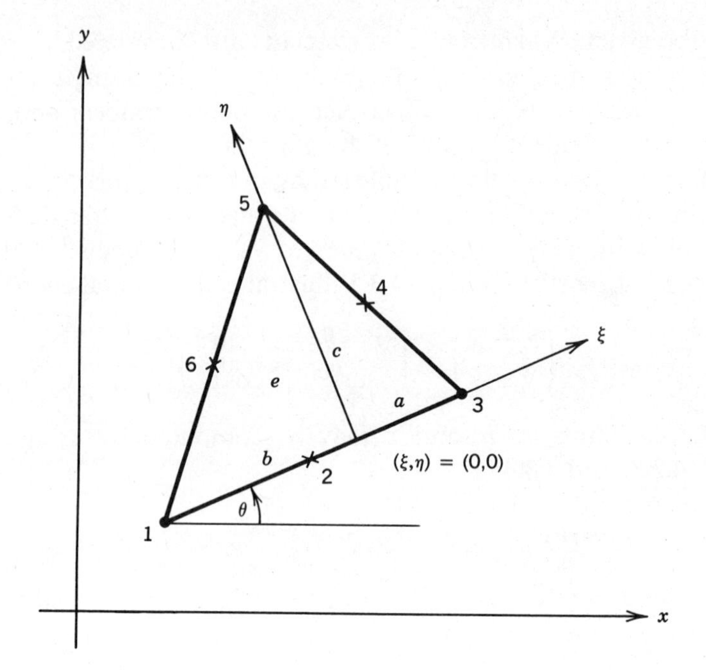

Figure 4.3.1

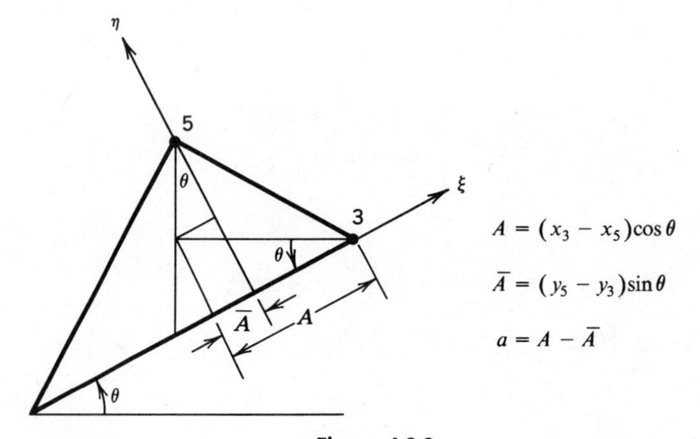

Figure 4.3.2

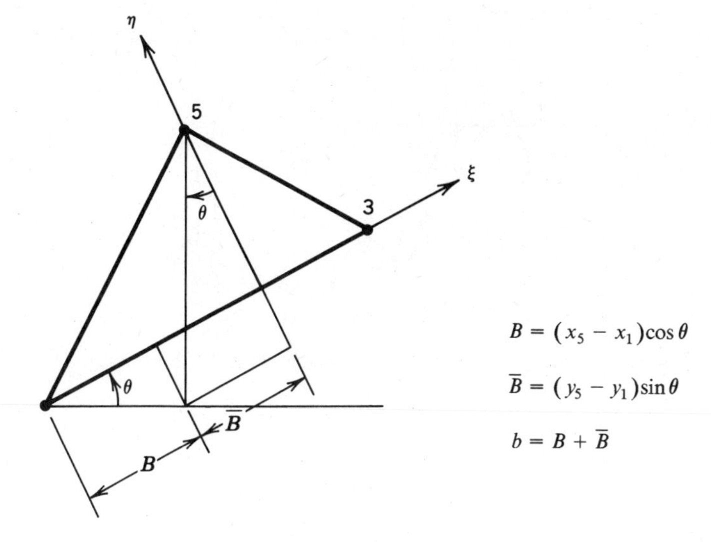

Figure 4.3.3

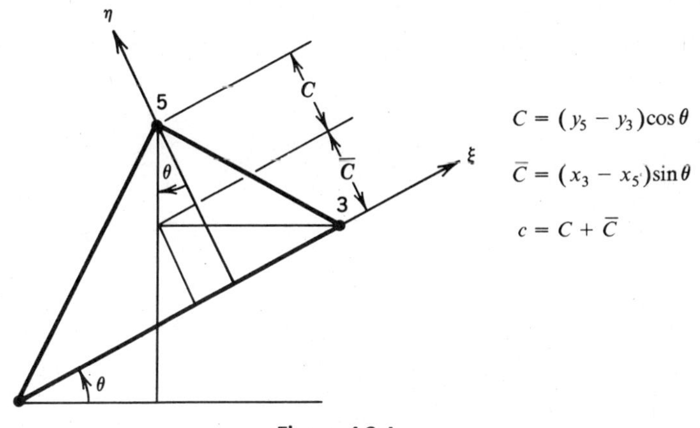

Figure 4.3.4

where m_i and n_i are defined in Table 4.3.1. Also in Table 4.3.1 are the (ξ, η) coordinates of the element nodes $1, \ldots, 6$. The six equations needed to compute α_i, $i = 1, \ldots, 6$ are found by evaluating (4.3.7) at these six element nodes:

$$A\boldsymbol{\alpha} = \mathbf{u}^e, \tag{4.3.8}$$

TABLE 4.3.1

i	m_i	n_i	(ξ, η)
1	0	0	$(-b, 0)$
2	1	0	$\left(\frac{a-b}{2}, 0\right)$
3	0	1	$(a, 0)$
4	2	0	$\left(\frac{a}{2}, \frac{c}{2}\right)$
5	1	1	$(0, c)$
6	0	2	$\left(\frac{-b}{2}, \frac{c}{2}\right)$

where $\boldsymbol{\alpha} = (\alpha_1, \ldots, \alpha_6)^{\mathrm{T}}$, $\mathbf{u}^e = (u_1^e, \ldots, u_6^e)^{\mathrm{T}}$, and

$$A \equiv \begin{pmatrix} 1 & -b & 0 & (-b)^2 & 0 & 0 \\ 1 & \dfrac{a-b}{2} & 0 & \left(\dfrac{a-b}{2}\right)^2 & 0 & 0 \\ 1 & a & 0 & a^2 & 0 & 0 \\ 1 & \dfrac{a}{2} & \dfrac{c}{2} & \left(\dfrac{a}{2}\right)^2 & \dfrac{ac}{4} & \left(\dfrac{c}{2}\right)^2 \\ 1 & 0 & c & 0 & 0 & c^2 \\ 1 & \dfrac{-b}{2} & \dfrac{c}{2} & \left(-\dfrac{b}{2}\right)^2 & \dfrac{-bc}{4} & \left(\dfrac{c}{2}\right)^2 \end{pmatrix}.$$

Consequently,

$$\boldsymbol{\alpha} = A^{-1}\mathbf{u}^e = B\mathbf{u}^e. \tag{4.3.9}$$

In practice, the matrix B, which depends on the element, will be computed numerically.

We shall develop the element matrix from the energy integral, written in (ξ, η) coordinates, for the sample problem

$$-\Delta u = f \quad \text{on } \Omega \subset \mathbb{R}^2, \tag{4.3.10}$$

$$u = 0 \quad \text{on } \partial\Omega, \tag{4.3.11}$$

with the energy integral

$$X(u) = \frac{1}{2} \iint_{\Omega} \left(u_x^2 + u_y^2 - 2uf \right). \tag{4.3.12}$$

The following proposition, which is analogous to line (4.2.3), is important since it implies the element matrix computed by using (ξ, η) coordinates equals the element matrix for the (x, y) coordinates.

Proposition 4.3.1. Let (x, y) and (ξ, η) be related by (4.3.2). Then

$$\begin{aligned} X(u^e) &= \frac{1}{2} \iint_{e} \left(u_x(x, y)^2 + u_y(x, y)^2 - 2u(x, y)f(x, y) \right) dx\, dy \\ &= \frac{1}{2} \iint_{e} \left(\bar{u}_\xi(\xi, \eta)^2 + \bar{u}_\eta(\xi, \eta)^2 - 2\bar{u}(\xi, \eta)\bar{f}(\xi, \eta) \right) d\xi\, d\eta, \end{aligned} \tag{4.3.13}$$

where $\bar{u}$ and $\bar{f}$ are defined by

$$\bar{u}(\xi, \eta) = u(\bar{x} + (\cos\theta)\xi - (\sin\theta)\eta, \bar{y} + (\sin\theta)\xi + (\cos\theta)\eta),$$

$$\bar{f}(\xi, \eta) = f(\bar{x} + (\cos\theta)\xi - (\sin\theta)\eta, \bar{y} + (\sin\theta)\xi + (\cos\theta)\eta).$$

Proof. The change-of-variable theorem states that

$$\iint H(x, y)\, dx\, dy = \iint \overline{H}(\xi, \eta) \det \begin{pmatrix} \dfrac{\partial x}{\partial \xi} & \dfrac{\partial x}{\partial \eta} \\ \dfrac{\partial y}{\partial \xi} & \dfrac{\partial y}{\partial \eta} \end{pmatrix} d\xi\, d\eta,$$

where $\overline{H}(\xi, \eta) \equiv H(\bar{x} + (\cos\theta)\xi - (\sin\theta)\eta,\ \bar{y} + (\sin\theta)\xi + (\cos\theta)\eta)$ and $\partial x/\partial \xi = \cos\theta$, $\partial x/\partial \eta = -\sin\theta$, $\partial y/\partial \xi = \sin\theta$,

$\partial y/\partial\eta = \cos\theta$. Thus, the determinant equals 1, and we let $H(x, y)$ be $\frac{1}{2}(u_x^2 + u_y^2 - 2uf)$. By the chain rule

$$\bar{u}_\xi = u_x \frac{\partial x}{\partial \xi} + u_y \frac{\partial y}{\partial \xi} = u_x \cos\theta + u_y \sin\theta, \qquad (4.3.14)$$

$$\bar{u}_\eta = u_x \frac{\partial x}{\partial \eta} + u_y \frac{\partial y}{\partial \eta} = u_x(-\sin\theta) + u_y \cos\theta. \qquad (4.3.15)$$

Lines (4.3.14) and (4.3.15) imply

$$\begin{aligned} \bar{u}_\xi^2 + \bar{u}_\eta^2 &= u_x^2 \cos^2\theta + u_y^2 \sin^2\theta + 2\cos\theta\sin\theta\, u_x u_y \\ &\quad + u_x^2 \sin^2\theta + u_y^2 \cos^2\theta - 2\cos\theta\sin\theta\, u_x u_y \\ &= u_x^2 + u_y^2 \end{aligned} \qquad (4.3.16)$$

Thus,

$$\bar{H}(\xi, \eta) \cdot \det \begin{pmatrix} \dfrac{\partial x}{\partial \xi} & \dfrac{\partial x}{\partial \eta} \\ \dfrac{\partial y}{\partial \xi} & \dfrac{\partial y}{\partial \eta} \end{pmatrix} = \left(\bar{u}_\xi^2 + \bar{u}_\eta^2 - 2\bar{u}\bar{f}\right)$$

$$= u_x^2 + u_y^2 - 2uf = H(x, y).$$

So, line (4.3.13) follows from the change-of-variable theorem.

In order to determine the element matrix, we must compute for $i = 1, \ldots, 6$

$$\frac{\partial X(u^e)}{\partial u_i^e}. \qquad (4.3.17)$$

By using (4.3.7) in (4.3.12) and using Proposition 4.3.1, we may compute $X(u^e)$ as

$$X(u^e) = \frac{1}{2} \iint_e \left[\left(\sum_{i=1}^{6} \alpha_i m_i \xi^{m_i - 1} \eta^{n_i} \right)^2 + \left(\sum_{i=1}^{6} \alpha_i n_i \xi^{m_i} \eta^{n_i - 1} \right)^2 - 2 \left(\sum_{i=1}^{6} \alpha_i \xi^{m_i} \eta^{m_i} \right) f(\xi, \eta) \right] d\xi \, d\eta$$

$$= \frac{1}{2} \iint_e \left[\sum_{i=1}^{6} \sum_{j=1}^{6} \alpha_i m_i \xi^{m_i - 1} \eta^{n_i} \alpha_j m_j \xi^{m_j - 1} \eta^{n_j - 1} + \sum_{i=1}^{6} \sum_{j=1}^{6} \alpha_i n_i \xi^{m_i} \eta^{n_i - 1} \alpha_j n_j \xi^{m_j} \eta^{n_j - 1} - 2 \sum_{i=1}^{6} \alpha_i \xi^{m_i} \eta^{n_i} f(\xi, \eta) \right] d\xi \, d\eta$$

$$= \frac{1}{2} \sum_{i=1}^{6} \sum_{j=1}^{6} \alpha_i \left(m_i m_j \iint_e \xi^{m_i + m_j - 2} \eta^{n_i + n_j} + n_i n_j \iint_e \xi^{m_i + m_j} \eta^{n_i + n_j - 2} \right) \alpha_j - \sum_{i=1}^{6} \alpha_i \iint_e \xi^{m_i} \eta^{n_i} f(\xi, \eta). \tag{4.3.18}$$

Line (4.3.18) may be written in matrix form when

$$g_{ij} = m_i m_j \iint_e \xi^{m_i + m_j - 2} \eta^{n_i + n_j} + n_i n_j \iint_e \xi^{m_i + m_j} \eta^{n_i + n_j - 2}, \tag{4.3.19}$$

$$G = (g_{ij}) \quad \text{is a } 6 \times 6 \text{ matrix}, \tag{4.3.20}$$

$$\mathbf{F} = (F_i) \quad \text{where } F_i = \iint_e \xi^{m_i} \eta^{n_i} f(\xi, \eta). \tag{4.3.21}$$

Line (4.3.18) becomes

$$
\begin{aligned}
X(u^e) &= \tfrac{1}{2}\boldsymbol{\alpha}^{\mathrm{T}} G \boldsymbol{\alpha} - \boldsymbol{\alpha}^{\mathrm{T}} \mathbf{F} \\
&= \tfrac{1}{2}\mathbf{u}^{e\mathrm{T}} B^{\mathrm{T}} G B \mathbf{u}^e - \mathbf{u}^{e\mathrm{T}} B^{\mathrm{T}} \mathbf{F} \qquad (4.3.22)
\end{aligned}
$$

As in the previous section for $\partial/\partial\mathbf{u}^e = (\partial/\partial u_1^e, \ldots, \partial/\partial u_6^e)^{\mathrm{T}}$, k^e is defined by

$$
\frac{\partial X(u^e)}{\partial \mathbf{u}^e} = k^e \mathbf{u}^e - B^{\mathrm{T}} \mathbf{F}. \qquad (4.3.23)
$$

Therefore, as $B^{\mathrm{T}} G B$ is symmetric, the element matrix is

$$
k^e = B^{\mathrm{T}} G B. \qquad (4.3.24)
$$

In order to compute G, we must be able to evaluate $\iint_e \xi^m \eta^n$ for different integers m and n. The integration is best done by dividing e into the left and right triangles.

Integration Formulas. Let e be a triangular element and a, b, and c defined from (4.3.4)–(4.3.6).

$$
\begin{aligned}
h(m,n) &\equiv \iint_e \xi^m \eta^n \\
&= \frac{c^{n+1}\left[a^{m+1} - (-b)^{m+1}\right] m!\,n!}{(m+n+2)!} \qquad (4.3.25)
\end{aligned}
$$

The m and n are determined by m_i, m_j, n_i, n_j in (4.3.19) for i, $j = 1, \ldots, 6$ and Table 4.3.1. This procedure applied to each of the 36 components gives

$$
G = \begin{pmatrix}
0 & 0 & 0 & 0 & 0 & 0 \\
0 & h(0,0) & 0 & 2h(1,0) & h(0,1) & 0 \\
0 & 0 & h(0,0) & 0 & h(1,0) & 2h(0,1) \\
0 & 2h(1,0) & 0 & 4h(2,0) & 2h(1,1) & 0 \\
0 & h(0,1) & h(1,0) & 2h(1,1) & h(0,2)+h(2,0) & 2h(1,1) \\
0 & 0 & 2h(0,1) & 0 & 2h(1,1) & 4h(0,2)
\end{pmatrix}
$$

(4.3.26)

For example, the g_{23} component is

$$\begin{aligned} g_{23} &= m_2 m_3 h(m_2 + m_3 - 2, n_2 + n_3) \\ &\quad + n_2 n_3 h(m_2 + m_3, n_2 + n_3 - 2) \\ &= 1 \cdot 0h(1 + 0 - 2, 0 + 1) + 0 \cdot 1h(1 + 0, 0 + 1 - 2) \\ &= 0. \end{aligned}$$

Example. Let a triangular element have nodes given by $(x_1, y_1) = (0,0)$, $(x_3, y_3) = (2,0)$, and $(x_5, y_5) = (1,1)$ (see Figure 4.3.1.). Then $a = b = c = 1$ and from (4.3.8) and (4.3.9) we have

$$A = \begin{pmatrix} 1 & -1 & 0 & 1 & 0 & 0 \\ 1 & 0 & 0 & 0 & 0 & 0 \\ 1 & 1 & 0 & 1 & 0 & 0 \\ 1 & \frac{1}{2} & \frac{1}{2} & \frac{1}{4} & \frac{1}{4} & \frac{1}{4} \\ 1 & 0 & 1 & 0 & 0 & 1 \\ 1 & -\frac{1}{2} & \frac{1}{2} & \frac{1}{4} & -\frac{1}{4} & \frac{1}{4} \end{pmatrix},$$

$$B = \frac{1}{2} \begin{pmatrix} 0 & 2 & 0 & 0 & 0 & 0 \\ -1 & 0 & 1 & 0 & 0 & 0 \\ -1 & -4 & -1 & 4 & -2 & 4 \\ 1 & -2 & 1 & 0 & 0 & 0 \\ 2 & 0 & -2 & 4 & 0 & -4 \\ 1 & 2 & 1 & -4 & 4 & -4 \end{pmatrix}.$$

From (4.3.25) and (4.3.26),

$$G = \frac{1}{3} \begin{pmatrix} 0 & 0 & 0 & 0 & 0 & 0 \\ 0 & 3 & 0 & 0 & 1 & 0 \\ 0 & 0 & 3 & 0 & 0 & 2 \\ 0 & 0 & 0 & 2 & 0 & 0 \\ 0 & 1 & 0 & 0 & 1 & 0 \\ 0 & 0 & 2 & 0 & 0 & 2 \end{pmatrix}.$$

The element matrix is then computed by (4.3.24):

$$k^e = \frac{1}{6}\begin{pmatrix} 3 & 0 & 0 & 0 & 1 & -4 \\ 0 & 16 & 0 & -8 & 0 & -8 \\ 0 & 0 & 3 & -4 & 1 & 0 \\ 0 & -8 & -4 & 16 & -4 & 0 \\ 1 & 0 & 1 & -4 & 6 & -4 \\ -4 & -8 & 0 & 0 & -4 & 16 \end{pmatrix}.$$

The preceding results are summarized in the next proposition.

Proposition 4.3.2. The element matrix for the problem (4.3.10), (4.3.11) is given by (4.3.24). B and G are computed by $B \equiv A^{-1}$ in line (4.3.8) and line (4.3.26). The values for a, b, c are given lines (4.3.4)–(4.3.6) and $h(m, n)$ are given by the formula in (4.3.25).

Remark. If the differential equation (4.3.10) changes, the G will be different and consequently k^e will be different. If the boundary conditions are of the form (2.5.3), then the integrals over $\partial\Omega_2$ reduce to those in Section 4.2.

4.4 BILINEAR SHAPE FUNCTIONS ON RECTANGULAR ELEMENTS

This section differs from the previous discussions about shape functions in two dimensions because the elements will be rectangular and not triangular. In this case there will be four element nodes for each element. The shape function will have the form

$$u^e(x, y) = \bar{\alpha}_1 + \bar{\alpha}_2 x + \bar{\alpha}_3 y + \bar{\alpha}_4 xy. \tag{4.4.1}$$

Consequently, the element matrix will be 4×4. As in Section 4.3 we shall use general coordinates (ξ, η), which are indicated in Figure 4.4.1. Note that Proposition 4.3.1 is still valid, and therefore, we may compute the element matrices in terms of the general coordinates. $u^e(x, y)$ is called bilinear because when either x or y is fixed $u^e(x, y)$ is linear in the other variable.

$u^e(x, y)$ may be written in terms of (ξ, η):

$$u^e(\xi, \eta) = \sum_{i=1}^{4} \alpha_i \xi^{m_i} \eta^{n_i}. \tag{4.4.2}$$

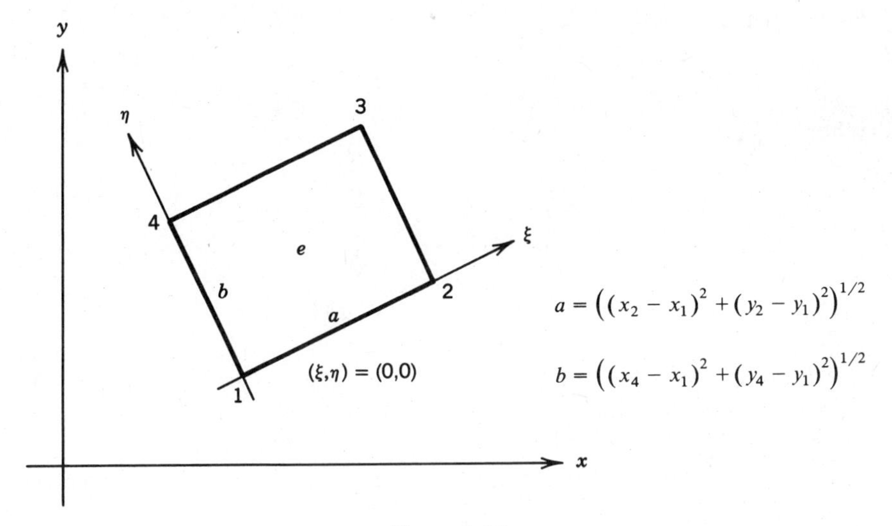

Figure 4.4.1

The values of m_i, n_i, and the (ξ, η) coordinates are given in Table 4.4.1. The four equations needed to compute α_i, $i = 1, \ldots, 4$, are found by evaluating (4.4.2) at each of the four element nodes.

$$A\boldsymbol{\alpha} = \mathbf{u}^e \tag{4.4.3}$$

where $\boldsymbol{\alpha} = (\alpha_1, \ldots, \alpha_4)^{\mathrm{T}}$, $\mathbf{u}^e = (u_1^e, \ldots, u_4^e)^{\mathrm{T}}$, and

$$A = \begin{pmatrix} 1 & 0 & 0 & 0 \\ 1 & a & 0 & 0 \\ 1 & a & b & ab \\ 1 & 0 & b & 0 \end{pmatrix}$$

TABLE 4.4.1

i	m_i	n_i	(ξ, η)
1	0	0	$(0,0)$
2	1	0	$(a,0)$
3	0	1	(a,b)
4	1	1	$(0,b)$

Consequently,

$$\boldsymbol{\alpha} = A^{-1}\mathbf{u}^e = B\mathbf{u}^e, \tag{4.4.4}$$

where

$$B = \begin{pmatrix} 1 & 0 & 0 & 0 \\ -1/a & 1/a & 0 & 0 \\ -1/b & 0 & 0 & 1/b \\ 1/(ab) & -1/(ab) & 1/(ab) & -1/(ab) \end{pmatrix}.$$

Again we consider problem (4.3.10), (4.3.11) as a sample problem. One derivation of the element matrix follows the same pattern as in lines (4.3.17)–(4.3.24). The only difference is the integrals $\iint_e \xi^{m_i}\eta^{n_i}$ are over a rectangular region and m_i and n_i are defined differently.

Integration Formulas. Let e be rectangular elements with sides of length a and b.

$$\begin{aligned} h_r(m,n) &= \iint_e \xi^m \eta^n \\ &= \frac{a^{m+1}}{m+1}\frac{b^{n+1}}{n+1}. \end{aligned} \tag{4.4.5}$$

The matrix G is easily computed to be

$$G = \begin{pmatrix} 0 & 0 & 0 & 0 \\ 0 & h_r(0,0) & 0 & h_r(0,1) \\ 0 & 0 & h_r(0,0) & h_r(1,0) \\ 0 & h_r(0,1) & h_r(1,0) & h_r(0,2) + h_r(2,0) \end{pmatrix}$$

$$= \begin{pmatrix} 0 & 0 & 0 & 0 \\ 0 & ab & 0 & \dfrac{ab^2}{2} \\ 0 & 0 & ab & \dfrac{a^2b}{2} \\ 0 & \dfrac{ab^2}{2} & \dfrac{a^2b}{2} & \dfrac{a^3b}{3} + \dfrac{b^3a}{3} \end{pmatrix}. \tag{4.4.6}$$

Proposition 4.4.1. Consider problem (4.3.10), (4.3.11). The 4×4 element matrix for the bilinear shape function (4.4.2) is $k^e = B^{\mathrm{T}}GB$, where B is from (4.4.4) and G is defined by (4.4.6). More precisely,

$$k^e = \begin{pmatrix} \frac{b}{3a} + \frac{a}{3b} & \frac{-b}{3a} + \frac{a}{6b} & \frac{-b}{6a} - \frac{a}{6b} & \frac{b}{6a} - \frac{a}{3b} \\ & \frac{b}{3a} + \frac{a}{3b} & \frac{b}{6a} - \frac{a}{3b} & \frac{-b}{6a} - \frac{a}{6b} \\ & & \frac{b}{3a} + \frac{a}{3b} & \frac{-b}{3a} + \frac{a}{6b} \\ \text{SYM} & & & \frac{b}{3a} + \frac{a}{3b} \end{pmatrix}. \tag{4.4.7}$$

Proof. Perform the multiplication $k^e = B^{\mathrm{T}}GB$.

The element matrix (4.4.7) can be derived without using the G matrix. The following discussion parallels the development used for the linear shape functions.

Define $X = (1, x, y, xy)$ and note that $u^e(x, y) = X\boldsymbol{\alpha}$, where $\boldsymbol{\alpha} = (\alpha_1, \alpha_2, \alpha_3, \alpha_4)^{\mathrm{T}}$. Evaluate $u^e(x, y)$ at each element node (x_i, y_i), $i = 1, \ldots, 4$.

$$\mathbf{u}^e = A_x\boldsymbol{\alpha}, \tag{4.4.8}$$

where

$$A_x = \begin{pmatrix} 1 & x_1 & y_1 & x_1y_1 \\ 1 & x_2 & y_2 & x_2y_2 \\ 1 & x_3 & y_3 & x_3y_3 \\ 1 & x_4 & y_4 & x_4y_4 \end{pmatrix}.$$

Then $\boldsymbol{\alpha} = A_x^{-1}\mathbf{u}^e$ and

$$u^e(x, y) = XA_x^{-1}\mathbf{u}^e. \tag{4.4.9}$$

Proposition 4.4.2. $XA_x^{-1} = \mathbf{N}^{\mathrm{T}}$, where

$$\mathbf{N}^{\mathrm{T}} = (N_1(x, y), N_2(x, y), N_3(x, y), N_4(x, y))$$

and

$$N_1(x, y) = L_1(x)L_1(y), \tag{4.4.10}$$

$$N_2(x, y) = L_2(x)L_1(y), \tag{4.4.11}$$

$$N_3(x, y) = L_2(x)L_2(y), \tag{4.4.12}$$

$$N_4(x, y) = L_1(x)L_2(y), \tag{4.4.13}$$

where

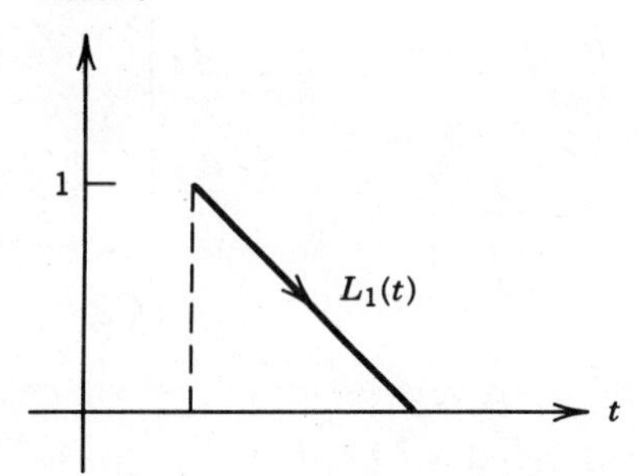

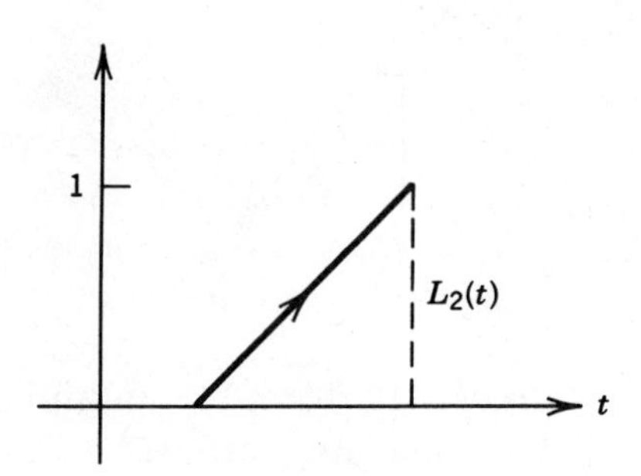

Proof. It suffices to show $X = \mathbf{N}^{\mathrm{T}} A_x$. This can be done by direct substitution and using the fact that e is a rectangle. For example, let us show

$$x = N_1(x, y)x_1 + N_2(x, y)x_2 + N_3(x, y)x_3 + N_4(x,y)x_4.$$

$$\begin{aligned}
N_1x_1 &+ N_2x_2 + N_3x_3 + N_4x_4 \\
&= L_1(x)L_1(y)x_1 + L_2(x)L_1(y)x_2 + L_2(x)L_2(y)x_3 \\
&\quad + L_1(x)L_2(y)x_4 \\
&= L_1(y)(L_1(x)x_1 + L_2(x)x_2) \\
&\quad + L_2(y)(L_2(x)x_3 + L_1(x)x_4) \\
&= L_1(y)\left(\frac{-(x - x_2)x_1 + (x - x_1)x_2}{x_2 - x_1}\right) \\
&\quad + L_2(y)\left(\frac{-(x - x_2)x_4 + (x - x_1)x_3}{x_2 - x_1}\right) \\
&= L_1(y) \cdot x + L_2(y)\left(\frac{-(x - x_2)x_4 + (x - x_1)x_3}{x_2 - x_1}\right).
\end{aligned} \tag{4.4.14}$$

Now use the rectangular shape and assume rotation has been done: $x_4 - x_3 = x_1 - x_2$, $x_2 = x_3$, and $x_1 = x_4$. The coefficient of $L_2(y)$ is then x, and (4.4.14) becomes

$$L_1(y)x + L_2(y)x = (L_1(y) + L_2(y))x = x.$$

This is what we wanted to show. The other three rows of $X = \mathbf{N}^{\mathrm{T}}A_x$ are similar.

Proposition (4.4.2) and equation (4.4.9) imply $u^e(x, y) = \sum_{i=1}^{4} u_i^e N_i(x, y)$. So one can compute $X(u^e)$ and $\partial X(u^e)/\partial u_i^e$ in just the same way as in the linear shape function cases. If one uses the fact that L_1 and L_2 are $N_1^{e_i}(t)$ and $N_2^{e_i}(t)$, then the computation of the element matrix (4.4.7) is straightforward.

4.5 COMPLETE CUBIC SHAPE FUNCTIONS ON TRIANGULAR ELEMENTS

In the previous examples the values of u^e were given at all nodes. Shape functions of this type are often called *Lagrangian*. If values and some of the function's partial derivatives are specified, then the shape function is called *Hermitian*. In higher-order problems, such as the clamped-plate problem, the weak formulation has second-order derivatives (see exercise 2-17). Therefore, we need the second-order derivative to be at least piecewise continuous. In this case the first-order derivative should be continuous. One attempt to obtain this desired property is to specify the first-order partial derivatives at each node of a triangular element.

The following discussion is very brief, and the interested reader should consult Norrie and deVries [13, Chapter 5]. We consider a triangular element with a general coordinate system as indicated in Figure 4.5.1. The shape function is a complete cubic given by

$$u^e(\xi, \eta) = \sum_{i=1}^{10} \alpha_i \xi^{m_i} \eta^{n_i}, \tag{4.5.1}$$

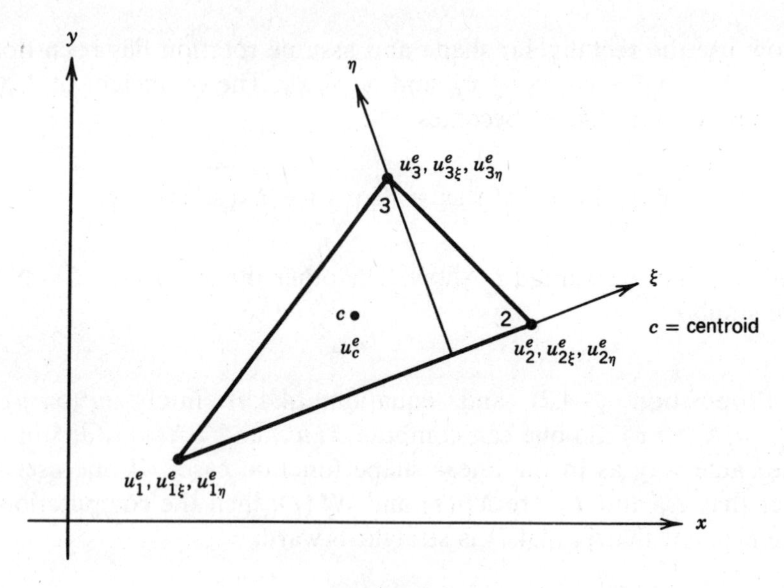

Figure 4.5.1

where

$$\begin{aligned} &m_1 = m_3 = m_6 = m_{10} = 0 = n_1 = n_2 = n_4 = n_7, \\ &m_2 = m_5 = m_9 = 1 = n_2 = n_5 = n_8, \\ &m_4 = m_8 = 2 = n_6 = n_9, \\ &m_7 = 3 = n_{10}. \end{aligned}$$

The partial derivatives of $u^e(\xi, \eta)$ are computed as

$$u_\xi^e(\xi, \eta) = \sum_{i=1}^{10} \alpha_i m_i \xi^{m_i - 1} \eta^{n_i}, \tag{4.5.2}$$

$$u_\eta^e(\xi, \eta) = \sum_{i=1}^{10} \alpha_i n_i \xi^{m_i} \eta^{n_i - 1}. \tag{4.5.3}$$

In order to determine all ten α_i, $i = 1, \ldots, 10$, we need ten equations. These are given by computing u^e, (4.5.1), at each of the four element nodes, three from the vertices and one from the centroid. The other six equations are u^e_ξ, (4.5.2), at each of the three vertex nodes, and u^e_η, (4.5.3), at each of the three vertex nodes. This gives

$$\mathbf{u}^e_{\mathrm{H}} = A\boldsymbol{\alpha}, \tag{4.5.4}$$

where A is a 10×10 matrix, $\boldsymbol{\alpha} = (\alpha_1, \ldots, \alpha_{10})^{\mathrm{T}}$, H represents Hermitian, and

$$\mathbf{u}^e_{\mathrm{H}} = \left(u^e_1, u^e_{1\xi}, u^e_{1\eta}, u^e_2, u^e_{2\xi}, u^e_{2\eta}, u^e_3, u^e_{3\xi}, u^e_{3\eta}, u^e_c\right)^{\mathrm{T}}.$$

From (4.5.4) one can compute the usual G matrix and write for a suitable energy integral

$$X(u^e_{\mathrm{H}}) = \tfrac{1}{2}\mathbf{u}^{e\mathrm{T}}_{\mathrm{H}} B^{\mathrm{T}} G B \mathbf{u}^e_{\mathrm{H}}. \tag{4.5.5}$$

Because u^e_{H} has partial derivatives, Proposition 4.3.1 does not hold and we cannot simply write $k^e = B^{\mathrm{T}}GB$. However, one can use the chain rule to obtain

$$\mathbf{u}^{e\mathrm{T}}_{\mathrm{L}} = R\mathbf{u}^e_{\mathrm{L},x,y}, \tag{4.5.6}$$

where

$$\mathbf{u}^e_{\mathrm{H},x,y} = \left(u^e_1, u^e_{1x}, u^e_{1y}, u^e_2, u^e_{2x}, u^e_{2y}, u^e_3, u^e_{3x}, u^e_{3y}, u^e_c\right)^{\mathrm{T}},$$

$$R = \begin{pmatrix} C & \phi & \phi & 0 \\ \phi & C & \phi & 0 \\ \phi & \phi & C & 0 \\ 0 & 0 & 0 & 1 \end{pmatrix}$$

is a 10×10 with

$$C = \begin{pmatrix} 1 & 0 & 0 \\ 0 & \cos\theta & \sin\theta \\ 0 & -\sin\theta & \cos\theta \end{pmatrix}$$

and

$$\phi = 3 \times 3 \text{ zero matrix}.$$

Thus, lines (4.5.5) and (4.5.6) give

$$X\left(u^e_{\mathrm{H},x,y}\right) = \tfrac{1}{2}\mathbf{u}^{e\mathrm{T}}_{\mathrm{H},x,y} R^{\mathrm{T}} B^{\mathrm{T}} G B R \mathbf{u}^e_{\mathrm{H},x,y}. \tag{4.5.7}$$

Since $R^{\mathrm{T}}B^{\mathrm{T}}GBR$ is symmetric, the 10×10 element matrix is

$$k^e = R^{\mathrm{T}}B^{\mathrm{T}}GBR. \tag{4.5.8}$$

4.6 OBSERVATIONS AND REFERENCES

There are many different shape functions and different shaped elements. We have mentioned only a few. In some problems a combination of different shape functions give good results. In Norrie and deVries [13, Chapter 9] and Bathe [2, Chapter 5], these and some other shape functions are discussed. When choosing shape functions for a problem that has k derivatives in its weak formulation, or its energy integral, one should make sure the shape functions give an approximate solution that is continuous, and has $(k - 1)$ continuous derivatives in at least some points along the element boundaries. Such shape functions are called *conforming*. Examples includes:

1. $-\Delta u = f$ where $k = 1$ with linear or quadratic shape functions.
2. $\Delta^2 u = f$ where $k = 2$ with cubic shape functions.

This topic is discussed in more detail in Norrie and deVries [13, Chapter 8].

Interesting exercises include 4-8, 4-10, 4-15, and 4-18.

EXERCISES

4-1 Prove Proposition 4.1.1.

4-2 In (4.1.6) verify the formula for the case $a = 1$, $b = 1$, $c = 0$, $d = 0$, and e the element with vertices $(0,0,0)$, $(1,0,0)$, $(0,1,0)$ and $(0,0,1)$.

4-3 Consider the problem (4.1.7), (4.1.8). Use Stokes's theorem (see Taylor and Mann [29]) to find the weak formulation. Verify that $X(u)$ given in (4.1.9) is admissible.

4-4 Prove Proposition 4.1.2. Compute

$$\overline{R}_i^e = \iiint f_I N_i^e$$

where f_I on e is

$$\sum_{j=1}^{4} f_j^e N_j^e \quad \text{and} \quad f_j^e = f(x_j, y_j, z_j).$$

4-5 Verify (4.2.6).

4-6 In (4.2.14) show $B^{\mathrm{T}}GB$ is symmetric, that is, $(B^{\mathrm{T}}GB)^{\mathrm{T}} = B^{\mathrm{T}}GB$. Also, for problem (4.2.8)–(4.2.10), compute R^e.

4-7 Complete the proof of Proposition 4.2.1.

4-8 Consider a variation of the problem (4.2.8)–(4.2.10). Let (4.2.8) be replaced by $-m\ddot{y} + ky = f(t)$. Find the element matrix and compute $\partial X(y^e)/\partial y_i^e$. Suppose $f(t) = t^2 + 1$, $m = k = 1$, and there are two equal elements. Find the solution of the resulting algebraic system and compare it with the results of exercises 1-1 and 1-9.

4-9 Verify the components of A in line (4.3.8).

4-10 Consider a variation of problem (4.3.10), (4.3.11). Let (4.3.10) be replaced by $-\Delta u + cu = f$. Find the element matrix for this problem when quadratic shape functions are used.

4-11 Verify the formula for $h(m, n)$ in (4.3.25).

4-12 Verify the components of G in (4.3.26).

4-13 Show G in (4.3.26) is symmetric and hence $B^{\mathrm{T}}GB$ in (4.3.22) is symmetric.

4-14 Verify the computations for the example that is given before Proposition 4.3.2. For this example, compute $B^{\mathrm{T}}\mathbf{F}$, where $\mathbf{F}$ is from line (4.3.21) with $f(\xi, \eta) = \xi + \eta$.

4-15 Find the 4×4 element matrix for the problem in 4-10 when bilinear shape functions on rectangular elements are used.

4-16 Complete the proof of Proposition 4.4.2.

4-17 Use the results of Proposition 4.4.2 and $\partial X(u^e)/\partial u_i^e$ for $i = 1, 2, 3, 4$ to compute the element matrix. It should be the same as in line (4.4.7).

4-18 Modify FEMI so that quadratic shape functions are used.

4-19 Modify FEMI so that bilinear shape functions are used.

5

ERROR ESTIMATES AND EXISTENCE

The primary objectives of this chapter are to develop for a model problem error estimates and existence of a solution. A simple estimate is the existence of a constant, C, which is independent of the mesh size h, such that

$$|u - u^h| \leq Ch,$$

where u is the continuum solution and u^h is the function associated with a finite element solution. The existence of a solution will be established by considering a sequence of functions u^l, such that $X(u^l) \downarrow d \equiv \inf_{v \in S} X(v)$, where $X(v)$ is an admissible energy integral and S is a set of "suitable" functions. The convergence of u^l to u where $X(u) = d$ will be proved using the completness property of the "suitable" set S.

In Sections 5.1 to 5.4 we develop some preliminary material. Section 5.5 contains the proof of equivalence of the classical, energy, and weak formulations. The error estimates are presented in Section 5.6. The existence of an energy solution is given in Section 5.7. Even though these results are for a simple model problem, the techniques may be used on more complicated problems (see the observations preceding the exercises).

5.1 DEFINITIONS OF $a(u, \Psi)$, $H_0^1(0, L)$

Throughout this chapter we consider the model problem (5.1.1)–(5.1.3):

$$-(m(x)\dot{u})^{\cdot} + k(x)u = f(x) \quad \text{on } [0, L], \tag{5.1.1}$$

$$u(0) = 0, \tag{5.1.2}$$

$$u(L) = 0, \tag{5.1.3}$$

where $m(x), \dot{m}(x), k(x), f(x) \in C[0, L]$ and $m, k \geq \delta > 0$ for all $x \in [0, L]$. The conditions on m, k and f may be relaxed or made stronger (see exercises 5-7 and 5-10 to 5-15). The preceding conditions on m, k, and f are used because they give an elementary approach to the topics of this chapter. The following example illustrates an important physical problem with a weaker condition on m.

Example. Consider the steady-state heat conduction in a thin rod with inhomogeneous material. In particular, assume the thermal conductivity $K(x) = m(x)$ has the form

$$K(x) = \begin{cases} k_1, & 0 \leq x \leq \frac{1}{2} \\ k_2, & \frac{1}{2} < x \leq 1. \end{cases}$$

Then the heat equation has the form

$$-(K(x)u_x)_x = f(x),$$

$$u(0) = 0,$$

$$u(1) = 0.$$

If $f(x) = 1$, $k_1 = 1$, and $k_2 = 2$, then the "solution" may be formed by solving for u when $x < \frac{1}{2}$, when $x > \frac{1}{2}$, and then demanding that u and Ku_x be continuous at $x = \frac{1}{2}$, that is, $u(\frac{1}{2}^-) = u(\frac{1}{2}^+)$ and

$(Ku_x)(\frac{1}{2}^-) = (Ku_x)(\frac{1}{2}^+)$. This gives

$$u(x) = \begin{cases} -\frac{1}{2}x^2 + \frac{5}{12}x, & 0 \le x \le \frac{1}{2} \\ -\frac{1}{4}x^2 + \frac{5}{24}x + \frac{1}{24}, & \frac{1}{2} < x \le 1. \end{cases}$$

Note u_x is not continuous at $x = \frac{1}{2}$, that is, $u_x(\frac{1}{2}^-) = -\frac{1}{12} \neq -\frac{1}{24} = u_x(\frac{1}{2}^+)$. Consequently, u is not a classical solution. However, as we shall see at the end of this section, u is a weak solution.

The smoothness of $K(x)$ (i.e., is $K(x) \in C[0, L]$, $C^1[0, L]$, or $C^2[0, L]$?) will determine the smoothness of the solution. For example, if

$$K(x) = \begin{cases} 1, & 0 \le x \le \frac{1}{2} \\ 2x, & \frac{1}{2} < x \le 1, \end{cases}$$

then $K \in C[0,1]$ and $K \notin C^1[0,1]$. The solution is

$$u(x) = \begin{cases} -\frac{1}{2}x^2 + ax, & 0 \le x \le \frac{1}{2}, \quad a = \dfrac{3}{4(1 + \ln 2)} \\ -\frac{1}{2}x + \frac{1}{2}a \ln x + \frac{1}{2}, & \frac{1}{2} < x \le 1 \end{cases}$$

and $u \in C^1[0,1]$, $u \notin C^2[0,1]$.

Another related example is given by

$$K(x) = \begin{cases} 1, & 0 \le x \le \frac{1}{2} \\ \left(x - \frac{1}{2}\right)^2 + 1, & \frac{1}{2} < x \le 1 \end{cases}$$

where $K \in C^1[0,1]$ and $K \notin C^2[0,1]$. The solution is

$$u(x) = \begin{cases} -\frac{1}{2}x^2 + ax, & 0 \le x \le \frac{1}{2} \\ -\frac{1}{2}\ln\left(\left(x - \frac{1}{2}\right)^2 + 1\right) & \\ \quad +\left(-\frac{1}{2} + a\right)\tan^{-1}\left(x - \frac{1}{2}\right) + b, & \frac{1}{2} < x \le 1, \end{cases}$$

where

$$a = 2b + \tfrac{1}{4},$$
$$b = \tfrac{1}{2}\ln\tfrac{5}{4} + \left(\tfrac{1}{2} - a\right)\tan^{-1}\tfrac{1}{2},$$
$$u \in C^2[0,1], \qquad u \notin C^3[0,1].$$

An admissible energy integral for (5.1.1)–(5.1.3) is

$$X(u) = \frac{1}{2}\int_0^L (m\dot{u}^2 + ku^2 - 2uf). \tag{5.1.4}$$

That $X(u)$ is admissible ($F'(0) = 0$ is the weak equation where $F(\lambda) \equiv X(u + \lambda\psi)$) follows from the weak formulation of (5.1.1)–(5.1.3). In particular, if $u \in C^2[0, L]$ is a classical solution of (5.1.1)–(5.1.3), then by multiplying (5.1.1) by Ψ and integrating by parts we obtain the weak equation

$$\int_0^L m(x)\dot{u}\dot{\Psi} + k(x)u\Psi = \int_0^L f(x)\Psi \tag{5.1.5}$$

for all $\Psi \in C^2[0, L]$, with $\Psi(0) = 0 = \Psi(L)$. In the previous discussions of a weak solution, we always stated that u and Ψ should be "suitable" functions such that all integrals exist. In this chapter we shall be more precise. In the following definitions we use the concept of a Lebesgue measurable function; the reader should either refer to Royden [19], or accept these functions and their integrals as axioms.

Definitions.

$$L_2(0, L) = \left\{u\colon [0, L] \to \mathbb{R} \,|\, u \text{ is the Lebesgue measurable and } \int_0^L u^2 < \infty\right\}.$$

$$\|u\|_{L_2} = \left(\int_0^L u^2\right)^{1/2} \text{ is called a } \textit{norm on } L_2(0, L).$$

$$S^h = \left\{u\colon [0, L] \to \mathbb{R} \,|\, u = \sum_{j=1}^{N-1} u_j\Psi_j,\ u_j \in \mathbb{R},\ h = L/N\right\}.$$

Ψ_j are the usual piecewise-linear shape functions.

$$H_0^1(0, L) = \{u\colon [0, L] \to \mathbb{R} \,|\, u,\ \dot{u} \in L_2(0, 1),\ u(0) = 0 = u(L)\}.$$

$$\|u\|_1 \equiv \left(\int_0^L \dot{u}^2 + u^2\right)^{1/2} \text{ is called the } \textit{norm on } H_0^1(0, L).$$

Remarks

1. Even if the reader is not familiar with Lebesgue-measurable functions, a good understanding of the error estimates may be obtained. Lebesgue measurable functions are used so that $L_2(0, L)$, $S^h \subset H_0^1(0, L)$, $H^1(0, L)$, and $H_0^1(0, L)$ are *complete spaces* (see Section 5.4 for the definition of a complete space). This important property is used to establish the existence theorem in Section 5.7.
2. In $H_0^1(0, L)$, $\dot{u}$ may not exist in a classical sense. In Section 5.4 we define the notion of a weak derivative in $L_2(0, L)$. Also, the value of u at given point needs to be defined. In Section 5.4 we show

$$H^1(0,L) \equiv \{u \in L_2(0,L) \mid \dot{u} \in L_2(0,L)\} \subset C[0,L]$$

and, consequently, we write

$$H_0^1(0,L) = \{u \in H^1(0,L) \mid u(0) = 0 = u(L)\}.$$

We introduce some notation that will ease our writing and generalizes to other more complicated problems.

Definitions. Let a: $H^1(0, L) \times H^1(0, L) \to \mathbb{R}$.

$$a(u,\Psi) \equiv \int_0^L m(x)\dot{u}\dot{\Psi} + k(x)u\Psi,$$

$$(f,\Psi) \equiv \int_0^L f(x)\Psi.$$

Line (5.1.5) may be rewritten

$$a(u,\Psi) = (f,\Psi). \tag{5.1.6}$$

Line (5.1.4) may be rewritten

$$X(u) = \tfrac{1}{2}a(u,u) - (f,u). \tag{5.1.7}$$

A useful inequality for functions in $L_2(0, L)$ is *Cauchy's inequality*, line (5.1.8).

Lemma . If $u, v \in L_2(0, L)$, then $\int_0^L uv$ is finite and, more precisely,

$$\left|\int_0^L uv\right| \le \|u\|_{L_2} \cdot \|v\|_{L_2} \tag{5.1.8}$$

Proof. Since $|uv| \le \frac{1}{2}(u^2 + v^2)$, $|\int_0^L uv| \le \int_0^L |uv| < \infty$. Let $(u, v) \equiv \int_0^L uv$ and note $(u, u) = \|u\|_{L_2}^2$. Define $f(t) = (u + tv, u + tv)$ and note $f(t) \ge 0$ and $f(t) = t^2(v, v) + 2t(u, v) + (u, u)$. When $t = t_1 = -(u, v)/(v, v)$, $0 \le f(t_1) = -(u, v)^2/(v, v) + (u, u)$. This is equivalent to line (5.1.8).

Proposition 5.1.1. If $u, \Psi \in H_0^1(0, L)$, then $a(u, \Psi)$ and (f, Ψ) are finite. Consequently, $X(u)$ is finite.

Proof. Since $m(x), k(x), f(x) \in C[0, L]$, $\max m(x) \le M_2 < \infty$, $\max k \le M_k < \infty$ and $\max|f| \le M_f < \infty$. By Cauchy's inequality, (5.1.8), and for $\Psi \in H^1(0, L)$

$$|(f, \Psi)| = \left|\int_0^L f\Psi\right| \le \|f\|_{L_2}\|\Psi\|_{L_2} \le LM_f\|\Psi\|_{L_2} < \infty.$$

By repeated use of (5.1.8),

$$\begin{aligned}
|a(u, \Psi)| &\le \int_0^L m(x)|\dot{u}|\,|\dot{\Psi}| + \int_0^L k(x)|u|\,|\Psi| \\
&\le M_m \int_0^L |\dot{u}|\,|\dot{\Psi}| + M_k \int_0^L |u|\,|\Psi| \\
&\le M_m\|\dot{u}\|_{L_2}\|\dot{\Psi}\|_{L_2} + M_k\|u\|_{L_2}\|\Psi\|_{L_2} \\
&\le M_m\|u\|_1\|\Psi\|_1 + M_k\|u\|_1\|\Psi\|_1 < \infty, \\
&\qquad\qquad u, \Psi \in H^1(0, L).
\end{aligned}$$

$X(u) = \frac{1}{2}(a(u, u) - 2(f, u))$ is finite because we may choose $\Psi = u$ and then use the previous results.

We are in a position to give a precise meaning to the classical, energy, and weak formulations.

Definitions. $u \in C^2[0, L]$ is called a *classical solution of* (5.1.1)–(5.1.3) if and only if (5.1.1)–(5.1.3) hold.

$u \in H_0^1(0, L)$ is called an *energy solution of* (5.1.1–5.1.3) if and only if $X(u) = \min_{v \in H_0^1(0, L)} X(v)$.

$u \in H_0^1(0, L)$ is called a *weak solution of* (5.1.1)–(5.1.3) if and only if for all $\Psi \in H_0^1(0, L)$, $a(u, \Psi) = (f, \Psi)$.

$u^h \in S^h$ is called a *variational finite element solution of* (5.1.1)–(5.1.3) if and only if $X(u^h) = \min_{v^h \in S^h} X(v^h)$.

Example. We return to the example at the beginning of this section. We wish to show that

$$u(x) = \begin{cases} -\frac{1}{2}x^2 + \frac{5}{12}x, & 0 \le x \le \frac{1}{2} \\ -\frac{1}{4}x^2 + \frac{5}{24}x + \frac{1}{24}, & \frac{1}{2} < x \le 1 \end{cases}$$

is a weak solution of (5.1.1)–(5.1.3), where

$$K(x) = \begin{cases} 1, & 0 \le x \le \frac{1}{2} \\ 2, & \frac{1}{2} < x \le 1. \end{cases}$$

We must show for all $\Psi \in H_0^1(0,1)$, $a(u, \Psi) = (f, \Psi)$:

$$\begin{aligned} a(u,\Psi) &= \int_0^1 K(x) u_x \Psi_x \, dx \\ &= \int_0^{1/2} K(x) u_x \Psi_x \, dx + \int_{1/2}^1 K(x) u_x \Psi_x \, dx \\ &= \int_0^{1/2} 1\left(-x + \tfrac{5}{12}\right)\Psi_x \, dx + \int_{1/2}^1 2\left(-\tfrac{1}{2}x + \tfrac{5}{24}\right)\Psi_x \, dx \\ &= \int_0^{1/2} 1\left(-x + \tfrac{5}{12}\right)\Psi_x \, dx + \int_{1/2}^1 \left(-x + \tfrac{5}{12}\right)\Psi_x \, dx \\ &= \int_0^1 \left(-x + \tfrac{5}{12}\right)\Psi_x \, dx \\ &= \left(-x + \tfrac{5}{12}\right)\Psi(x)\Big|_{x=0}^{x=1} - \int_0^1 (-1)\Psi \, dx. \end{aligned}$$

Since $\Psi_{(0)} = 0 = \Psi(1)$, we have

$$a(u, \Psi) = \int_0^1 1\Psi \, dx = (f, \Psi).$$

5.2 LINEAR SPACES OF REAL-VALUED FUNCTIONS

Before proceeding toward the error estimates, we want to examine some other spaces of functions and their relationship to each other. The following definition makes the term linear space of function precise.

Definition. Let S be a set of real-valued functions. S is called a *space* or *linear space* or a *vector space* of functions if and only if

(i) $u \in S$ and $\alpha \in \mathbb{R}$ imply $\alpha u \in S$,

(ii) $u, v \in S$ imply $u + v \in S$.

A set S that is *not linear* is given by

$$S = \{u \in H_0^1(0, L) | u(\tfrac{1}{2}) = 4\}.$$

Note that neither (i) nor (ii) holds, for example, if $u(\frac{1}{2}) = 4$, then $2u(\frac{1}{2}) = 8 \neq 4$ and so $2u \notin S$.

The following inequality, (5.2.1), is known as *Minkowski's inequality*, and it is a generalization of the triangle inequality for the absolute-value function.

Lemma . If $u, v \in L_2(0, L)$, then $u + v \in L_2(0, L)$ and, more precisely,

$$\|u + v\|_{L_2} \leq \|u\|_{L_2} + \|v\|_{L_2} \tag{5.2.1}$$

Proof. If suffices to show $\|u + v\|_{L_2}^2 \leq (\|u\|_{L_2} + \|v\|_{L_2})^2$.

$$\begin{aligned} \|u + v\|_{L_2}^2 &= \int_0^L (u + v)^2 \\ &= \int_0^L u^2 + 2\int_0^L uv + \int_0^L v^2 \\ &\leq \int_0^L u^2 + 2\|u\|_{L_2}\|v\|_{L_2} + \int_0^L v^2, \quad \text{by Cauchy's inequality.} \end{aligned}$$

$$\|u + v\|_{L_2}^2 \leq (\|u\|_{L_2} + \|v\|_{L_2})^2.$$

Proposition 5.2.1. $L_2(0, L)$, S^h, $H^1(0, L)$ and $H_0^1(0, L)$ are linear spaces.

Proof. We give only the proof that $H^1(0, L)$ is a linear space. Let $\alpha \in \mathbb{R}$ and $u \in H^1(0, L)$. Since $L_2(0, L)$ is linear, $\alpha u \in L_2(0, L)$ and $(\alpha u)^{\cdot} = \alpha \dot{u} \in L_2(0, L)$. Let $u, v \in H^1(0, L)$. Then $u, v, \dot{u}, \dot{v} \in L_2(0, L)$. Since $L_2(0, L)$ is linear, $u + v \in L_2(0, L)$ and $(u + v)^{\cdot} = \dot{u} + \dot{v} \in L_2(0, L)$. (If the derivative is interpreted as a weak derivative, then $(\alpha u)^{\cdot} = \alpha \dot{u}$ and $(u + v)^{\cdot} = \dot{u} + \dot{v}$ must be proved; see exercise 5-5.)

One can show that $\|\cdot\|_{L_2}$ and $\|\cdot\|_1$ have the following properties, which also hold for the absolute-value function:

(i) $\|u\| \geq 0$; $\|u\| = 0$ if and only if $u = 0$.

(ii) $\|\alpha u\| = |\alpha|\,\|u\|$ for all real α.

(iii) $\|u + v\| \leq \|u\| + \|v\|$.

Let S be a linear space of functions. Any function $\|u\| \in \mathbb{R}$ where $u \in S$ is called a *norm of S* if (i), (ii), and (iii) hold (see exercise 5-5).

Other examples of linear spaces are given in Figures 5.2.1 and 5.2.2. Note that these are a nested sequences of sets with the smallest sets consisting of those functions that are the smoothest.

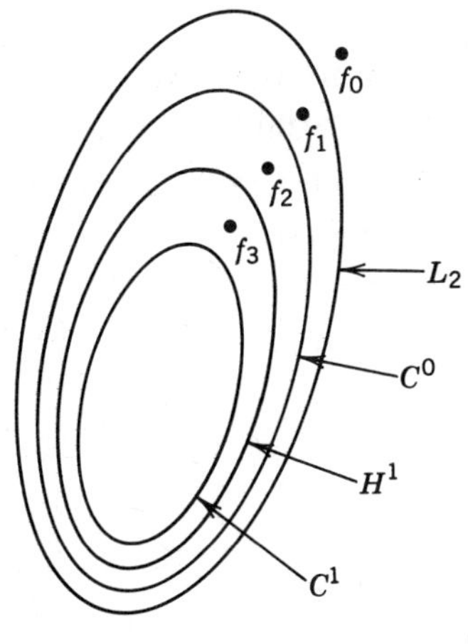

$f_0 \notin L_2(0,1)$, $\quad f_0(x) = x^{-1/2}$

$f_1 \notin C^0[0,1]$, $\quad f_1 \in L_2(0,1)$

$f_1(x) = x^{-1/4}$

$f_2 \notin H^1(0,1)$, $\quad f_2 \in C^0[0,1]$

$f_2(x) = x^{1/2}$

$f_3 \notin C^1[0,1]$, $\quad f_3 \in H^1(0,1)$

$f_3(x) = x^{3/4}$ where

$H^1(0,1) \equiv \{u\colon (0,1) \to \mathbb{R} \mid u, \dot{u} \in L_2(0,1)\}$

Figure 5.2.1

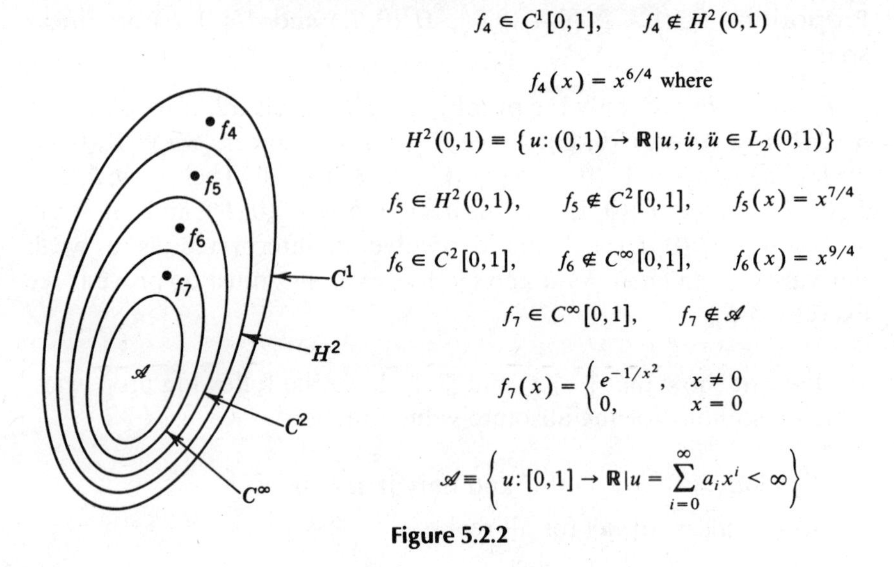

Figure 5.2.2

5.3 PROPERTIES OF $a(u, \Psi)$

The two propositions in this section state some routine but useful properties of a: $H^1(0, L) \times H^1(0, L) \to \mathbb{R}$.

Proposition 5.3.1 (Coercive and Bounded). Let $u, \Psi \in H_0^1(0, L)$ and $a(u, \Psi)$ be defined from (5.1.1)–(5.1.3).

(a) $a(u, \Psi)$ is a symmetric bilinear form, that is, $a(u, \Psi) = a(\Psi, u)$ and $a(u, \Psi)$ is linear in each variable.

(b) There exist constants C_1 and C_2, which are independent of the choice of $u, \Psi \in H_0^1(0, L)$, such that

$$C_1 \|u - \Psi\|_1^2 \leq a(u - \Psi, u - \Psi), \qquad (5.3.1)$$

$$a(u - \Psi, u - \Psi) \leq C_2 \|u - \Psi\|_1^2. \qquad (5.3.2)$$

If (5.3.1) holds, $a(u, \Psi)$ is called *coercive on* $H_0^1(0, L)$. If (5.3.1) holds, $a(u, \Psi)$ is called *bounded*.

Proof

(a)
$$a(u, \Psi) = \int_0^L m\dot{u}\dot{\Psi} + ku\Psi$$
$$= \int_0^L m\dot{\Psi}\dot{u} + k\Psi u$$
$$= a(\Psi, u).$$

Since $a(u, \Psi)$ is symmetric, it suffices to show $a(u, \Psi)$ is linear in Ψ, that is, for all $\alpha_1, \alpha_2 \in \mathbb{R}$

$$a(u, \alpha_1\Psi_1 + \alpha_2\Psi_2) = \alpha_1 a(u, \Psi) + \alpha_2 a(u, \Psi_2).$$

Note, by Proposition 5.2.1, $\alpha_1\Psi_1 + \alpha_2\Psi_2 \in H_0^1(0, L)$.

$$a(u, \alpha_1\Psi_1 + \alpha_2\Psi_2) = \int_0^L m\dot{u}(\alpha_1\dot{\Psi}_1 + \alpha_2\dot{\Psi}_2) + ku(\alpha_1\Psi_1 + \alpha_2\Psi_2)$$
$$= \alpha_1\int_0^L m\dot{u}\dot{\Psi}_1 + \alpha_2\int_0^L m\dot{u}\dot{\Psi}_2 + \alpha_1\int_0^L ku\Psi_1 + \alpha_2\int_0^L ku\Psi_2$$
$$= \alpha_1 a(u, \Psi_1) + \alpha_2 a(u, \Psi_2).$$

(b) In order to show inequality (5.3.1) holds, let $m_m = \min m(x)$ and $m_k = \min k(x)$, and note $0 < \min(m_m, m_k)$. Also, by Proposition 5.2.1, $u - \psi \in H_0^1(0, L)$.

$$a(u - \Psi, u - \Psi) = \int_0^L m(\dot{u} - \dot{\Psi})^2 + k(u - \Psi)^2$$
$$\geq \min(m_m, m_k)\int_0^L (\dot{u} - \dot{\Psi})^2 + (u - \Psi)^2$$
$$= \min(m_m, m_k)\|u - \Psi\|_1^2$$

Therefore, choose $C_1 = \min(m_m, m_k)$. In order to show inequality

(5.3.2) holds, let $M_m = \max m(x)$ and $M_k = \max k(x)$.

$$a(u - \Psi, u - \Psi) = \int_0^L m(\dot{u} - \dot{\Psi})^2 + k(u - \dot{\Psi})^2$$

$$\leq M_m \int_0^L (\dot{u} - \dot{\Psi})^2 + M_k \int_0^L (u - \Psi)^2$$

$$\leq \max(M_m, M_k) \|u - \Psi\|_1^2.$$

Therefore, choose $C_2 = \max(M_m, M_k)$.

In the next proposition we assume that the energy solution, u, and the variational finite element solution, u^h, exist.

Proposition 5.3.2. Let $u \in H_0^1(0, L)$ and $u^h \in S^h$ satisfy $X(u) = \min_{v \in H_0^1(0, L)} X(v)$ and $X(u^h) = \min_{v^h \in S^h} X(v^h)$. The following hold:

(a) $$a(u, \Psi) = (f, \Psi) \quad \text{for all } \Psi \in H_0^1(0, L). \tag{5.3.3}$$

(b) $$a(u^h, v^h) = (f, v^h) \quad \text{for all } v^h \in S^h. \tag{5.3.4}$$

(c) $$a(u - u^h, v^h) = 0 \quad \text{for all } v^h \in S^h. \tag{5.3.5}$$

(d) $$a(u - u^h, u - u^h) \leq a(u - u^h - v^h, u - u^h - v^h) \quad \text{for all } v^h \in S^h. \tag{5.3.6}$$

Proof. **(a)** and **(b)** Equations (5.3.3) and (5.3.4) are derived in the usual way by defining

$$F(\lambda) = X(u + \lambda\Psi), \qquad u + \lambda\Psi \in H_0^1(0, 1),$$

$$F^h(\lambda) = X(u^h + \lambda v^h), \qquad u^h + \lambda v^h \in S^h.$$

As $F(0)$ and $F^h(0)$ are minimum values, $F'(0) = 0$ and $F^{h'}(0) = 0$. These equations imply (5.3.3) and (5.3.4).

(c) Equation (5.3.5) is derived by letting $\Psi = v^h$ in line (5.3.3). Then we have

$$a(u, v^h) = (f, v^h),$$

$$a(u^h, v^h) = (f, v^h).$$

By subtracting these two and using the linearity of part (a) in Proposition 5.3.1 we have

$$a(u, v^h) - a(u^h, v^h) = (f, v^h) - (f, v^h),$$

$$a(u - u^h, v^h) = 0$$

(d) In order to prove the inequality (5.3.6), use the linearity of (a) in Proposition 5.3.1

$$a((u - u^h) - v^h, (u - u^h) - v^h) = a(u - u^h, u - u^h) - 2a(u - u^h, v^h) + a(v^h, v^h)$$

Since $a(u - u^h, v^h) = 0$ and $a(v^h, v^h) \geq 0$,

$$a(u - u^h - v^h, u - u^h - v^h) \geq a(u - u^h, u - u^h).$$

5.4 INTERPOLATION, COMPLETENESS, AND CONTINUITY OF FUNCTIONS IN $H_0^1(0, L)$

The interpolation, or approximation, of a function in $H_0^1(0, L)$ is studied. The concept of a weak derivative is introduced and used to show $H^1(0, L)$ is complete. We conclude the section by showing that any function in $H^1(0, L)$ is continuous.

Definition. Let $u \in C[0, L]$. The *linear interpolation* of u, u_I, is

$$u_I(x) = \sum_{j=0}^{N} u\left(\frac{jL}{N}\right)\Psi_j(x).$$

Proposition 5.4.1 (Interpolation). If $u \in C^2[0, L]$, then

(a) $|u - u_I| \leq h^2 \max_{(0, L)}|\ddot{u}|, \qquad h = L/N.$ (5.4.1)

(b) $|\dot{u} - \dot{u}_I| \leq h \max_{(0, L)}|\ddot{u}|.$ (5.4.2)

(c) There exists a C_3, which is independent of h, such that

$$\|u - u_I\|_1^2 \leq C_3 h^2 \left(\max_{(0, L)}|\ddot{u}|\right)^2. \tag{5.4.3}$$

Proof. **(a)** and **(b)** Consider $u - u_I$ on each interval (x_i, x_{i+1}). As $(u - u_I)(x_i) = 0$ and by the fundamental theorem of calculus,

$$(u - u_I)(x) = \int_{x_i}^{x} (\dot{u} - \dot{u}_I)\,dx. \tag{5.4.4}$$

Since $\dot{u} - \dot{u}_I \in C^1[0, L]$ and $(u - u_I)(x_i) = 0 = (u - u_I)(x_{i+1})$, the mean value theorem gives a $z \in (x_i, x_{i+1})$ such that $(\dot{u} - \dot{u}_I)(z) = 0$. Again by the fundamental theorem of calculus

$$\begin{aligned}(\dot{u} - \dot{u}_I)(x) &= \int_{z}^{x} (\ddot{u} - \ddot{u}_I)\,dx, \quad z, x \in (x_i, x_{i+1}) \\ &= \int_{z}^{x} \ddot{u}\,dx, \qquad\qquad u_I \text{ is linear on } (x_i, x_{i+1}).\end{aligned} \tag{5.4.5}$$

Since $u \in C^2[x_i, x_{i+1}]$, $\max_{(x_i, x_{i+1})}|\ddot{u}| \le \max_{(0, L)}|\ddot{u}| < \infty$. Thus, since $z, x \in (x_i, x_{i+1})$, line (5.4.5) yields (5.4.2). Line (5.4.1) is obtained by inserting (5.4.2) into (5.4.4) and using the fact that $x \in (x_i, x_{i+1})$.

(c) In order to derive (5.4.3), use (5.4.1) and (5.4.2) in the definition of $\|u - u_I\|_1^2$.

$$\begin{aligned}\|u - u_I\|_1^2 &= \int_0^L (\dot{u} - \dot{u}_I)^2 + (u - u_I)^2 \\ &\le \int_0^L \Big(h \max_{(0, L)}|\ddot{u}|\Big)^2 + \Big(h^2 \max_{(0, L)}|\ddot{u}|\Big)^2 \\ &\le (h^2 + h^4)\int_0^L \Big(\max_{(0, L)}|\ddot{u}|\Big)^2 \\ &\le h^2 \cdot L(1 + L^2) \cdot \Big(\max_{(0, L)}|\ddot{u}|\Big)^2\end{aligned}$$

Therefore, for this interval $(0, L)$ choose $C_3 = (1 + L^2)L$.

We shall now precisely define the meaning of a derivative for any function in $L_2(0, L)$. This definition gives more insight to the definition of a weak solution, and it will be used in the study of completeness of $H^1(0, L)$.

Definition. A *weak derivative of* $u \in L_2(0, L)$ *in* $L_2(0, L)$ is a function $v \in L_2(0, L)$ such that for all $\Psi \in C_c^\infty(0, L) =$

$$\{\Psi \in C^\infty(0, L) | \Psi \equiv 0 \text{ on } (0, L)\backslash(\epsilon, L - \epsilon) \text{ for some } \epsilon > 0\}$$

$$-\int_0^L u\dot{\Psi} = \int_0^L v\Psi. \tag{5.4.6}$$

The motivation for this definition comes from the property that any function $u \in C^1(0, L)$, with a classical derivative, has

$$\begin{aligned}\int_0^L u\dot{\Psi} &= u\Psi|_0^L - \int_0^1 \dot{u}\Psi \\ &= 0 - \int_0^L \dot{u}\Psi.\end{aligned}$$

Consequently, if $u \in C^1(0, L)$, then $v = \dot{u}$, and so a function with a classical derivative also has a weak derivative. The next example illustrates a function with a weak derivative and no classical derivative.

Example. $u = |x - \frac{1}{2}|$, $L = 1.0$. We claim that

$$v = \begin{cases} -1, & x < \frac{1}{2} \\ 1, & x > \frac{1}{2} \end{cases}$$

is a weak derivative of u. We must show (5.4.6) holds:

$$\begin{aligned}-\int_0^1 u\dot{\Psi} &= -\int_0^{1/2} -(x - \tfrac{1}{2})\dot{\Psi} - \int_{1/2}^1 (x - \tfrac{1}{2})\dot{\Psi} \\ &= -(-(x - \tfrac{1}{2}))\Psi(x)|_0^{1/2} + \int_0^{1/2} (-1)\Psi \\ &\quad -((x - \tfrac{1}{2})\Psi(x))|_{1/2}^1 + \int_{1/2}^1 (+1)\Psi \\ &= \int_0^{1/2} (-1)\Psi + \int_{1/2}^1 (+1)\Psi = \int_0^1 v\Psi\end{aligned}$$

by using $\Psi(0) = 0 = \Psi(L)$. So, even though u does not have a classical derivative, it does have a weak derivative.

Remark. If the derivatives in (5.1.1) are viewed as weak derivatives, then this solution will be the weak solution, that is, $a(u, \Psi) = (f, \Psi)$ for all $\Psi \in H_0^1(0, L)$. In particular, the reader should consider the first example in Section 5.1, where $k = 0$, $m = K(x) = 1$ for $0 \le x \le \frac{1}{2}$, and $K(x) = 2$ for $\frac{1}{2} < x \le 1$.

As previously mentioned, the proof of existence is given by finding a sequence $u^l \in H_0^1(0, L)$ such that $X(u^l) \downarrow d$ as $l \to \infty$ where $d = \inf_{v \in S} X(v)$ and $S = H_0^1(0, L)$. We hope that u^l will "converge" to $u \in H_0^1(0, L)$ and $d = X(u)$. The definition of "convergence" in $H_0^1(0, L)$ is as follows.

Definition. Let $u^l, u \in H_0^1(0, L)$. $u^l \to u$ *converges in* $H_0^1(0, L)$ if and only if $\|u^l - u\|_1 \to 0$ as $l \to \infty$.

The difficulty with this concept is that in order to show convergence, one must have both the limit u and the sequence u^l. In our case the limit is the would-be solution whose existence we are trying to establish! Fortunately, we may extend the concept of completeness of the real line $\mathbb{R}$ to the completeness of $H_0^1(0, L)$. The completeness of the real line characterizes convergence as follows:

> Let $x^l, x \in \mathbb{R}$. x^l converges to x in $\mathbb{R}$ if and only if x^l is a Cauchy sequence.
>
> x^l is defined to be *Cauchy* if and only if for all $\epsilon > 0$ there exists an N such that if $l, m \ge N$, then $|x^l - x^m| < \epsilon$.

The proof of this characterization may be found in Taylor and Mann [29], chapter 16. Note, the Cauchy criterion does not require the limit x. This property is very useful and is illustrated for some elementary numerical methods in Chapter 7.

The generalization of completeness of $\mathbb{R}$ to $L_2(0, L)$ is discussed in Royden [19], chapters 1–6. In this case $\mathbb{R}$ is replaced by $L_2(0, L)$ and the distance $|x^l - x^m|$ is replaced by $\|u^l - u^m\|_{L_2}$. We assume the reader is convinced that $L_2(0, L)$ is complete with respect to $\|\cdot\|_{L_2}$. The generalization of $\mathbb{R}$ to $H^1(0, L)$ (or $H_0^1(0, L)$) is as follows and is proved in Proposition 5.4.2. The reader should note how the concept of a weak derivative is used in the proof.

Definition. $H^1(0, L)$ (or $H_0^1(0, L)$) is *complete with respect to* $\|\cdot\|_1$ if and only if convergence is characterized by a sequence being Cauchy. $u^l \in H^1(0, L)$ (or $H_0^1(0, L)$) is *Cauchy with respect to* $\|\cdot\|_1$ if and only if for all $\epsilon > 0$ there exists an N such that if $l, m \geq N$, then $\|u^l - u^m\|_1 < \epsilon$.

Proposition 5.4.2. (Completeness) $H_0^1(0, L)$ and $H^1(0, L)$ are complete with respect to $\|\cdot\|_1$.

Proof. First, let us show $H^1(0, L)$ is complete. Let $u^l \in H^1(0, L)$ with respect to $\|\cdot\|_1$. By definition of $\|\cdot\|_1$, u^l and $(u^l)^{\cdot}$ are Cauchy with respect to $\|\cdot\|_{L_2}$. Since $L_2(0, L)$ is complete, u^l and $(u^l)^{\cdot}$ converge in $L_2(0, L)$ to u and v, respectively. It remains to show $\dot{u} = v$, that is, line (5.4.6) holds. By definition,

$$-\int_0^L u^l \dot{\Psi} = \int_0^L (u^l)^{\cdot}\, \Psi.$$

By Cauchy's inequality,

$$\left|\int_0^L u^l \dot{\Psi} - \int_0^L u \dot{\Psi}\right| = \left|\int_0^L (u^l - u) \dot{\Psi}\right|$$

$$\leq \|u^l - u\|_{L_2(0, L)} \|\dot{\Psi}\|_{L_2(0, L)}$$

and

$$\left|\int_0^L (u^l)^{\cdot}\, \Psi - \int_0^L v \Psi\right| \leq \left|\int_0^L ((u^l)^{\cdot} - v) \Psi\right|$$

$$\leq \|(u^l)^{\cdot} - v\|_{L_2(0, L)} \|\Psi\|_{L_2(0, L)}.$$

Consequently,

$$-\int_0^L u^l \dot{\Psi} \to -\int_0^L u \dot{\Psi},$$

$$\int_0^L (u^l)^{\cdot}\, \Psi \to \int_0^L v \Psi.$$

Thus

$$-\int_0^L u\dot{\Psi} = \int_0^L v\Psi \quad \text{for all } \Psi \in C_c^\infty(0, L).$$

This shows that the weak derivative of u is $v \in L_2(0, L)$ and hence $u \in H^1(0, L)$.

In order to show $H_0^1(0, L)$ is complete we need part (b) of the next proposition when $H^1(0, L)$ is used. We leave the proof to the reader.

Remarks

1. $C[0, L]$ is not complete with respect to $\|\cdot\|_{L_2}$. For example,

$$u_n = \begin{cases} n^{1/4}, & 0 \le x \le 1/n \\ x^{-1/4}, & 1/n < x \le 1 \end{cases}$$

is in $C[0,1]$ and converges in $L_2(0,1)$ to $u = x^{-1/4} \in L_2(0,1)$. But u is not continuous at $x = 0$ and so $u \notin C[0,1]$. Also, $C^1[0, L]$ is not complete with respect to $\|\cdot\|_1$. To see this, just consider $v_n \equiv \int_0^x u_n$ and $v \equiv \int_0^x u$.
2. $H_0^1(0, L)$ could be viewed as a completion of $C_c^1(0, L) = \{u \in C^1(0, L) | u(0) = 0 = u(L)\}$ with respect to the $\|\cdot\|_1$ norm.

The next proposition simply states that any function in $H_0^1(0, L)$ is also continuous. Its generalization to $\Omega \subset \mathbb{R}^2$ or $\mathbb{R}^3$ is known as Sobolev's embedding lemma (See Friedman [11, chapter I.11]).

Proposition 5.4.3 (Continuity). Let $u \in H^1(0, L)$.

(a) $H^1(0, L) \subset C[0, L]$ and $|u(y) - u(x)| \le |y - x|^{1/2} \|\dot{u}\|_{L_2}$.

(b) There is a constant C_4, which is independent of $u \in H^1(0, L)$, such that

$$|u| \le C_4 \|u\|_1. \tag{5.4.7}$$

Consequently, if $u^l \to u$ in $H^1(0, L)$, then $u^l \to u$ uniformly in $C[0, L]$.

Proof. **(a)** By Cauchy's inequality

$$
\begin{aligned}
|u(y) - u(x)| &= \left|\int_x^y 1\dot{u}\right| \\
&\le |y - x|^{1/2}\left(\int_x^y (\dot{u})^2\right)^{1/2} \\
&\le |y - x|^{1/2}\left(\int_0^L (\dot{u})^2\right)^{1/2} \\
&= |y - x|^{1/2}\|\dot{u}\|_{L_2}. \qquad (5.4.8)
\end{aligned}
$$

(b) Since u is continuous on $[0, L]$, there exists an $x_0 \in [0, L]$ such that $u(x_0)^2 = \min u(x)^2 \le u(x)^2$. By integrating from 0 to L we obtain

$$u(x_0)^2 \le L^{-1}\int_0^L u^2.$$

Also, $u(x) = \int_{x_0}^x \dot{u} + u(x_0)$, and hence

$$
\begin{aligned}
|u(x)| &\le \left|\int_{x_0}^x 1\dot{u}\right| + |u(x_0)| \\
&\le L^{1/2}\|\dot{u}\|_{L_2} + L^{-1/2}\|u\|_{L_2} \\
&\le (L^{1/2} + L^{-1/2})\|u\|_1.
\end{aligned}
$$

Define $C_4 = L^{1/2} + L^{-1/2}$ to obtain (5.4.7). In order to show $u^l \to u$ uniformly when $u^l \to u$ in $H^1(0, L)$, simply apply (5.4.7) with u replaced by $u^l - u$

$$|u^l(x) - u(x)| \le C_4\|u^l - u\|_1. \qquad (5.4.9)$$

Proposition 5.4.3 allows one to show that the norm $\|u\|_{①} \equiv \|\dot{u}\|_{L_2}$ on $H_0^1[0, L]$ is *equivalent to the norm* $\|u\|_1$ on $H_0^1[0, L]$, that is, there are constants K_1, K_2, which are independent of $u \in H_0^1(0, L)$, such that

$$K_1\|u\|_{①} \le \|u\|_1 \le K_2\|u\|_{①}.$$

In order to prove this, note by (5.4.8) with $x = 0$

$$\begin{aligned}\|u\|_{①}^2 = \|\dot{u}\|_{L_2}^2 \le \|u\|_1^2 &= \|\dot{u}\|_{L_2}^2 + \|u\|_{L_2}^2 \\ &= \|\dot{u}\|_{L_2}^2 + L\|\dot{u}\|_{L_2}^2 \\ &= (1 + L)\|u\|_{①}^2.\end{aligned}$$

Thus, choose $K_1 = 1$ and $K_2 = (1 + L)^{1/2}$. As a consequence, the results of this chapter for problem (5.1.1)–(5.1.3) hold when $k(x) \ge 0$, which is less restrictive than $k(x) \ge \delta > 0$ (see exercise 5-7).

5.5 EQUIVALENCE OF CLASSICAL, ENERGY, AND WEAK FORMULATIONS

Before proving the equivalence theorem, we need to show the weak solution $u \in C^2[0, L]$. This will allow us to use the interpolation results of Proposition 5.4.1. The condition that $m \in C^1[0, L]$ is essential as the examples in Section 5.1 have shown.

Proposition 5.5.1 (Smoothness). Let $u \in H_0^1(0, L)$ be a weak solution of (5.1.1)–(5.1.3). Then $u \in C^2[0, L]$.

Proof. By the continuity proposition (Proposition 5.4.3), $u \in C[0, L]$. Since u is a weak solution, $-(m(x)\dot{u})^{\cdot} + k(x)u = f(x)$ where weak derivatives are used. Thus, as $k, f \in C[0, L]$,

$$-(m(x)\dot{u})^{\cdot} = f(x) - k(x) \cdot u(x)$$

is continuous, and consequently $-m(x)\dot{u} \in C^1[0, L]$. Since $0 < m(x) \in C[0, L]$, $\dot{u}$ must be continuous, that is, $u \in C^1[0, L]$. Since $m \in C^1[0, L]$, $-(m\dot{u})^{\cdot} = -\dot{m}\dot{u} - m\ddot{u} = f - ku$. Thus, $\ddot{u} = (f - ku + \dot{m}\dot{u})/(-m)$ is continuous, that is, $u \in C^2(0, L]$.

Lemma. Let $f,\ g \in C[0, L]$. If for all $\Psi \in C_c[0, L] = \{u \in C[0, L] | \Psi = 0 \text{ on } (0, 1)\setminus(\epsilon, L - \epsilon) \text{ for some } \epsilon > 0\}$

$$\int_0^L g\Psi = \int_0^L f\Psi,$$

then $g = f$.

Proof. If $f \neq g$, then there is an $x_0 \in (0, L)$ such that $f(x_0) \neq g(x_0)$. Suppose $f(x_0) - g(x_0) \geq \delta > 0$. Since $(f - g)(x)$ is continuous at x_0, there is an $\epsilon > 0$ such that for all $x \in (x_0 - \epsilon, x_0 + \epsilon)$, $f(x) - g(x) \geq \delta/2$. Choose $\Psi \in C[0, L]$ such that $\Psi(x) = 0$ for $x \notin (x_0 - \epsilon, x_0 + \epsilon)$, $\Psi(x) \geq 0$ for $x \in (x_0 - \epsilon, x_0 + \epsilon)$, and $\Psi(x_0) > 0$. Then $\int_0^L (f - g)\Psi > 0$, which is a contradiction of the assumption. Hence, $f = g$.

Theorem 5.5.1. Consider (5.1.1)–(5.1.3) with $m \in C^1[0, L]$ and $k, f \in C[0, L]$. The classical, energy, and weak formulations are equivalent. Moreover, the solution is unique (if it exists).

Proof. The uniqueness follows just as in the proof of Theorem 1.4.1. Also, that any classical and energy solutions are the unique weak solution follows in the same way. It remains to show that any weak solution is also a classical solution. Proposition 5.5.1 shows that a weak solution is in $C^2[0, L]$. Now, integrate backward the weak equation for all $\Psi \in H_0^1(0, L)$:

$$a(u, \Psi) = (f, \Psi),$$

$$\int_0^L m(x)\dot{u}\dot{\Psi} + k(x)u\Psi = \int_0^L f(x)\Psi,$$

$$\int_0^L (-m(x)\dot{u})^{\cdot} + k(x)u)\Psi = \int_0^L f(x)\Psi.$$

Since $-(m\dot{u})^{\cdot} + ku$ and f are continuous, we may apply the above lemma with $g = -(m\dot{u})^{\cdot} + ku$ and $f = f$ to conclude $-(m(x)\dot{u})^{\cdot} + k(x)u = f(x)$.

5.6 ERROR ESTIMATES

Finally, we are in a position to derive the simplest of error estimates given in line (5.6.1). If more smoothness of m, k, and f is assumed, then higher-order estimates given in lines (5.6.9) and (5.6.10) may be established.

Theorem 5.6.1. Consider problem (5.1.1)–(5.1.3) with $m \in C^1[0, L]$. $u^h \in H_0^1(0, L)$ be a variational finite element solution of (5.1.1)–(5.1.3) (or, a Galerkin finite element solution of (5.1.1)–(5.1.3), that is, u^h satisfies 5.3.4). Then there exists a constant C, which is independent of h, such that

$$|u - u^h| \leq Ch. \tag{5.6.1}$$

Proof. By the smoothness proposition (Proposition 5.5.1), $u \in C^2[0, L]$. By the interpolation proposition (Proposition 5.4.1, part (c)),

$$\|u - u_{\mathrm{I}}\|_1^2 \leq C_3 h^2 (\max|\ddot{u}|)^2 \tag{5.6.2}$$

By the coercive portion of Proposition 5.3.1, part (b), with $\Psi = u^h$,

$$C_1 \|u - u^h\|_1^2 \leq a(u - u^h, u - u^h). \tag{5.6.3}$$

By Proposition 5.3.2, part (d), with $v^h = -u^h + u_{\mathrm{I}} \in S^h$,

$$a(u - u^h, u - u^h) \leq a(u - u_{\mathrm{I}}, u - u_{\mathrm{I}}). \tag{5.6.4}$$

By the bounded portion of Proposition 5.3.1, part (b),

$$a(u - u_{\mathrm{I}}, u - u_{\mathrm{I}}) \leq C_2 \|u - u_{\mathrm{I}}\|_1^2 \tag{5.6.5}$$

Combine lines (5.6.2)–(5.6.5) to get

$$C_1 \|u - u^h\|_1^2 \leq C_2 C_3 h^2 (\max|\ddot{u}|)^2. \tag{5.6.6}$$

By the continuity proposition (Proposition 5.4.3, part (b)), with $u = u - u^h$,

$$|u - u^h| \leq C_4 \|u - u^h\|_1, \tag{5.6.7}$$

$$|u - u^h| \leq C_4 \left(\frac{C_2 C_3}{C_1} (\max|\ddot{u}|)^2 \right)^{1/2} h. \tag{5.6.8}$$

Therefore, choose

$$C = C_4 \left(\frac{C_2 C_3}{C_1} (\max|\ddot{u}|)^2 \right)^{1/2}$$

to get (5.6.1).

Corollary 1. If $f, k \in C^2[0, L]$ and $m \in C^3[0, L]$, then there exists a constant $\bar{C}$ such that

$$|u - u^h| \leq \bar{C}h^2 \tag{5.6.9}$$

Corollary 2. If $f, k \in C^2[0, L]$, $m \in C^3[0, L]$ and u^h is a finite element solution of $-(m(x)\dot{u})^{\dot{}} \underline{\underline{+}} k(x) \cdot u = f_{\mathrm{I}}(x)$, $u(0) = 0 = u(L)$, then there exists a constant $\bar{\bar{C}}$ such that, for u a solution of $-(m\dot{u})^{\dot{}} + ku = f$ and $u(0) = 0 = u(L)$,

$$|u - u^h| \leq \bar{\bar{C}}h^2 \tag{5.6.10}$$

5.7 EXISTENCE

In this section we establish existence of $u \in H_0^1(0, L)$ and $u^h \in S^h$ that are the energy and variational finite element solutions of (5.1.1)–(5.1.3). The crucial step in the argument is that $H_0^1(0, L)$ is complete with respect to $\|\cdot\|_1$ (see Proposition 5.4.2 and the preceding definition). This enables us to demonstrate convergence of $u^l \in H_0^1(0, L)$ without having the limit $u \in H_0^1(0, L)$.

Theorem 5.7.1. There exists $u \in H_0^1(0, L)$ and $u^h \in S^h$ that are energy and variational finite element solutions of (5.1.1)–(5.1.3).

Proof. The proof has three parts. First, we establish that there exists a number $\Gamma > -\infty$ such that for all $u \in H_0^1(0, L)$,

$$X(u) \geq \Gamma. \tag{5.7.1}$$

Now, we let $u^l \in H_0^1(0, L)$ be such that

$$\Gamma \leq d = \inf_{v \in H_0^1(0, L)} X(v) \downarrow X(u^l) \tag{5.7.2}$$

Note that d is finite from (5.7.1). Second, we show u^l is Cauchy with respect to $\|\cdot\|_1$. By the completeness proposition (Proposition 5.4.2), u^l has a limit $u \in H_0^1(0, L)$. The third part is to show $d = X(u)$.

Part one. Recall $X(u) = \frac{1}{2}a(u,u) - (f,u)$. By the Cauchy inequality (5.1.8),

$$(f,u) = \int_0^L fu \le \|f\|_{L_2}\|u\|_{L_2} \le \|f\|_{L_2}\|u\|_1. \tag{5.7.3}$$

By the coercive property of $a(u,u)$ in Proposition 5.3.1,

$$\frac{1}{2}a(u,u) \ge \frac{C_1}{2}\|u\|_1^2. \tag{5.7.4}$$

Therefore, (5.7.3) and (5.7.4) imply

$$\begin{aligned} X(u) &\ge \frac{C_1}{2}\|u\|_1^2 - \|f\|_{L_2}\|u\|_1 \\ &= \frac{C_1}{2}\left(\|u\|_1^2 - 2\left(\frac{\|f\|_{L_2}}{C_1}\right)\|u\|_1\right) \\ &= \frac{C_1}{2}\left(\|u\|_1^2 - 2\alpha\|u\|_1 + \alpha^2\right) - \frac{C_1}{2}\alpha^2, \qquad \alpha = \frac{\|f\|_{L_2}}{C_1}, \end{aligned}$$

and so

$$X(u) \ge -\frac{C_1}{2}\alpha^2.$$

So, choose $\Gamma = -(C_1/2)\alpha^2$ and line (5.7.1) holds.

Part two. Because $a(u,\Psi)$ in symmetric and bilinear (see Proposition 5.3.1),

$$\begin{aligned} a\left(\frac{u^l+u^m}{2}, \frac{u^l+u^m}{2}\right) + a\left(\frac{u^l-u^m}{2}, \frac{u^l-u^m}{2}\right) \\ = \frac{1}{2}a(u^l,u^l) + \frac{1}{2}a(u^m,u^m). \end{aligned} \tag{5.7.5}$$

Now, subtract

$$2\left(f, \frac{u^l+u^m}{2}\right) = (f,u^l) + (f,u^m)$$

from both sides of (5.7.5):

$$a\left(\frac{u^l + u^m}{2}, \frac{u^l + u^m}{2}\right) - 2\left(f, \frac{u^l + u^m}{2}\right) + a\left(\frac{u^l - u^m}{2}, \frac{u^l - u^m}{2}\right)$$

$$= \frac{1}{2}a(u^l, u^l) - (f, u^l) + \frac{1}{2}a(u^m, u^m) - (f, u^m). \tag{5.7.6}$$

In terms of $X(u)$, (5.7.6) is

$$2X\left(\frac{u^l + u^m}{2}\right) + a\left(\frac{u^l - u^m}{2}, \frac{u^l - u^m}{2}\right) = X(u^l) + X(u^m). \tag{5.7.7}$$

Now, choose $\overline{N}$ so that if $l, m \geq \overline{N}$, then

$$d \leq X(u^l) \leq d + \epsilon, \tag{5.7.8}$$

$$d \leq X(u^m) \leq d + \epsilon. \tag{5.7.9}$$

Also, note

$$d \leq X\left(\frac{u^l + u^m}{2}\right). \tag{5.7.10}$$

By placing (5.7.8)–(5.7.10) into (5.7.7), we have

$$a\left(\frac{u^l - u^m}{2}, \frac{u^l - u^m}{2}\right) \leq (d + \epsilon) + (d + \epsilon) - 2d = 2\epsilon. \tag{5.7.11}$$

Again use the coercive property in Proposition 5.3.1 and line (5.7.11)

$$C_1\left\|\frac{u^l - u^m}{2}\right\|_1^2 \leq a\left(\frac{u^l - u^m}{2}, \frac{u^l - u^m}{2}\right) \leq 2\epsilon. \tag{5.7.12}$$

Since

$$\left\|\frac{u^l - u^m}{2}\right\|_1^2 = \frac{1}{4}\|u^l - u^m\|_1^2,$$

line (5.7.12) yields

$$\|u^l - u^m\|_1^2 \le (2\epsilon \cdot 4)/C_1 \tag{5.7.13}$$

when $l, m \ge \overline{N}$. Thus, u^l is Cauchy and so there is a $u \in H_0^1(0, L)$ such that $u^l \to u$ in $H_0^1(0, L)$.

Part three. In order to show $X(u) = d = \inf_{v \in H_0^1(0,L)} X(v)$, it suffices to show $X(u^l) \to X(u)$. Consider the terms of $X(u) = \frac{1}{2}a(u, u) - (f, u)$.

$$\begin{aligned} |(f,u) - (f,u^l)| &= \left| \int_0^L fu - \int_0^L fu^l \right| \\ &= \left| \int_0^L f(u - u^l) \right| \\ &\le \|f\|_{L_2} \|u - u^l\|_{L_2} \quad \text{by Cauchy's inequality.} \end{aligned}$$

Therefore,

$$|(f,u) - (f,u^l)| \le \|f\|_{L_2} \|u - u^l\|_1. \tag{5.7.14}$$

$$\begin{aligned} |a(u,u) - a(u^l,u^l)| &= \left| \int_0^L m\dot{u}^2 + ku^2 - \int_0^L m\dot{u}^{l^2} + k\dot{u}^{l^2} \right| \\ &\le \left| \int_0^L m(\dot{u}^2 - \dot{u}^{l^2}) \right| + \left| \int_0^1 k(u^2 - u^{l^2}) \right| \\ &= \left| \int_0^L m(\dot{u} - \dot{u}^l)(\dot{u} + \dot{u}^l) \right| \\ &\quad + \left| \int_0^L k(u - u^l)(u + u^l) \right| \\ &\le M_m \left| \int_0^L (\dot{u} - \dot{u}^l)(\dot{u} + \dot{u}^l) \right| \\ &\quad + M_k \left| \int_0^L (u - u^l)(u + u^l) \right|, \end{aligned} \tag{5.7.15}$$

where $M_m = \max m$ and $M_k = \max k$. Again use Cauchy's inequality in (5.7.15).

$$\begin{aligned}|a(u,u) - a(u^l,u^l)| &\le M_m\|\dot{u} - \dot{u}^l\|_{L_2}\|\dot{u} + \dot{u}^l\|_{L_2} \\ &\quad + M_k\|u - u^l\|_{L_2}\|u + u^l\|_{L_2} \\ &\le M_m\|u - u^l\|_1 \cdot \|u + u^l\|_1 \\ &\quad + M_k\|u - u^l\|_1 \cdot \|u + u^l\|_1. \qquad (5.7.16)\end{aligned}$$

Since u^l converges in $H_0^1(0, L)$, there is a constant M, which is independent of l, such that

$$\|u + u^l\|_1 \le M. \qquad (5.7.17)$$

Lines (5.7.16) and (5.7.17) combine to yield

$$|a(u,u) - a(u^l,u^l)| \le (M_m M + M_k M)\|u - u^l\|_1 \qquad (5.7.18)$$

Thus, by the definition of $X(u)$, lines (5.7.14) and (5.7.18) imply $X(u^l) \to X(u)$. The theorem is proved for the energy solution part.

In order to prove the variational finite element segment, note S^h is closed in $H_0^1(0, L)$, that is, $u^l \in S^h$ and $u^l \to u$ in $H_0^1(0, L)$ implies $u \in S^h$. In order to show $u \in S^h$, note that $u^l = \Sigma_j u_j^l \Psi_j$ and by Proposition 5.4.3, part (b), $u^l(x_j) = u_j^l \to u(x_j)$. Thus, $u(x) = \Sigma_j u_j \Psi_j$, where $u_j = u(x_j)$ and so $u \in S^h$. Then it follows that S^h is complete with respect to $\|\cdot\|_1$, that is, if $u^l \in S^h$ is Cauchy, then $u^l \to u$ in $H_0^1(0, L)$ and $u \in S^h$. The proof for the variational finite element case proceeds in the same three parts once $H_0^1(0, L)$ is replaced by S^h.

5.8 OBSERVATIONS AND REFERENCES

The error estimates that we have derived have been in terms of each point evaluation. Somewhat weaker estimates may be derived for mean square norm, $\|\cdot\|_{L_2}$. In particular, an elementary proof is given in Strang and Fix [28, chapter 1] for problem (5.1.1)–(5.1.3) with $m = 1$ and $k = 0$, that

$$\|u - u^h\|_{L_2} \leq Ch^2.$$

Note that the norm $\|\cdot\|_{L_2}$ is weaker than absolute value, but the order is now 2 and not 1. Also, in higher space dimensions the $\|\cdot\|_{L_2}$ norm is easier to use than absolute value.

If one uses higher-order shape functions, then the interpolation estimate may be improved and, hence, the order of the error estimate improved. For example, in Ortega [15], quadratic shape functions give the estimate

$$\|u - u_{\mathrm{I}}\|_1^2 \leq Ch^4.$$

$H^1(0, L)$ is a special case of sequences of Sobolev spaces for higher-order derivatives, higher space dimensions, and other L_p spaces of function. All these are complete with respect to certain norms. Some examples and their uses are given in Friedman [11].

The existence theorem in Section 5.7 may be generalized for Hilbert spaces and has many applications to more complicated problems. Also, the minimizing set does not have to be the entire space; it need only be a closed convex subset of the space. This, in general, gives rise to variational inequalities (in place of the weak equation). These will be discussed in Chapter 9. Interesting books on this subject are Elliott and Ockendon [9] and Stampacchia [27].

EXERCISES

5-1 Consider the three problems given in the example at the beginning of Section 5.1. Verify that they are weak solutions.

5-2 Consider

$$-(K(x)u_x)_x = x,$$

$$u(0) = 0 = u(1),$$

where

$$K(x) = \begin{cases} 2, & 0 \le x \le \frac{1}{2} \\ 3, & \frac{1}{2} < x \le 1. \end{cases}$$

Find a solution and verify that it is a weak solution.

5-3 Consider the problem in 5-2 with $u(1) = 0$ replaced by $u_x(1) = 4 - u(1)$. Find the solution and verify that it is a weak solution.

5-4 Show that the weak derivative of

$$u(x) = \begin{cases} -2x, & 0 \le x \le \frac{1}{2} \\ x - \frac{3}{2}, & \frac{1}{2} < x \le 1 \end{cases}$$

is

$$v(x) = \begin{cases} -2, & 0 \le x \le \frac{1}{2} \\ 1, & \frac{1}{2} < x \le 1. \end{cases}$$

5-5 For weak derivatives show the following properties:

(a) $(\alpha u)^{\dot{}} = \alpha \dot{u}$.

(b) $(u + v)^{\dot{}} = \dot{u} + \dot{v}$.

(c) $\|\cdot\|_1$ is a norm on $H^1(0, L)$.

(d) Show that the weak derivative of any continuous function is unique. Use the Lemma that follows Proposition 5.5.1.

(e) Show that any weak derivative that is continuous must also be a classical derivative. This is used in the proof of Proposition 5.5.1; can you find where it is used?

(f) Show that the weak derivative of $\int_{x_0}^{x} u(x)\, dx + C$ is $u(x)$ where $u \in L_2(0, L)$.

(g) Show that $u =$ constant when the weak derivative is assumed to be zero and u is continuous.

(h) In the Lemma following Proposition 5.5.1 and parts (d), (e), and (g) continuity was assumed. Do these results hold if continuity is replaced by the function being in $L_2(0, L)$?

5-6 Complete the proof of Proposition 5.4.2 by showing that $H_0^1(0, L)$ is complete.

5-7 Prove the results of the chapter when $k(x) \geq 0$ of problem (5.1.1)–(5.1.3). See the paragraph following the proof of Proposition 5.4.3.

5-8 In the context of Theorem 5.5.1 prove Theorem 1.4.1. See the first two sentences of the proof of Theorem 5.5.1.

5-9 Consider the following problem:

$$-(m(x)\dot{u})^{\cdot} + k(x)u = f(x), \tag{1}$$

$$u(0) = 0, \tag{2}$$

$$\dot{u}(L) = s(u_s - u(L)). \tag{3}$$

Let $\bar{a}(u, \psi)$ and $(\bar{f}, \psi)$ be as indicated in (1.4.9), let $X(u)$ be as indicated in (1.4.10), and let $H_0^1(0, L)$ be replaced by

$$\overline{H}^1(0, L) = \left\{u\colon [0, L] \to \mathbb{R} \,|\, u(0) = 0, \int_0^L (\dot{u}^2 + u^2) < \infty, \right.$$

$$\left. u \text{ is Lebesgue measurable}\right\}.$$

Prove all the results of this chapter for (1)–(3). Find the least restrictive assumptions on $m(x)$, $k(x)$, $f(x)$, s, and u_s so that the analogs of these results hold (see exercise 5-15(c)).

5-10 Consider Proposition 5.5.1. Assume $k, f \in C^2[0, L]$ and $m \in C^3[0, L]$. Show $u \in C^4[0, L]$.

5-11 Consider Corollary 1 of Theorem 5.6.1. Use the results of 5-10 to prove the error estimate. (The reader may wish to consult Section 1.6 in Strang and Fix [28].) Must u^h be quadratic?

5-12 Prove Corollary 2 of Theorem 5.6.1 (see exercises 5-10 and 5-11).

5-13 Consider (5.1.1)–(5.1.3) with nonzero boundary conditions $u(0) = a$ and $u(L) = b$. Let

$$w = u - \left(\frac{(b-a)}{L} x + a \right)$$

and show w has the form of (5.1.1)–(5.1.3) where $w(0) = 0 = w(L)$. This gives the existence of the solution u!

5-14 Consider 5-13 but establish existence by the following steps:

(a) Consider Theorem 5.7.1 where $H_0^1(0, L)$ is replaced by $K = \{u \in H^1(0, L) | u(0) = a,\ u(L) = b\}$ and S^h is replaced by $K^h = \{v^h = \Sigma_0^N v_j \Psi_j | v_0 = a,\ v_N = b\}$.

(b) Show K and K^h are convex. (A set S is called *convex* if and only if $u, v \in S$ imply $\lambda u + (1 - \lambda)v \in S$ for all $0 \leq \lambda \leq 1$.)

(c) Show K and K^h are closed in $H^1(0, L)$. (A set S is *closed in* $H^1(0, L)$ if and only if $u^l \in S$ and u^l converges to $u \in H^1(0, L)$ implies $u \in S$.)

(d) $a(u, \psi)$ is coercive on $K - K$ and $K^h - K^h$.

(e) Show there exist $u \in K$ and $u^h \in K$ such that

$$X(u) = \inf_{v \in K} X(v) \quad \text{and} \quad X(u^h) = \inf_{v^h \in K^h} X(v^h).$$

5-15 Consider Theorem 5.7.1 with a weaker condition on f. Namely, assume (f, Ψ) is a *linear form* which means (f, Ψ) is a real number for every $\Psi \in H_0^1(0, L)$ such that

(i) $(f, \alpha_1 \Psi_1 + \alpha_2 \Psi_2) = \alpha_1 (f, \Psi_1) + a_2 (f, \Psi_2)$,

(ii) $\Psi^l \to \Psi$ in $H_0^1(0, L)$ implies $(f, \Psi^l) \to (f, \Psi)$.

(a) Show $f \in L_2(0, L)$ with $(f, \Psi) \equiv \int_0^L f \Psi$ satisfies (i) and (ii).

(b) Show $f = \delta(x - x_0)$, that is $(f, \Psi) \equiv \Psi(x_0)$, satisfies (i) and (ii)

(c) Show $(\bar{f}, \Psi) \equiv \int_0^L f\Psi + su_s\Psi(L)$ satisfies (i) and (ii) (see exercise 5-9).

(d) Prove Theorem 5.7.1 with assumptions (i) and (ii) on (f, Ψ).

5-16 Consider the bilinear forms $a(u, \Psi)$ and $\bar{a}(u, \Psi)$ given by line 5.1.6 and exercise 5-9. Show they are *continuous*, that is, there exists a constant, C, such that $|a(u, \Psi)| \leq C\|u\|_1 \|\Psi\|_1$. How is this used in the proof of Theorem 5.7.1?

6

TIME-DEPENDENT PROBLEMS

In Section 6.1 we describe a model time-dependent heat-conduction problem. Section 6.2 contains a review of three basic finite difference methods (explicit, implicit, and Crank–Nicolson). In Section 6.3 we give a brief description of stability and the Lax equivalence theorem. Both the matrix and the von Neumann stability methods are presented. The finite element method for an implicit time discretization of a one-space-variable problem is given in Section 6.4. The Crank–Nicolson method is discussed in Section 6.5. Extensions of the FEM to several space variables are given in Section 6.6.

6.1 A SAMPLE PROBLEM

Throughout this chapter we shall refer to the sample heat-conduction problem that is discussed in this section:

$$\rho c u_t - (K u_x)_x = 0, \qquad (x,t) \in (0,1) \times (0,T), \tag{6.1.1}$$

$$u(0,t) = 100, \qquad t \in (0,T), \tag{6.1.2}$$

$$u_x(1,t) = 0, \qquad t \in (0,T), \tag{6.1.3}$$

$$u(x,0) = u^0(x), \qquad x \in (0,1). \tag{6.1.4}$$

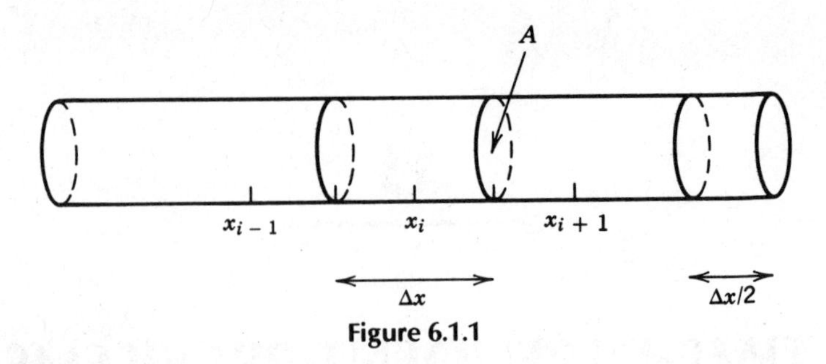

Figure 6.1.1

It is useful to review the derivation of (6.1.1)–(6.1.4). In order to do this, we must recall the emperical law that governs heat conduction.

Fourier's Heat Law

(a) Heat flows from hot to cold.

(b) For "small" changes in time Δt, the amount of heat flowing across a given cross-sectional area A is proportional to the product of (i) the temperature gradient in a direction normal to the given cross sectional area, (ii) A, and (iii) Δt.

Derivation of (6.1.1) – (6.1.4) for a Thin Rod. Consider a thin rod whose cylinderical sides are insulated, whose right end is insulated, and whose left side is held at a temperature of 100°. See Figure 6.1.1. The important physical parameters are the temperature $u(x, t)$, the density ρ, the specific heat c, and the thermal conductivity K (the proportionality constant in part (b) of Fourier's heat law). Over a small change in time, Δt, we have the change in heat energy for a small volume

$$\rho c(u(x, t + \Delta t) - u(x, t)) \cdot A \cdot \Delta x$$

$$= \text{heat in} - \text{heat out}$$

$$\cong (-Ku_x(x, t) - (-Ku_x(x + \Delta x, t)))A \cdot \Delta t.$$

Thus,

$$\rho c \frac{u(x,t+\Delta t) - u(x,t)}{\Delta t} \cong \frac{Ku_x(x+\Delta x,t) - Ku_x(x,t)}{\Delta x}. \tag{6.1.5}$$

By letting Δt, $\Delta x \to 0$, (6.1.5) implies (6.1.1), the continuum model of heat conduction.

For later use, we note that the exact solution of (6.1.1)–(6.1.4) may be obtained by separation of variables and by Fourier series.

Exact Solution of (6.1.1) – (6.1.4). Let $u(x,0) = 0$.

$$u(x,t) = 100 + \sum_{n=1}^{\infty} a_n e^{-(K/\rho c)(n\pi - \pi/2)^2 t} \sin\left(n\pi - \frac{\pi}{2}\right)x,$$

$$a_n = \left(\int_0^1 (-100)\sin\left(n\pi - \frac{\pi}{2}\right)x\,dx\right) \Big/ \int_0^1 \sin^2\left(n\pi - \frac{\pi}{2}\right)x\,dx. \tag{6.1.6}$$

6.2 FINITE DIFFERENCE SCHEMES

In this section we give a very brief review of the explicit, implicit, and Crank–Nicolson finite difference methods. We use the following notation:

$$u^k \cong u^k(x) = u(x, k\Delta t), \qquad \Delta t = T/\overline{K},$$

$$u_i \cong u_i(t) = u(i\Delta x, t), \qquad \Delta x = 1/N,$$

$$u_i^k \cong u(i\Delta x, k\Delta t),$$

$$u_{\text{exn}} = \left(u_i^k\right) \in \mathbb{R}^M \quad \text{where } M = N\overline{K}$$

$$= \text{the exact numerical solution (no roundoff error)},$$

$$u_{\text{ex}} = \left(u(i\Delta x, k\Delta t)\right) \in \mathbb{R}^M$$

$$= \text{the exact continuum solution at } (x,t) = (i\Delta x, k\Delta t).$$

In equation (6.1.1) the time derivative is approximated by

$$u_t \to \frac{u^{k+1} - u^k}{\Delta t}.$$

In order to approximate the space derivative in (6.1.1), we use the derivation of the heat equation as given in line (6.1.5). Note how we have used the boundary condition (6.1.3) in line (6.1.5) when $i = N$.

$$(Ku_x)_x \to \delta(K\delta u_i)$$
$$= \begin{cases} \left(K\dfrac{u_{i+1} - u_i}{\Delta x} - K\dfrac{u_i - u_{i-1}}{\Delta x}\right)/\Delta x, & 1 \le i < N \\ \left(0 - K\dfrac{u_N - u_{N-1}}{\Delta x}\right)/(\Delta x/2), & i = N. \end{cases}$$

In $\delta(K\delta u_i)$ we may choose to evaluate u_i as u_i^k or u_i^{k+1} or $(u_i^k + u_i^{k+1})/2$. These choices give the three methods.

Definitions. Consider problems (6.1.1)–(6.1.4). The initial condition is given by $u_i^0 = u^0(i\Delta x)$ for $i = 0, \ldots, N$. The left boundary condition gives $u_0^k = 0$ for $k = 1, \ldots, \overline{K}$. The partial differential equation (6.1.1) and the right boundary condition (6.1.3) give the three methods:

Explicit FDM: $$\rho c\frac{u_i^{k+1} - u_i^k}{\Delta t} - \delta(K\delta u_i^k) = 0. \tag{6.2.1}$$

Implicit FDM: $$\rho c\frac{u_i^{k+1} - u_i^k}{\Delta t} - \delta(K\delta u_i^{k+1}) = 0. \tag{6.2.2}$$

Crank–Nicolson FDM: $$\rho c\frac{u_i^{k+1} - u_i^k}{\Delta t} - \frac{1}{2}\left(\delta(K\delta u_i^{k+1}) + \delta(K\delta u_i^k)\right) = 0. \tag{6.2.3}$$

Each of these methods may also be written in the form

$$F(u_{\text{exn}}) = 0, \tag{6.2.4}$$

where the i, k component of $F(u_{\text{exn}})$ is given by one of the lines (6.2.1)–(6.2.3). Another representation is to use a matrix to go from one time step to the next

$$u_{\text{exn}}^{k+1} = Bu_{\text{exn}}^{k} + f^{k}, \tag{6.2.5}$$

where $u_{\text{exn}}^{k} \in \mathbb{R}^{N}$, $(u_{\text{exn}}^{k}) = (u_{i}^{k})$, k is fixed, and u_{i}^{k} is the exact numerical solution. The forms of B and f^{k} are easily illustrated for $N = 4$.

Explicit FDM: $B = I + C$, $f^{k} = (100d, 0, 0, 0)^{\mathrm{T}}$, $d = (K/\rho c)(\Delta t/\Delta x^{2})$,

$$C = d\begin{pmatrix} -2 & 1 & 0 & 0 \\ 1 & -2 & 1 & 0 \\ 0 & 1 & -2 & 1 \\ 0 & 0 & -2 & 2 \end{pmatrix}.$$

Implicit FDM: $B = (I - C)^{-1}$, $f^{k} = (100d, 0, 0, 0)^{\mathrm{T}}$.

Crank–Nicolson FDM: $B = (I - \frac{1}{2}C)^{-1}(I + \frac{1}{2}C)$, $f^{k} = (100d, 0, 0, 0)^{\mathrm{T}}$.

The explicit method does not require the solution of an algebraic system. Both the implicit method and the Crank–Nicolson method do require the solution of an algebraic problem. In one space dimension the problem has the form $Au = d$, where A is tridiagonal. Fortunately, they can easily be solved by using the tridiagonal (or Thomas) algorithm, which is described at the end of this section.

The explicit method is not so good as it might first appear. A few numerical experiments with variable Δt and fixed $\Delta x = \frac{1}{4}$ and $K = \rho = c = 1$ show this. If $\Delta t = \frac{1}{30}$, the numerical solution will blow up after several time steps. If $\Delta t = \frac{1}{40}$, then the scheme gives good results. If one changes the physical parameters or Δx, then the critical value of Δt will change. This important topic will be discussed in detail in Section 6.3. We summarize those results.

Stability

Explicit FDM: Stable if $(K/\rho c)(2\Delta t/\Delta x^2) < 1$.
Implicit and Crank–Nicolson FDM: Always stable.

Let us return to the problem of solving the algebraic system of the form

$$Au = d, \tag{6.2.6}$$

where A is an $N \times N$ matrix and u, $d \in \mathbb{R}^N$. In the implicit method and the Crank–Nicolson method, A has the form $(I - C)$ and $(I - \frac{1}{2}C)$, respectively. Note A has the following properties:

(i) A is tridiagonal.

$$-a_i u_{i-1} + b_i u_i - c_i u_{i+1} = d_i, \quad i = 1, \ldots, N. \tag{6.2.7}$$

(ii) $a_i, b_i, c_i \geq 0$. (6.2.8)
(iii) $b_i > a_i + c_i$. (6.2.9)

Under these conditions the following algorithm will give a unique solution of (6.2.6).

As motivation consider the simple 3×3 system given in (1.7.5). Note that after the first part of the Gaussian elimination method the matrix equation has the form

$$\begin{pmatrix} \cdot & \cdot & 0 \\ 0 & \cdot & \cdot \\ 0 & 0 & \cdot \end{pmatrix} \begin{pmatrix} u_1 \\ u_2 \\ u_3 \end{pmatrix} = \begin{pmatrix} \cdot \\ \cdot \\ \cdot \end{pmatrix}. \tag{6.2.10}$$

Therefore, there are numbers A_i and D_i such that

$$u_{i-1} = A_i u_i + D_i. \tag{6.2.11}$$

If A satisfies (6.2.7)–(6.2.9), then one can always find such A_i and D_i.

Tridiagonal Algorithm (Thomas) for $Au = d$

$$A_{i+1} = \frac{c_i}{b_i - a_i A_i}, \qquad A_1 = 0, \quad 1 \le i \le N - 1, \quad (6.2.12)$$

$$D_{i+1} = \frac{d_i + a_i D_i}{b_i - a_i A_i}, \qquad D_1 = 0, \quad 1 \le i \le N, \quad (6.2.13)$$

$$u_{i-1} = A_i u_i + D_i, \qquad u_N = D_{N+1}, \quad 1 < i \le N. \quad (6.2.14)$$

Remarks

1. If A does not satisfy (6.2.7)–(6.2.9), the algorithm may or may not work. In that case the denominators in (6.2.12) and (6.2.13) may be zero. For example, consider $N = 2$ with

$$\begin{pmatrix} 2 & -2 \\ -2 & 2 \end{pmatrix}\begin{pmatrix} x_1 \\ x_2 \end{pmatrix} = \begin{pmatrix} 1 \\ 1 \end{pmatrix}.$$

Then $b_1 = b_2 = 2$, $c_1 = 2$, $a_2 = 2$, and $d_1 = d_2 = 1$.

$$A_2 = \frac{2}{2 - 0.0} = 1,$$

$$D_2 = \frac{1 + 0.0}{2 - 0.0} = \frac{1}{2},$$

$$D_3 = \frac{1 + 0 \cdot \frac{1}{2}}{2 - 2 \cdot 1} = \text{not defined.}$$

2. This algorithm is important because it does not require storing the full matrix, and the number of computations is small.
3. If a_i, b_i, and c_i are replaced by blocks that are invertible, then under certain conditions the tridiagonal algorithm generalizes to a useful algorithm. The following derivation can be generalized to block tridiagonal matrices. See exercise 6-1.

Derivation of Tridiagonal Algorithm. Place

$$u_{i-1} = A_i u_i + D_i \quad \text{and} \quad u_i = A_{i+1} u_{i+1} + D_{i+1}$$

into

$$-a_i u_{i-1} + b_i u_i - c_i u_{i+1} = d_i$$

to get

$$-a_i(A_i(A_{i+1}u_{i+1} + D_{i+1}) + D_i) + b_i(A_{i+1}u_{i+1} + D_{i+1}) \\ -c_i u_{i+1} = d_i. \tag{6.2.15}$$

(6.2.15) may be rewritten

$$u_{i+1}(-a_i A_i A_{i+1} + b_i A_{i+1} - c_i) - a_i(A_i D_{i+1} + D_i) + b_i D_{i+1} = d_i. \tag{6.2.16}$$

Since (6.2.16) must hold for all i, a_i, b_i, and c_i, the coefficient of u_{i+1} in (6.2.16), must be zero. Hence,

$$-a_i A_i A_{i+1} + b_i A_{i+1} - c_i = 0, \tag{6.2.17}$$

$$-a_i(A_i D_{i+1} + D_i) + b_i D_{i+1} = d_i. \tag{6.2.18}$$

(6.2.17) gives the formula for A_{i+1} in (6.2.12), and (6.2.18) gives the formula for D_{i+1} in (6.2.13). In order to show $A_1 = D_1 = 0$, note that for $i = 1$, $a_1 = 0$ and, consequently,

$$b_1 u_1 - c_1 u_2 = d_1, \tag{6.2.19}$$

$$u_1 = A_2 u_2 + D_2. \tag{6.2.20}$$

Thus, $A_2 = c_1/b_1$ and $D_2 = d_1/b_1$. The only way to ensure this in formulas (6.2.12) and (6.2.13) is to demand $A_1 = D_1 = 0$. In order to show $u_N = D_{N+1}$, note that for $i = N$, $c_N = 0$, and so, $A_{N+1} = 0$. Thus, $u_N = A_{N+1}u_N + D_{N+1} = D_{N+1}$.

6.3 STABILITY AND THE LAX EQUIVALENCE THEOREM

The convergence of a numerical solution to the solution of a time-dependent continuum problem is studied. We limit our discussion to *well-posed problems*, which include those (1) that have a solution, (2) whose solution is unique, and (3) whose solution depends continuously on the data. We shall show convergence of the numerical scheme by establishing the scheme's stability and consistency. This approach does not require the explicit formula for the exact solution. For the explicit FDM, we illustrate Lax's equivalence theorem, which gives convergence when the scheme is stable and consistent. In order to define convergence, consistency, and stability, we shall need the general concept of the distance between two column vectors.

The traditional method of measuring distance between u and v ($u, v \in \mathbb{R}^M$) is the *Euclidean* distance

$$\|u - v\|_2 \equiv \left(\sum_{i=1}^{M} (u_i - v_i)^2 \right)^{1/2},$$

where $u = (u_i)$ and $v = (v_i)$. This function $\|u\|_2$ gives a real number, and it satisfies the following properties:

(i) $\|u\|_2 \geq 0$; $\|u\|_2 = 0$ if and only if $u = 0$. (6.3.1)

(ii) $\|\alpha u\|_2 = |\alpha| \, \|u\|_2$ for all $\alpha \in \mathbb{R}$. (6.3.2)

(iii) $\|u + v\|_2 \leq \|u\|_2 + \|v\|_2$. (6.3.3)

Definition. Let $u, v \in \mathbb{R}^M$. Any real-valued function $\|u\|$ that satisfies (6.3.1)–(6.3.3) when $\|\cdot\|_2$ is replaced by $\|\cdot\|$ is called a *norm* on $\mathbb{R}^M$. The *distance* between u and v is $\|u - v\|$.

Another important norm is

$$\|u\|_\infty \equiv \max_i |u_i|. \tag{6.3.4}$$

For example, if $M = 2$, $u = (-4, 1)^T$, then $\|u\|_\infty = \max(|-4|, |1|) = 4$. In addition to being a norm, $\|u\|_\infty$ has several other important properties.

Proposition 6.3.1. Let $\|u\|_\infty$ be defined by (6.3.4). Let B be an $M \times M$ matrix and u, u^{k+1}, u^k, $f^k \in \mathbb{R}^M$.

(a) $\|u\|_\infty$ is a norm.

(b) $\|Bu\|_\infty \le \|B\|_\infty \|u\|_\infty$ where

$$\|B\|_\infty \equiv \max_i \sum_j |b_{ij}| \text{ (called a } \textit{matrix norm}\text{)}$$

$$B = (b_{ij}), \qquad i, j = 1, \ldots, M.$$

(c) Define $u^{k+1} = Bu^k + f^k$.

$$\|u^{k+1}\|_\infty \le \|B\|_\infty^{k+1} \|u^0\|_\infty + \|B\|_\infty^k \|f^0\|_\infty + \|B\|_\infty^{k-1} \|f^1\|_\infty + \cdots + \|f^k\|_\infty. \quad (6.3.5)$$

Proof. The proof of (a) follows from the definition (6.3.4) and the analogous properties of the absolute-value norm on $\mathbb{R}^1$.

In order to prove part (b), use the component form of matrix multiplication $Bu = (\Sigma_j b_{ij} u_j)$:

$$\begin{aligned} \|Bu\|_\infty &= \max_j \left| \sum_j b_{ij} u_j \right| \\ &\le \max_j \sum_j |b_{ij}|\,|u_j| \\ &\le \max_i \sum_j |b_{ij}| \left(\max_j |u_j| \right) \\ &= \max_i \sum_j |b_{ij}| (\|u\|_\infty) \\ &= \|B\|_\infty \|u\|_\infty. \end{aligned}$$

The proof of (c) follows from working backward with the definition of u^{k+1}:

$$\begin{aligned} u^{k+1} &= Bu^k + f^k \\ &= B\left(Bu^{k-1} + f^{k-1}\right) + f^k \\ &= B^2 u^{k-1} + Bf^{k-1} + f^k \\ &\vdots \\ &= B^{k+1} u^0 + B^k f^0 + B^{k-1} f^1 + \cdots + f^k. \end{aligned}$$

By repeated use of (6.3.3) with (6.3.4), and use of part (b), this gives the estimate

$$\|u^{k+1}\|_\infty \le \|B^{k+1}u^0\|_\infty + \|B^k f^0\|_\infty + \|B^{k-1}f^1\|_\infty + \cdots + \|f^k\|_\infty$$

$$\le \|B\|_\infty^{k+1}\|u^0\|_\infty + \|B\|_\infty^k\|f^0\|_\infty + \|B\|_\infty^{k-1}\|f^1\|_\infty$$

$$+ \cdots + \|f^k\|_\infty.$$

Example. Let $M = 4$, and from the explicit FDM as given after line (6.2.5), let

$$B = \begin{pmatrix} 1-2d & d & 0 & 0 \\ d & 1-2d & d & 0 \\ 0 & d & 1-2d & d \\ 0 & 0 & 2d & 1-2d \end{pmatrix}.$$

Then

$$\sum_j |b_{1j}| = |1-2d| + |d|,$$

$$\sum_j |b_{2j}| = \sum_j |b_{3j}| = |d| + |1-2d| + |d|,$$

$$\sum_j |b_{4j}| = |2d| + |1-2d|.$$

If $0 < 2d \le 1$, then $\|B\|_\infty = \max(1-d, 1) = 1$. Also, $\|f^k\|_\infty = 100d \le 50$.

As we shall see, it will often be useful if one can find, or at least know that there exists, a norm such that $\|B\| < 1$. One such norm is derived from $\|\cdot\|_\infty$ and an invertible matrix E:

$$\|u\|_E \equiv \|Eu\|_\infty. \tag{6.3.6}$$

The reader should show that Proposition 6.3.1 holds when $\|\cdot\|_\infty$ is replaced by $\|\cdot\|_E$ (see exercise 6-2). The only modification is in part (b), where $\|B\|_E$ must be defined as follows:

$$\begin{aligned}\|Bu\|_E &= \|EBu\|_\infty \\ &= \|(EBE^{-1})Eu\|_\infty \\ &\le \|EBE^{-1}\|_\infty \|Eu\|_\infty.\end{aligned}$$

Therefore, replace $\|B\|_\infty$ by $\|B\|_E \equiv \|EBE^{-1}\|_\infty$.

Example. Consider the above example of B with E of the form

$$E = \begin{pmatrix} 1 & 0 & 0 & 0 \\ 0 & a & 0 & 0 \\ 0 & 0 & b & 0 \\ 0 & 0 & 0 & c \end{pmatrix}.$$

Then

$$EBE^{-1} = \begin{pmatrix} 1-2d & \frac{d}{a} & 0 & 0 \\ ad & 1-2d & \frac{a}{b}d & 0 \\ 0 & \frac{b}{a}d & 1-2d & \frac{b}{c}d \\ 0 & 0 & \frac{c}{b}2d & 1-2d \end{pmatrix}.$$

By assuming $2d \le 1$ and with the proper choice of a, b, c, we hope $\|B\|_E < 1$. In particular, if $a = \frac{12}{20}$, $b = \frac{10}{20}$, and $c = \frac{9}{20}$, then $\|B\|_E = 1 - d/18 < 1$.

Remarks

1. Consider any matrix $B \ge 0$ with $\|B\|_\infty = 1$ and at least one $i = i_0$ such that $\Sigma_j |b_{i_0 j}| < 1$. Then there exists an invertible diagonal matrix E with $\|B\|_E < 1$ (see exercise 6-3). Assume $b_{ii} > 0$.

2. If E is an invertible diagonal matrix, then it is not difficult to show that there are constants C_1 and C_2, which are independent of u, such that

$$C_1\|u\|_E \le \|u\|_\infty \le C_2\|u\|_E. \tag{6.3.7}$$

In fact, this is true for any norm on $\mathbb{R}^M$ (see Ortega [15, Chapter 1]).

3. Given a vector norm, $\|\cdot\|$, one can always generate a matrix norm by

$$\|B\| \equiv \sup_{u \neq 0} \frac{\|Bu\|}{\|u\|}.$$

Consequently, $\|B\| \geq \|Bu\|/\|u\|$ and $\|Bu\| \leq \|B\|\,\|u\|$. Then part (c) of Proposition 6.3.1 holds for any norm. Examples include:

(i) $\|u\|_\infty$, where one can show $\|B\| = \|B\|_\infty = \max_i \Sigma_j |b_{ij}|$.
(ii) $\|u\|_E$, where one can show $\|B\| = \|B\|_E = \|EBE^{-1}\|_\infty$.

In order to describe the numerical methods, we shall use either $F(u_{\text{exn}}) = 0$ in line (6.2.4) or $u_{\text{exn}}^{k+1} = Bu_{\text{exn}}^k + f^k$ in line (6.2.5).

Notation

$$M = N\bar{K}, \qquad 1 \leq i \leq N,$$

$$\Delta x = 1/N, \qquad 1 \leq k \leq \bar{K} \quad \text{and} \quad \Delta t = T/\bar{K}.$$

$u_{\text{exn}}, u_{\text{ex}}, u_{\text{num}} \in \mathbb{R}^M$ where

$u_{\text{exn}} = \left(u_i^k\right) =$ *exact numerical* solution

$u_{\text{ex}} = \left(u(i\Delta x, k\Delta t)\right) =$ *exact* solution

$u_{\text{num}} =$ *computed or numerical* solution

$u_{\text{exn}} = u_{\text{num}} +$ *global round-off* error

$u_{\text{exn}}^k, u_{\text{ex}}^k, u_{\text{num}}^k \in \mathbb{R}^N$ are the vectors associated with $u_{\text{exn}}, u_{\text{ex}}, u_{\text{num}}$, respectively, when k is fixed.

$r^k =$ *local round-off* error
$=$ the error resulting from just the computations at the k time step

$$u_{\text{num}}^{k+1} = Bu_{\text{num}}^k + f^k + r^k.$$

The analysis of the finite difference scheme may be broken into two parts.

$$u_{\mathrm{ex}} - u_{\mathrm{num}} = (u_{\mathrm{ex}} - u_{\mathrm{exn}}) + (u_{\mathrm{exn}} - u_{\mathrm{num}}) = E_{\mathrm{d}} + E_{\mathrm{r}}.$$

$E_{\mathrm{d}} = u_{\mathrm{ex}} - u_{\mathrm{exn}}$ is called the *discretization error*, and one hopes that it will go to zero as $\Delta t, \Delta x$ go to zero. $E_{\mathrm{r}} = u_{\mathrm{exn}} - u_{\mathrm{num}}$ is called the *global round-off error*, and it is dependent on the machine's accuracy. One hopes that as k increases, it will remain bounded. And, if the local round-off errors are uniformly small, then the global round-off error should also be small. As we shall see in Propositions 6.3.2 and 6.3.3, the following definitions are relevant to the preceding considerations.

Definitions. Let $F(u_{\mathrm{exn}}) = 0$ represent a numerical scheme.

$$\|u_{ex} - u_{exn}\| = \textit{discretization}\ \text{error}$$

$$\|F(u_{ex}) - F(u_{exn})\| = \textit{truncation}\ \text{error}.$$

The scheme is called *consistent* if and only if the truncation error goes to zero as $\Delta t, \Delta x$ go to zero. The scheme is called *stable* if and only if $\|u_{\mathrm{exn}}^k\|$ is uniformly bounded with respect to k and Δt. u_{exn} *converges* to u_{ex} as $\Delta t, \Delta x$ go to zero if and only if the discretization error goes to zero.

As an example we consider the explicit FDM given in line (6.2.1) for the model problem (6.1.1)–(6.1.4). The properties listed above are established in the proof of the next proposition.

Remark. Stability may also defined by a uniformly bounded global round-off error $= \|E_r^{k+1}\|$. For additional discussion see Proposition 6.3.3 and the observations which appear before the exercises.

Proposition 6.3.2. Consider the explicit FDM given by (6.2.1) and the problem (6.1.1)–(6.1.4). Assume $0 < 2d \leq 1$ where $d =$

$(K/\rho c)(\Delta t/\Delta x^2)$. Then we have

(a) The scheme is consistent.

(b) The scheme is stable.

(c) The scheme converges. More precisely, there is a constant C, which is independent of Δt and Δx, such that

$$\|u_{\text{exn}} - u_{\text{ex}}\| \le C(\Delta t + \Delta x^2).$$

Proof. In order to show consistency, consider the i, k component of $F(u_{\text{exn}})$, given in line (6.3.8) where $1 \le i < N$,

$$\rho c \frac{u_i^{k+1} - u_i^k}{\Delta t} - K \frac{u_{i+1}^k - 2u_i^k - u_{i-1}^k}{\Delta x^2} = 0 \qquad (6.3.8)$$

and the i, k component of $F(u_{\text{ex}})$, given in line (6.3.9), where $1 \le i < N$,

$$\rho c \frac{u(i\Delta x, (k+1)\Delta t) - u(i\Delta x, k\Delta t)}{\Delta t}$$
$$- K \frac{u((i+1)\Delta x, k\Delta t) - 2u(i\Delta x, k\Delta t) + u((i-1)\Delta x, k\Delta t)}{\Delta x^2}. \qquad (6.3.9)$$

In order to estimate the terms in $F(u_{\text{ex}})$, we use Taylor's theorem. (We use the notation $A = 0(\alpha)$, which means $|A| \le C\alpha$, where C is independent of α.) Since $u_{tt}(x, t)$ and $u_{xxxx}(x, t)$ are continuous on $[0, 1] \times [0, T]$, we have for all i, k

$$u(i\Delta x, (k+1)\Delta t) = u(i\Delta x, k\Delta t) + u_t(i\Delta x, k\Delta t)\frac{\Delta t}{1!} + 0(\Delta t^2), \qquad (6.3.10)$$

$$u((i \pm 1)\Delta x, k\Delta t) = u(i\Delta x, k\Delta t) + u_x(i\Delta x, k\Delta t)\frac{(\pm\Delta x)}{1!} + u_{xx}(i\Delta x, k\Delta t)\frac{(\pm\Delta x)^2}{2!} + u_{xxx}(i\Delta x, k\Delta t)\frac{(\pm\Delta x)^3}{3!} + 0(\Delta x^4). \qquad (6.3.11)$$

Because of the continuity of u_{tt} and u_{xxxx}, the constants in $0(\Delta t^2)$ and $0(\Delta x^4)$ are independent of i, k. By placing (6.3.10) and (6.3.11) into (6.3.9) and using (6.3.8), we obtain for $1 \le i < N$ the i, k component of $|F(u_{\text{exn}}) - F(u_{\text{ex}})|$

$$\begin{aligned} &\left|0 - \left[\rho c\left(u_t(i\Delta x, k\Delta t) + 0(\Delta t)\right) - K\left(u_{xx}(i\Delta x, k\Delta t) + 0(\Delta x^2)\right)\right]\right| \\ &\qquad = |(\rho c u_t - K u_{xx})(i\Delta x, k\Delta t)| + 0(\Delta t + \Delta x^2)| \\ &\qquad = 0 + 0(\Delta t + \Delta x^2). \end{aligned}$$

The term for $i = N$ is similar. Thus, for $\|\cdot\| = \|\cdot\|_\infty$,

$$\|(F(u_{\text{ex}}) - F(u_{\text{exn}})\|_\infty = 0(\Delta t + \Delta x^2),$$

and we have shown the scheme is consistent.

In order to show the scheme is stable, we use the matrix description $u_{\text{exn}}^{k+1} = Bu_{\text{exn}}^{k+1} + f^k$. B and f^k have the form as given after line (6.2.5). As shown after the proof of Proposition 6.3.1, $\|B\|_\infty = 1$ and $\|f^k\|_\infty = 100d \le 50$. Moreover, that discussion gives the existence of a norm $\|\cdot\|_E$ such that $\|B\|_E = \delta < 1$, and using (6.3.7) $\|f^k\|_E \le 50C_2$. Apply part (c) of Proposition 6.3.1 with the substitutions $u^{k+1} \to u_{\text{exn}}^{k+1}$, $f^k \to f^k$ and $\|\cdot\| \to \|\cdot\|_E$.

$$\begin{aligned} \|u_{\text{exn}}^{k+1}\|_E &\le \|B\|_E^{k+1}\|u_{\text{exn}}^0\|_E + \|B\|_E^k\|f^0\|_E + \cdots + \|f^k\|_E \\ &\le \delta^{k+1}\|u_{\text{exn}}^0\|_E + (\delta^k + \delta^{k-1} + \cdots + 1)50C_2 \\ &\le \|u_{\text{exn}}^0\|_E + \frac{1}{1-\delta}50C_2 \quad \text{by the geometric series.} \end{aligned}$$

Therefore,

$$\|u_{\text{exn}}^{k+1}\|_E \le \frac{1}{C_1}\|u_{\text{exn}}^0\|_\infty + \frac{1}{1-\delta}50C_2. \tag{6.3.12}$$

The right-hand side is independent of Δt and k, but the C_1 and C_2 may depend on $\Delta x = 1/N$. Thus the scheme is stable.

The convergence of the discretization error $\|u_{\text{ex}} - u_{\text{exn}}\|_\infty$ to zero may be deduced from the stability and the consistency. Let $E_{\text{d}}^{k+1} = u_{\text{ex}}^{k+1} - u_{\text{exn}}^{k+1}$, and note for $F(u)$ restricted to $\mathbb{R}^N$ (an abuse of notation), we have

$$\frac{\Delta t}{\rho c} F\left(u_{\text{ex}}^{k+1}\right) = u_{\text{ex}}^{k+1} - B u_{\text{ex}}^{k} - f^k,$$

$$\frac{\Delta t}{\rho c} F\left(u_{\text{exn}}^{k+1}\right) = u_{\text{exn}}^{k+1} - B u_{\text{exn}}^{k} - f^k.$$

Thus

$$\begin{aligned} E_{\text{d}}^{k+1} &= u_{\text{ex}}^{k+1} - u_{\text{exn}}^{k+1} \\ &= \frac{\Delta t}{\rho c}\left(F\left(u_{\text{ex}}^{k+1}\right) + B u_{\text{ex}}^{k} + f^k\right) \\ &\quad - \frac{\Delta t}{\rho c}\left(F\left(u_{\text{exn}}^{k+1}\right) + B u_{\text{exn}}^{k} + f^k\right) \\ &= B E_{\text{d}}^{k} + \frac{\Delta t}{\rho c}\left(F\left(u_{\text{ex}}^{k}\right) - F\left(u_{\text{exn}}^{k}\right)\right). \end{aligned} \tag{6.3.13}$$

Apply part (c) of Proposition 6.3.1 with the substitutions $u^{k+1} \to E_{\text{d}}^{k+1}$,

$$f^k \to (\Delta t/\rho c)\left(F\left(u_{\text{ex}}^{k}\right) - F\left(u_{\text{exn}}^{k}\right)\right) \equiv \Delta t \bar{f}^k, \quad \text{and} \quad \|\cdot\| \to \|\cdot\|_\infty.$$

Thus

$$\|E_{\text{d}}^{k+1}\|_\infty \le \|B\|_\infty^{k+1} \|E_{\text{d}}^{0}\|_\infty + \Delta t\left(\|B\|_\infty^{k} \|\bar{f}^0\|_\infty + \cdots + \|\bar{f}^k\|_\infty\right). \tag{6.3.14}$$

Since $0 < 2d \le 1$, $\|B_\infty\| = 1$ and the scheme is stable. Since the scheme is consistent, and more precisely, the truncation error is $0(\Delta t + \Delta x^2)$, we have $\|\bar{f}^k\|_\infty \le \bar{C}(\Delta t + \Delta x^2)$, where $\bar{C}$ is independent of i, k, Δt, and Δx. Thus (6.3.14) gives

$$\|E_{\text{d}}^{k+1}\|_\infty \le \|E_{\text{d}}^{0}\|_\infty + \Delta t\, k \bar{C}(\Delta t + \Delta x^2). \tag{6.3.15}$$

Since there is no round-off error, $E_{\mathrm{d}}^{0} = 0$. Also, $k\,\Delta t \le T$, and hence, for $C = \bar{C}T$,

$$\|E_{\mathrm{d}}^{k+1}\|_{\infty} \le C(\Delta t + \Delta x^2).$$

This proves the convergence and the error estimate.

Remarks

1. Note how the truncation error estimate gives rise to the discretization error estimate. The truncation error estimate was obtained from the Taylor series and the finite difference scheme; one does not need an explicit formula for the exact solution.

2. Also, the stability condition does not require an explicit formula for the solution.

The consistency and stability condition implied convergence. This is very useful because these properties may be checked before one tries to use a proposed scheme. This is in general true, and the following important theorem contains some of these results (see Richtmeyer [18, Chapter 3]).

Lax Equivalence Theorem. Consider a well-posed time-dependent problem of the form $du/dt = Bu$. Let a finite difference scheme be consistent. Then stability and convergence are equivalent.

Before returning to some other methods of establishing stability, we discuss global and local round-off errors.

$u_{\mathrm{ex}} - u_{\mathrm{num}} =$ *global round-off* error
$E_{\mathrm{r}}^{k} = u_{\mathrm{exn}}^{k} - u_{\mathrm{num}}^{k} =$ *global round-off* error at time $k\Delta t$
$r^{k} =$ *local round-off* error given by

$$u_{\mathrm{num}}^{k+1} = Bu_{\mathrm{num}}^{k} + f^{k} + r^{k}$$

Then

$$\begin{aligned} E_{\mathrm{r}}^{k+1} &= u_{\mathrm{exn}}^{k+1} - u_{\mathrm{num}}^{k+1} \\ &= \left(Bu_{\mathrm{exn}}^{k} + f^{k}\right) - \left(Bu_{\mathrm{num}}^{k} + f^{k} + r^{k}\right) \\ &= BE_{\mathrm{r}}^{k} - r^{k}. \end{aligned}$$

By the arguments in part (c) of Proposition 6.3.1, we may write the global round-off errors in terms of the previous local round-off errors

$$E_{\mathrm{r}}^{k+1} = B^{k+1}E_{\mathrm{r}}^{0} - (B^k r^0 + \cdots + r^k)$$

or

$$\|E_{\mathrm{r}}^{k+1}\| \le \|B\|^{k+1}\|E_{\mathrm{r}}^{0}\| + \|B\|^k\|r^0\| + \cdots + \|r^k\|. \quad (6.3.16)$$

If stability is taken to mean $\|B\| = \delta < 1$ for some norm, then the effects of local round-off errors in (6.3.16) go to zero as k increases. Moreover, the global round-off will remain bounded, provided the local round-off errors are uniformly bounded, that is, $\|r^k\| \le C$. To see this, simply apply $\|B\| \le \delta < 1$ and the geometric series to (6.3.16):

$$\|E_{\mathrm{r}}^{k+1}\| \le \delta^{k+1}\|E_{\mathrm{r}}^{0}\| + \frac{1-\delta^{k+1}}{1-\delta}C.$$

This result is summarized in the next proposition.

Proposition 6.3.3. Let $u_{\mathrm{exn}}^{k+1} = Bu_{\mathrm{exn}}^{k} + f^k$ be a numerical scheme for a time-dependent problem. Let r^k be the local round-off error where $u_{\mathrm{num}}^{k+1} = Bu_{\mathrm{num}}^{k} + f^k + r^k$. If the scheme is asymtotically stable, that is, $\|B\| < 1$ for some norm, and if the round-off error is uniformly bounded, that is, $\|r^k\| \le C$ for all k, then the global round-off error is uniformly bounded. Moreover, if the local round-off error is uniformly small, then the global round-off error will be small.

There are two methods frequently used to test for stability, the matrix (or eigenvalue) method and the von Neumann (or Fourier-series) method. The technique we used for the explicit FDM was the matrix method. We simply found a norm such that $\|B\| < 1$, where the scheme was given by $u^{k+1} = Bu^k + f^k$. It is not always so simple to find such a norm. Fortunately, the following proposition, whose proof may be found in Ortega [15, Chapter 1], characterizes this important problem.

Definitions. A nonnull vector u such that $Bu = \lambda u$ is called an *eigenvector of* B, and $\lambda \in \mathbb{C}$ is called an *eigenvalue of* B:

$$\rho(B) \equiv \max_{\lambda} \{|\lambda| \,|Bu = \lambda u\}$$

$$= \textit{spectral radius of } B.$$

Example. Let $M = 2$, and let

$$B = \begin{pmatrix} 1 - 2d & d \\ 2d & 1 - 2d \end{pmatrix}.$$

For $u \neq 0$, $Bu = \lambda u$ if and only if $\det(B - \lambda I) = 0$:

$$\det\begin{pmatrix} 1 - 2d - \lambda & d \\ 2d & 1 - 2d - \lambda \end{pmatrix} = (1 - 2d - \lambda)^2 - 2d^2 = 0.$$

So, either $(1 - 2d - \lambda) + \sqrt{2}\,d = 0$ or $(1 - 2d - \lambda) - \sqrt{2}\,d = 0$. Then $\lambda_1 = 1 - 2d + \sqrt{2}\,d$ and $\lambda_2 = 1 - 2d - \sqrt{2}\,d$. If $0 < 2d \leq 1$, then $0 < \lambda_1 = 1 + (-2 + \sqrt{2})d < 1$ and $-1 < 1 - d(2 + \sqrt{2}) = \lambda_2 < 1$. Thus, if $2d \leq 1$, then $\rho(B) < 1$.

Proposition 6.3.4. For all $\epsilon > 0$, there exists a norm $\|\cdot\|$ such that $\rho(B) \leq \|B\| \leq \rho(B) + \epsilon$. In particular, $\rho(B) < 1$ if and only if there exists a norm such that $\|B\| < 1$.

There are several useful methods for estimating $\rho(B)$, and some of these are summarized in Smith [16, Chapter 3]. Consequently, the matrix method has two forms.

Matrix (or Eigenvalue) Method for Stability of $u^{k+1} = Bu^k + f^k$. Either find a norm such that $\|B\| < 1$ or find an estimate such that $\rho(B) < 1$.

The second stability method is based on the Fourier-series expansion for the error (exact solution minus numerical solution). Any Fourier-series expansion has terms of the form (see line (6.1.6))

$$A_{k,n} \exp\left(\sqrt{-1}\, C_n i \,\Delta x\right)$$

where n is the nth term or mode, $A_{k,n}$ the time-dependent coefficient, for example $a_n \exp[-(K/\rho c)(n\pi - \pi/2)^2 k\,\Delta t]$, and C_n the associated spatial eigenvalues, for example, $(n\pi - \pi/2)$. Von Neumann's idea was that each node, or nth term, of the Fourier series ought to remain bounded as k increases. If e_i^k is any term or mode of the series for the error, then e_i^k will be bounded when $|e_i^{k+1}| \le |e_i^k|$ for all k.

von Neumann (or Fourier-Series) Method for Stability

1. Let $e_i^k = A_k \exp(\sqrt{-1}\ Ci\,\Delta x)$.
2. Substitute e_i^k into the algorithm and compute G_i from $e_i^{k+1} = G_i e_i^k$.
3. Demand $|G_i| \le 1$ for each i.

Examples. Consider the explicit FDM given by (6.2.1). Let $1 \le i < N$. Then

$$e_i^k = A_k \exp(\sqrt{-1}\ Ci\,\Delta x),$$

$$e_i^{k+1} = A_{k+1} \exp(\sqrt{-1}\ Ci\,\Delta x),$$

$$e_{i\pm1}^k = A_k \exp\big(\sqrt{-1}\ C(i \pm 1)\,\Delta x\big).$$

Line (6.2.1) for $1 \le i < N$ becomes

$$\frac{\rho c}{\Delta t}\big(A_{k+1}\exp(\sqrt{-1}\ Ci\,\Delta x) - A_k \exp(\sqrt{-1}\ Ci\,\Delta x)\big)$$

$$= \frac{K}{\Delta x^2}\big(A_k \exp(\sqrt{-1}\,C(i+1)\,\Delta x)$$

$$-2A_k \exp(\sqrt{-1}\ i\,\Delta x) + A_k \exp\big(\sqrt{-1}\ C(i-1)\,\Delta x\big)\big).$$

If $\theta = C\,\Delta x$, $\cos\theta = (\exp(\sqrt{-1}\ C\,\Delta x) + \exp(-\sqrt{-1}\ C\,\Delta x))/2$ and $d = (K/\rho c)\Delta t/\Delta x^2$, then $e_i^{k+1} = (1 + 2d(\cos\theta - 1))e_i^k$. So,

$$G_i = 1 + 2d(\cos\theta - 1),$$

and consequently, for $0 < 2d \leq 1$, $|G_i| \leq 1$. One should check the last case $i = N$.

The implicit FDM, line (6.2.2), gives for $1 \leq i < N$,

$$G_i = \frac{1}{1 - 2d(\cos\theta - 1)},$$

and so $|G_i| \leq 1$ with no condition on d. Also, the Crank–Nicolson method has no stability condition; for $1 \leq i < N$,

$$G_i = \frac{1 + d(\cos\theta - 1)}{1 - d(\cos\theta - 1)}.$$

The last two examples indicate that the von Neumann method may be easier to use than the matrix method. In particular, B for the implicit (or Crank–Nicolson) method is $B = (I - C))^{-1}$ (or $(I - \frac{1}{2}C)^{-1}(I + \frac{1}{2}C))$, and it appears $\rho(B)$ may be more difficult to estimate (see Smith [26], where the pros and cons of each method are more carefully discussed).

6.4 FEM FOR IMPLICIT TIME DISCRETIZATION

We shall see that the previous three methods for finite difference methods have a similar form when the finite element method is used. An implicit time discretization of (6.1.1)–(6.1.4) yields a sequence of elliptic boundary-value problems of the form that we have studied in the previous chapters:

$$\rho c \frac{u^{k+1}(x) - u^k(x)}{\Delta t} - K u_{xx}^{k+1}(x) = 0, \tag{6.4.1}$$

$$u^{k+1}(0) = 100, \tag{6.4.2}$$

$$u_x^{k+1}(1) = 0. \tag{6.4.3}$$

Thus, at each time step an elliptic problem of the form (6.4.4)–(6.4.6) must be solved:

$$-mu_{xx} + Cu = g, \tag{6.4.4}$$

$$u(0) = u_0, \tag{6.4.5}$$

$$u_x(1) = 0. \tag{6.4.6}$$

In (6.4.4)–(6.4.6) we make the following substitutions:

$$u \mapsto u^{k+1}, \qquad g \mapsto \frac{\rho c}{\Delta t} u^k,$$

$$m \mapsto K, \qquad u_0 \mapsto 100,$$

$$C \mapsto \frac{\rho c}{\Delta t}.$$

The energy integral for (6.4.4)–(6.4.6) is

$$X(u) = \frac{1}{2}\int_0^1 m u_x^2 + \frac{1}{2}\int_0^1 C u^2 - \int_0^1 g u$$

$$= X_\Delta(u) + X_c(u) - X_g(u). \tag{6.4.7}$$

Let $u^e = u_1^e N_1^e + u_2^e N_2^e$ and $\mathbf{u}^e = (u_1^e, u_2^e)^{\mathrm{T}}$. The element matrices are given by

$$\begin{pmatrix} \dfrac{\partial X(u^e)}{\partial u_1^e} \\ \dfrac{\partial X(u^e)}{\partial u_2^e} \end{pmatrix} = (k^\Delta + k^c)\mathbf{u}^e - \mathbf{R}^e, \tag{6.4.8}$$

where

$$k^\Delta = K\begin{pmatrix} 1/h & -1/h \\ -1/h & 1/h \end{pmatrix},$$

$$k^c = C\begin{pmatrix} h/3 & h/6 \\ h/6 & h/3 \end{pmatrix},$$

$$\mathbf{R}^e = \begin{pmatrix} \displaystyle\int_0^1 g N_1^e \\ \displaystyle\int_0^1 g N_2^e \end{pmatrix}.$$

In the case of (6.4.1)–(6.4.3), $g = (\rho c/\Delta t) u^{k,e}$ and $u^{k,e} = u_1^{k,e} N_1^e +$

$u_2^{k,e}N_2^e$. Thus,

$$\mathbf{R}^e = \frac{\rho c}{\Delta t}\begin{pmatrix} \int_0^1 \left(u_1^{k,e}N_1^e + u_2^{k,e}N_2^e\right)N_1^e \\ \int_0^1 \left(u_1^{k,e}N_1^e + u_2^{k,e}N_2^e\right)N_2^e \end{pmatrix}$$

$$= \frac{\rho c}{\Delta t}\begin{pmatrix} h/3u_1^{k,e} + h/6u_2^{k,e} \\ h/6u_1^{k,e} + h/3u_2^{k,e} \end{pmatrix}$$

$$= k^c\mathbf{u}^{k,e}, \qquad C = \frac{\rho c}{\Delta t}. \tag{6.4.9}$$

Thus, for (6.4.1)–(6.4.3), (6.4.8) may be rewritten

$$(k^\Delta + k^c)\mathbf{u}^{k+1,e} - k^c\mathbf{u}^{k,e}. \tag{6.4.10}$$

The system matrix may now be constructed by assembly by elements:

$$\overline{C} = \text{“sum” of } k^c,$$

$$\overline{A_\Delta} = \text{“sum” of } k^\Delta,$$

$$\overline{R} = \text{“sum” of } \mathbf{R}^e = \overline{C}\mathbf{u}^{k,e},$$

$$\overline{\mathrm{SM}} = \overline{C} + \overline{A_\Delta}.$$

Now, insert the boundary condition into $\overline{C}$, $\overline{A_\Delta}$, $\overline{R}$, and $\overline{\mathrm{SM}}$ to get C, A_Δ, R, and SM, respectively. Thus, (6.4.10) gives the algorithm for the *finite element method with implicit time discretization*:

$$\mathrm{SM}\,\mathbf{u}^{k+1} = (C + A_\Delta)\mathbf{u}^{k+1} = C\mathbf{u}^k = R \tag{6.4.11}$$

or

$$\mathbf{u}^{k+1} = (C + A_\Delta)^{-1}C\mathbf{u}^k. \tag{6.4.12}$$

Example. In (6.4.1)–(6.4.3) let $h = \Delta x = \frac{1}{4}$, $N = 4$. The reader should verify (6.4.13) and compare it with the system for the implicit

FDM, line (6.2.2), where the system matrix is $(I - C)$. Let

$$a = \frac{\rho c}{\Delta t}\frac{h}{6} - \frac{K}{h} \quad \text{and} \quad b = \frac{\rho c}{\Delta t}\frac{h}{3} + \frac{K}{h}.$$

Then

$$\begin{pmatrix} 1 & 0 & 0 & 0 & 0 \\ a & 2b & a & 0 & 0 \\ 0 & a & 2b & a & 0 \\ 0 & 0 & a & 2b & a \\ 0 & 0 & 0 & a & b \end{pmatrix} \begin{pmatrix} u_0^{k+1} \\ u_1^{k+1} \\ u_2^{k+1} \\ u_3^{k+1} \\ u_4^{k+1} \end{pmatrix} = \frac{\rho c}{\Delta t}\frac{h}{6} \begin{pmatrix} 100\dfrac{\Delta t}{\rho c}\dfrac{6}{h} \\ u_0^k + 4u_1^k + u_2^k \\ u_1^k + 4u_2^k + u_3^k \\ u_2^k + 4u_3^k + u_4^k \\ u_3^k + 2u_4^k \end{pmatrix}. \tag{6.4.13}$$

6.5 FEM FOR ONE SPACE VARIABLE

The Crank–Nicolson method for the finite element method is very similar to the finite difference case. Let $0 \le \lambda \le 1$, and consider the implicit and explicit time discretizations of (6.1.1). Multiply the implicit equation by λ and the explicit equation by $(1 - \lambda)$ and then add the two equations:

$$\rho c \lambda \frac{u^{k+1} - u^k}{\Delta t} = \lambda K u_{xx}^{k+1}, \tag{6.5.1}$$

$$\rho c (1 - \lambda) \frac{u^{k+1} - u^k}{\Delta t} = (1 - \lambda) K u_{xx}^k, \tag{6.5.2}$$

$$\rho c \frac{u^{k+1} - u^k}{\Delta t} = \lambda K u_{xx}^{k+1} + (1 - \lambda) K u_{xx}^k. \tag{6.5.3}$$

As in Section 6.4, we make the following substitutions:

$$\frac{\rho c}{\Delta t} u^{k+1} \to k^c \mathbf{u}^{k+1,e}, \qquad -K u_{xx}^{k+1} \to k^\Delta \mathbf{u}^{k+1,e},$$

$$\frac{\rho c}{\Delta t} u^{k} \to k^c \mathbf{u}^{k,e}, \qquad -K u_{xx}^{k} \to k^\Delta \mathbf{u}^{k,e}.$$

(6.5.3) gives the following expression for each element matrix

$$(k^c + \lambda k^{\Delta})\mathbf{u}^{k+1,e} = \left(k^c\mathbf{u}^{k,e} - (1-\lambda)k^{\Delta}\mathbf{u}^{k,e}\right). \qquad (6.5.4)$$

After "summing" over all elements and inserting the boundary condition, (6.5.4) gives (6.5.5), where C, A_{Δ} are the same as in Section 6.4:

$$(C + \lambda A_{\Delta})\mathbf{u}^{k+1} = C\mathbf{u}^k - (1-\lambda)A_{\Delta}\mathbf{u}^k \qquad (6.5.5)$$

or

$$\mathbf{u}^{k+1} = (C + \lambda A_{\Delta})^{-1}(C - (1-\lambda)A_{\Delta})\mathbf{u}^k. \qquad (6.5.6)$$

Finite Element Algorithms for (6.1.1) – (6.1.3)

$\lambda = 1$: *implicit* time discretization—(6.5.6)
$\lambda = \frac{1}{2}$: *Crank–Nicolson* method—(6.5.6)
$\lambda = 0$: *explicit* time discretization—(6.5.6)
$\lambda = 0$ and $C \mapsto \rho c/\Delta t\ D$ where $D = \mathrm{diag}(\int_0^L \Psi_i)$: *modified explicit*—(6.5.7)

$$\frac{\rho c}{\Delta t}D\mathbf{u}^{k+1} = \left(\frac{\rho c}{\Delta t}D - A_{\Delta}\right)\mathbf{u}^k. \qquad (6.5.7)$$

The explicit finite element method requires solving a linear algebraic system. The modified explicit does not. Both versions of the explicit method require a stability condition on Δt. This is similar to the finite difference method (see exercise 6-14).

6.6 FEM FOR TWO SPACE VARIABLES

In this section we consider the problem (6.6.1)–(6.6.3). We shall restrict our discussion to two space variables, triangular elements, and linear shape functions. The analysis for other elements, shape functions, or boundary conditions follows a similar format.

$$\rho c u_t - K\Delta u = f(x, y, t), \qquad \Omega \subset \mathbb{R}^2, \tag{6.6.1}$$

$$u = g_1(x, y), \qquad \partial\Omega_1 \times (0, \infty), \tag{6.6.2}$$

$$\frac{du}{d\nu} = 0, \qquad \partial\Omega_3 \times (0, \infty),\ \partial\Omega_3 = \partial\Omega \setminus \partial\Omega_1. \tag{6.6.3}$$

The Crank–Nicolson scheme may be derived as in Section 6.5. Note the slight changes because $f(x, y, t) \neq 0$. Let

$$f^{k+1} = f(x, y, (k+1)\Delta t) \quad \text{and} \quad f^k = f(x, y, k\,\Delta t).$$

$$\lambda \rho c \frac{u^{k+1} - u^k}{\Delta t} - \lambda K \Delta u^{k+1} = \lambda f^{k+1}, \tag{6.6.4}$$

$$(1-\lambda)\rho c \frac{u^{k+1} - u^k}{\Delta t} - (1-\lambda) K \Delta u^k = (1-\lambda) f^k, \tag{6.6.5}$$

$$\frac{\rho c}{\Delta t} u^{k+1} - \lambda K \Delta u^{k+1} = \lambda f^{k+1} + (1-\lambda) f^k + \frac{\rho c}{\Delta t} u^k + (1-\lambda) K \Delta u^k. \tag{6.6.6}$$

The element matrices k^Δ and k^c are now 3×3, and from Chapter 2 we have

$$k^\Delta = K\left(\frac{b_i b_j + c_i c_j}{4\Delta}\right),$$

where $i, j = 1, 2, 3$, and

$$k^c = \frac{\rho c}{\Delta t}\begin{pmatrix} \Delta/6 & \Delta/12 & \Delta/12 \\ \Delta/12 & \Delta/6 & \Delta/12 \\ \Delta/12 & \Delta/12 & \Delta/6 \end{pmatrix}.$$

The expression for each element matrix with $u^e = u_1^e N_1^e + u_2^e N_2^e +$

$u_3^e N_3^e$ is

$$(k^\Delta + k^c)\mathbf{u}^e - \mathbf{R}^e, \tag{6.6.7}$$

where

$$R_i^e = \iint_e g N_i^e$$

and

$$g = \lambda f^{k+1,e} + (1-\lambda) f^{k,e} + \frac{\rho c}{\Delta t} u^{k,e} + (1-\lambda) K \Delta u^{k,e}.$$

If the integration for $f^{k+1,e}$ and $f^{k,e}$ is done by approximating $f^{k+1,e}$ and $f^{k,e}$ by their linear interpolations, f_{I}^{k+1} and f_{I}^{k}, then

$$\mathbf{R}^e \simeq \lambda \frac{\Delta t}{\rho c} k^c \mathbf{f}^{k+1,e} + (1-\lambda) \frac{\Delta t}{\rho c} k^c \mathbf{f}^{k,e} + k^c \mathbf{u}^{k,e} - (1-\lambda) k^\Delta \mathbf{u}^{k,e} \tag{6.6.8}$$

where

$$\mathbf{f}^{k+1,e} = \big(f(x_1^e, y_1^e, (k+1)\Delta t), f(x_2^e, y_2^e, (k+1)\Delta t),$$
$$f(x_3^e, y_3^e, (k+1)\Delta t)\big)^{\mathrm{T}},$$

$$\mathbf{f}^{k,e} = \big(f(x_1^e, y_1^e, k\Delta t), f(x_2^e, y_2^e, k\Delta t), f(x_3^e, y_3^e, k\Delta t)\big)^{\mathrm{T}}.$$

Let C and A be the matrices that are formed from assembly by elements, that is,

$$C = \text{“sum” of } k^c \text{ with boundary conditions inserted,}$$

$$A_\Delta = \text{“sum” of } k^\Delta \text{ with boundary conditions inserted.}$$

Lines (6.6.7) and (6.6.8) give

$$(C + \lambda A_\Delta)\mathbf{u}^{k+1} = C\mathbf{u}^k + \frac{\Delta t}{\rho c} C\big(\lambda \mathbf{f}^{k+1} + (1-\lambda)\mathbf{f}^k\big) - (1-\lambda) A_\Delta \mathbf{u}^k \tag{6.6.9}$$

or

$$\mathbf{u}^{k+1} = (C + \lambda A_\Delta)^{-1}$$
$$\times\left[(C - (1-\lambda)A_\Delta)\mathbf{u}^k + \frac{\Delta t}{\rho c}C(\lambda \mathbf{f}^{k+1} + (1-\lambda)\mathbf{f}^k)\right]. \tag{6.6.10}$$

Finite Element Algorithms for (6.6.1) – (6.6.3)

$\lambda = 1$: *implicit* time discretization—(6.6.10)
$\lambda = \frac{1}{2}$: *Crank–Nicolson* method—(6.6.10)
$\lambda = 0$: *explicit* time discretization—(6.6.10)
$\lambda = 0$ and $D = \text{diag}(\int\int \Psi_i)$: *modified explicit*—(6.7.11)

$$\frac{\rho c}{\Delta t} D\mathbf{u}^{k+1} = \left(\frac{\rho c}{\Delta t}D - A_\Delta\right)\mathbf{u}^k + \frac{\Delta t}{\rho c}C\mathbf{f}^k. \tag{6.6.11}$$

The matrices C and A_Δ in (6.6.9) and (6.6.10) are not, in practice, computed and stored. The matrix $C + \lambda A_\Delta$ may be stored in the $N \times N$ part of SM, the system matrix. Note the $N \times N$ portion of SM is destroyed when IMAT is called. Therefore, in order to avoid computing $C + \lambda A_\Delta$ at each time step, one should store this matrix. Also, in order to compute the right-hand side at each time step, only DELTA for each element needs to be known. These remarks are valid only when ρ, c, and K are independent of time!

6.7 OBSERVATIONS AND REFERENCES

Although for the finite element method we did not discuss stability of the three methods, criteria similar to the finite difference method may be established. The reader should experiment with this in exercise 6-14. In exercise 6-14 the tridiagonal algorithm should be used. Exercise 6-15 is the time-dependent version of the problem in exercise 2-14. If Gaussian elimination is used, such as IMAT in FEMI, the system matrix will be destroyed at each time step.

Exercise 6-11 is useful in one's attempt to understand stability, consistency, and convergence.

Let us consider stability for the scheme given by $u_{\text{exn}}^{k+1} = Bu_{\text{exn}}^{k}$. This is stable if u_{exn}^{k} is bounded. By the results in chapter 1 of Ortega [15], this is characterized by either $\rho(B) < 1$, or $\rho(B) = 1$ and B is of class M. If $\rho(B) = 1$, then as indicated by line (6.3.16) the effect of a given local round-off error may or may not go to zero as k increases. Consequently, one may wish to use the stronger concept of *asymptotic stability*, which is defined by $\rho(B) < 1$. For example, when $2d \leq 1$, the explicit FDM is asymptotically stable.

An important related result is given by the following characterization of $\rho(B) < 1$. See Ortega [15, Chapters 1 and 7].

Proposition 6.7.1. Let B be an $M \times M$ matrix and $u^{k+1}, u^k, F \in \mathbb{R}^M$. The following are equivalent:

(a) $\rho(B) < 1$.

(b) There exists a norm $\|\cdot\|$ such that $\|B\| < 1$.

(c) $B^{k+1} \to 0$ as $k \to \infty$.

(d) Let $u^{k+1} \equiv Bu^k + f$. Then $u^{k+1} \to u$ as $k \to \infty$ and $u = Bu + f$.

Property (iv) has several important applications. Suppose $u^{k+1} = Bu^k + f$ represents a numerical scheme for a time-dependent problem. Then the limit u corresponds to the numerical solution of the steady-state problem. Another application of (iv) is to iterative solution of $Au = d$. For example, the Gauss–Seidel algorithm may be written in this form when $B = (D - L)^{-1}U$, $f = (D - L)^{-1}d$, and $A = (D - L) - U$.

EXERCISES

6-1 Derive a tridiagonal algorithm when the components are replaced by matrices.

6-2 Show $\|\cdot\|_\infty$ and $\|\cdot\|_E$ are norms. Prove Proposition 6.3.1 holds for any norm where $\|B\|$ is defined by $\sup_{u \neq 0}\|Bu\|/\|u\|$.

6-3 Let $B \geq 0$ be an $N \times N$ matrix with $\|B\|_\infty = 1$, and for at least one $i = i_0$, $\Sigma_j |b_{i_0, j}| < 1$. Show there exists an invertible diagonal matrix E such that $\|B\|_E < 1$. (*Hint*: Let E be a product of diagonal matrices such that each gives one more row $i = i'_0$ with $\Sigma_j |b_{i'_0, j}| < 1$.) Assume $b_{ii} > 0$.

6-4 Complete the proof of part (a) in Proposition 6.3.2 for $i = N$.

6-5 Consider the explicit method for the problem (6.1.1)–(6.1.4). For $i = N$, complete the von Neumann stability analysis.

6-6 Repeat exercise 6-4 when (6.1.3) is replaced by $Ku_x(1, t) = s(u_s - u(1, t))$.

6-7 Consider the problem (6.1.1)–(6.1.4). Use the von Neumann stability analysis to show that both the implicit and the Crank–Nicolson FDMs have no stability condition. What happens for the problem in exercise 6-9 when $a > 0$?

6-8 Use the matrix method to establish stability for the implicit FDM applied to problem (6.1.1)–(6.1.4).

6-9 Consider

$$\rho c u_t + a u_x - (K u_x)_x = 0, \qquad a \geq 0,$$

$$u(0, t) = 100,$$

$$u(1, t) = 0,$$

$$u(x, 0) = 0.$$

Use an explicit method and approximate au_x by $a(u_i^k - u_{i-1}^k)/\Delta x$. Use the von Neumann stability analysis to find a stability criterion on Δt.

6-10 Write a program to solve the problem in exercise 6-6 by using the explicit method. Observe the stability criterion.

6-11 Consider the problem in exercise 6-9. Prove a version of Proposition 6.3.2.

6-12 Write a subroutine for the tridiagonal algorithm. Use this subroutine in a program for the implicit or Crank–Nicolson methods (use $0 < \lambda \leq 1$) for the problem in exercise 6-6.

6-13 Consider the problem in exercise 6-6. For $N = 4$, use the implicit time discretization and the finite element method to determine the matrix equation that is analogous to (6.4.13).

6-14 Consider

$$\rho c u_t - K u_{xx} + a u = f(x,t), \qquad (0,1) \times (0,T),$$

$$u(0,t) = b, \qquad x = 0, \quad t \in (0,T),$$

$$K u_x(1,t) = s(u_s - u(1,t)), \qquad x = 1,$$

$$t \in (0,T),$$

$$u(x,0) = g(x), \qquad t = 0, \quad x \in (0,1).$$

Write a program for the implicit, Crank–Nicolson, explicit finite element methods as given in (6.5.6). Note because in the one-space-variable problem a given node has only two surrounding nodes, it is not really necessary to use the computer to assemble the system matrix. Observe a stability criterion; how do a and s change stability? In order to test your program, let $\lambda = 1$ and compute the exact solution for the first time step, and compute the exact steady-state solution.

6-15 Consider the two-space-variable problem

$$\rho c u_t - K \Delta u = f(x,y,t), \qquad \Omega x (0,T),$$

$$u(x,y,t) = g_1(x,y), \qquad \partial\Omega_1 \times (0,T),$$

$$K \frac{du}{dn} + su = g_2(x,y), \qquad \partial\Omega_2 \times (0,T),$$

$$K \frac{du}{dn} = 0, \qquad \partial\Omega_3 \times (0,T),$$

$$u(x,y,0) = g_0(x,y), \qquad \Omega.$$

Modify FEMI so that the methods in lines (6.6.10) and (6.6.11) can be used to approximate the solution. In order to test your program, let $\lambda = K = \rho = c = \Delta t = f = s = g_2 = 1$ and $g_0 = 0$. Use the results of exercise 2-14 to compare with the first time step. For this test case, find an explicit formula for the steady-state solution. Then show that after about ten time steps the numerical solution will have converged to the steady-state solution. Finally, use this program to solve a problem that is interesting to you.

7

NUMERICAL SOLUTION OF NONLINEAR ALGEBRAIC SYSTEMS

This chapter deals with a number of techniques for approximating solutions of nonlinear algebraic systems. Section 7.1 contains a discussion of nonlinearity and some problems that give nonlinear algebraic systems. Section 7.2 contains a description of some one-variable methods, including the bisection, Picard iteration, and Newton methods. The very important Newton's method for N unknowns is described in Section 7.3. Variations of Newton's method, called Gauss–Seidel–Newton and Newton–SOR methods, are presented in Section 7.4. The quasi-Newton method (Broyden updates) is discussed in Section 7.5. In Section 7.6 continuation (homotopy) techniques are studied. Finally, in Section 7.7 several versions of the general nonlinear Gauss–Seidel method are presented.

7.1 MOTIVATING EXAMPLES

Let V be a subset of either real-valued functions or column vectors. V is called a *linear space* if and only if $\alpha u + \beta v \in V$ for all $\alpha, \beta \in \mathbb{R}$ and $u, v \in V$. A mapping $L\colon V \to V$ is called a *linear operator* if and only if $L(\alpha u + \beta v) = \alpha L(u) + \beta L(v)$ for all $\alpha, \beta \in \mathbb{R}$ and $u, v \in V$. $Lu = f \in V$ is defined to be a *linear problem* if and only if L is a linear operator. All the problems in Chapters 1–6 are linear, and all the problems in Chapters 7–9 are nonlinear. Some interesting nonlinear problems follow.

Example 1. (Eigenvalues) Consider the following eigenvalue problem:

$$-u'' = \lambda u, \tag{7.1.1}$$

$$u(0) = 0, \tag{7.1.2}$$

$$u'(1) + u(1) = 0. \tag{7.1.3}$$

If $\lambda < 0$ or $\lambda = 0$, then it is easy to show $u \equiv 0$. If $\lambda > 0$, then (7.1.1) implies $u = A \sin\sqrt{\lambda}\, x + B \cos\sqrt{\lambda}\, x$. (7.1.2) implies $B = 0$. The last condition (7.1.3) gives

$$A\sqrt{\lambda} \cos\sqrt{\lambda} + A \sin\sqrt{\lambda} = 0 \tag{7.1.4}$$

Hence, either $A = 0$, in which case $u \equiv 0$, or λ must satisfy

$$\sqrt{\lambda} + \tan\sqrt{\lambda} = 0. \tag{7.1.5}$$

Figure 7.1.1 indicates that there are a countable number of such solutions.

Example 2. (Systems of Ordinary Equations) Consider the system of ordinary equations given by (7.1.6)–(7.1.7):

$$\dot{x}(t) = f(t, x), \tag{7.1.6}$$

$$x(0) = x_0, \tag{7.1.7}$$

where $x(t) = (x_1(t), x_2(t))^{\mathrm{T}}$ and $f(t, x) = (f_1(t, x_1, x_2), f_2(t, x_1, x_2))^{\mathrm{T}}$. A simple numerical method is an implicit Euler method in which $\dot{x}(t)$ is approximated by $(x(t + \Delta t) - x(t))/\Delta t$, and f_1, f_2 are evaluated at time equal to $t + \Delta t$. If $x_i^{k+1} = x_i(t + \Delta t)$ and

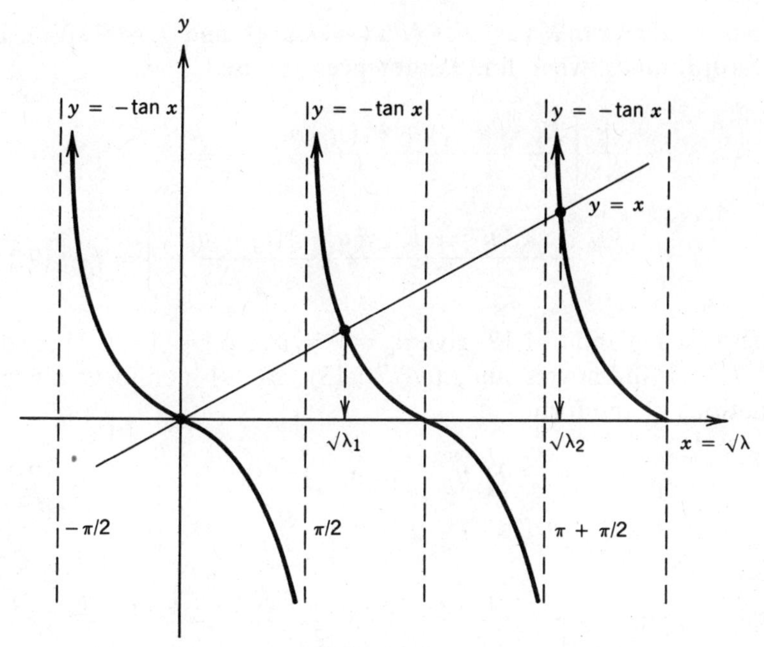

Figure 7.1.1

$x_i^k = x_i(t)$, $i = 1, 2$, then (7.1.6) gives

$$x_1^{k+1} = \Delta t f_1\left(t^{k+1}, x_1^{k+1}, x_2^{k+1}\right) + x_1^k, \tag{7.1.8}$$

$$x_2^{k+1} = \Delta t f_2\left(t^{k+1}, x_1^{k+1}, x_2^{k+1}\right) + x_2^k, \tag{7.1.9}$$

where $(x_1^0, x_2^0)^{\mathrm{T}} = x^0$. Thus, at each time step we have two unknowns, x_1^{k+1} and x_2^{k+1}, and two nonlinear algebraic equations, (7.1.8) and (7.1.9).

Example 3. (A Single Nonlinear Ordinary Differential Equation) Suppose we are considering a steady-state heat-conduction problem with a nonlinear thermal conductivity, $K(x, u)$. This may take the form, with u the temperature,

$$-\left(K(x, u)u_x\right)_x = f(x), \tag{7.1.10}$$

$$u(0) = 0, \tag{7.1.11}$$

$$u(1) = 0. \tag{7.1.12}$$

Let $u_i = u(i\,\Delta x)$, $K_i(u_i) = K(i\,\Delta x, u(i\,\Delta x))$, and $f_i = f(i\,\Delta x)$. Line (7.1.10) implies, when finite differences are used,

$$-\frac{1}{\Delta x}\left[\frac{K_{i+1}(u_{i+1}) + K_i(u_i)}{2}\,\frac{u_{i+1} - u_i}{\Delta x} - \frac{K_i(u_i) + K_{i-1}(u_{i-1})}{2}\,\frac{u_i - u_{i-1}}{\Delta x}\right] = f_i. \quad (7.1.13)$$

Lines (7.1.11) and (7.1.12) give $u_0 = 0 = u_N$, $\Delta x = 1/N$. Thus, there are $N - 1$ unknowns and, by (7.1.13), $N - 1$ nonlinear algebraic equations of the form

$$F_i(u_{i-1}, u_i, u_{i+1}) = 0, \quad (7.1.14)$$

where

$$F_i(u_{i-1}, u_i, u_{i+1}) = -\frac{1}{\Delta x}\left[\frac{K_{i+1}(u_{i+1}) + K_i(u_i)}{2}\,\frac{u_{i+1} - u_i}{\Delta x} - \frac{K_i(u_i) + K_{i-1}(u_{i-1})}{2}\,\frac{u_i - u_{i-1}}{\Delta x}\right] - f_i.$$

Example 4. (A Semilinear Ordinary Differential Equation) Consider an ordinary differential equation with a nonlinear forcing term, $f(x, u)$. For example, when $f(x, u)$ represents heat lost due to radiation it has the form $\pi\, d\epsilon\, \sigma_{SB}(u_s^4 - u^4)$ with $\pi, d\epsilon, \sigma_{SB}, u_s$ are constants.

$$-(K(x)u_x)_x = f(x, u), \quad (7.1.15)$$

$$u(0) = 0, \quad (7.1.16)$$

$$u(1) = 0. \quad (7.1.17)$$

Line (1.7.15) implies

$$-\frac{u_{i+1} - 2u_i + u_{i-1}}{\Delta x^2} = f_i(u_i) \quad (7.1.18)$$

with $u_0 = 0$, $u_N = 0$, and $\Delta x = 1/N$. This may also be written in matrix form

$$Au = (f_i(u_i)), \tag{7.1.19}$$

where $u = (u_1, \ldots, u_{N-1})^{\mathrm{T}}$, $(f_i(u_i)) \in \mathbb{R}^{N-1}$, and $A \in \mathbb{R}^{N-1} \times \mathbb{R}^{N-1}$:

$$A = \frac{1}{\Delta x^2}\begin{pmatrix} 2 & -1 & 0 & \cdot & \cdot & \cdot & 0 \\ -1 & 2 & -1 & & & & \cdot \\ \cdot & & & & & & \cdot \\ \cdot & & & & & & \cdot \\ \cdot & & & & & & \cdot \\ \cdot & & & & & & \cdot \\ \cdot & & & & -1 & 2 & -1 \\ 0 & \cdot & \cdot & \cdot & 0 & -1 & 2 \end{pmatrix}.$$

Example 5. (Burger's Equation)

$$u_t + uu_x - \mu u_{xx} = 0. \tag{7.1.20}$$

An implicit time discretization of (1.7.20) with finite differences and with upwind differences on uu_x is

$$\frac{u_i^{k+1} - u_i^k}{\Delta t} + u_i^{k+1}\delta u_i^{k+1} - \mu \frac{u_{i+1}^{k+1} - 2u_i^{k+1} + u_{i-1}^{k+1}}{\Delta x^2} = 0, \tag{7.1.21}$$

where δu_i refers to the upwind differencing

$$\delta u_i \equiv \begin{cases} \dfrac{u_i - u_{i-1}}{\Delta x} \equiv \delta^- u_i, & u_i \geq 0 \\ \dfrac{u_{i+1} - u_i}{\Delta x} \equiv \delta^+ u_i, & u_i < 0. \end{cases} \tag{7.1.22}$$

The $u_i^{k+1}\delta u_i^{k+1}$ term in (7.1.21) makes this equation nonlinear.

Example 6. (Navier–Stokes Equations) We consider isothermal, incompressible, viscous fluid flow. Let (u, v) be the normalized velocity and P the normalized pressure. The equations of x momentum, y momentum, and conservation of mass are, respectively,

$$u_t + uu_x + vu_y - \mu\,\Delta u = f_1 - \mathrm{P}_x, \tag{7.1.23}$$

$$v_t + uv_x + vv_y - \mu\,\Delta v = f_2 - \mathrm{P}_y, \tag{7.1.24}$$

$$u_x + v_y = 0. \tag{7.1.25}$$

f_1 and f_2 are forcing functions and $\mu = 1/\mathrm{Re}$, where Re is the Reynolds number LU^∞/ν, L the characteristic length, U^∞ the free stream speed, and $0 < \nu$ the kinematic viscosity. There are a number of techniques for approximating solutions to (1.7.23)–(1.7.25). For our present purposes, let us note that the nonlinearities arise from the terms $uu_x + vu_y$ and $uv_x + vv_y$. Upwind differencing as in Burger's equation may be used on these terms.

In Chapter 8 we shall implement the techniques of this chapter to solve some of these and other problems.

7.2 ONE-VARIABLE METHODS

This section contains three methods for approximating the solution of

$$f(x) = 0 \tag{7.2.1}$$

The simplest is the bisection method, which does not easily generalize to N variables. The Picard iteration and Newton methods generalize to N variables with N equations. Some theoretical results are presented for these two methods. Namely, the attractor and the quadratic convergence properties of Newton's method for functions of one variable are established. These two important properties hold for Newton's method for suitable problems with N unknowns and N equations.

Bisection Method. In order to justify the steps of this method, we shall need the following theorem of calculus.

Intermediate-Value Theorem. Let $f: [a, b] \to \mathbb{R}$ be continuous. If $f(a) > 0$ and $f(b) < 0$, then there is a $a < \bar{x} < b$ such that $f(\bar{x}) = 0$.

The theorem does not directly tell us what $\bar{x}$ equals, but the intermediate-value theorem may be used repeatedly to approximate $\bar{x}$.

Bisection Algorithm. Let $x^0 = a$, $x^1 = b$, and $f(a) > 0$, $f(b) < 0$ (or $f(a) < 0$ and $f(b) > 0$).

1. x^{k+1} is the midpoint of the interval that has the root.
2. Compute $f(x^{k-1})$, $f(x^k)$, and $f(x^{k+1})$.
3. Use the intermediate-value theorem to choose the interval with the root: either given by x^{k-1} and x^k or given by x^k and x^{k+1}.
4. Increase k by 1 and return to step 1.

Example. Let $f(x) = x + \tan x$ for $\pi/2 + \epsilon \le x \le \pi/2 + \pi - \epsilon$, where ϵ is "small" (see Figure 7.1.1). Note $f(x)$ is continuous, $f(\pi/2 + \epsilon) > 0$, and $f(\pi/2 + \pi - \epsilon) < 0$. Let $\epsilon = 0.1$.

$$x^0 = \frac{\pi}{2} + 0.1, \qquad f(x^0) = -8.296,$$

$$x^1 = \frac{\pi}{2} + \pi - 0.1, \qquad f(x^1) = 14.579,$$

$$\begin{aligned} x^2 &= (x^1 - x^0)/2 + x^1 \\ &= (x^1 + x^0)/2 \\ &= \pi, \qquad f(x^2) = \pi. \end{aligned}$$

Therefore, by the intermediate-value theorem, there is an $\bar{x}$ between x^0 and x^2 such that $f(\bar{x}) = 0$. By the algorithm

$$x^3 = (x^2 - x^0)/2 + x^0 = (x^2 + x^0)/2 \qquad f(x^3) = 1.502$$
$$= 2.406,$$
$$x_4 = (x^3 - x^0)/2 + x^0 = (x^3 + x^0)/2 \qquad f(x^4) = 0.058$$
$$= 2.038,$$
$$x^5 = (x^4 - x^0)/2 + x^0 = (x^4 + x^0)/2 \qquad f(x^5) = -1.573$$
$$= 1.855,$$
$$x^6 = (x^4 - x^5)/2 + x^5 = (x^4 + x^5)/2 \qquad f(x^6) = -0.588$$
$$= 1.946,$$

At this point in the calculations we know by the intermediate-value theorem that there is an $\bar{x}$ between 1.946 and 2.038 such that $f(\bar{x}) = 0$. One can continue this scheme until $\bar{x}$ is found to be in a suitably small interval.

Picard Iteration Method. This iteration scheme attempts to approximate a fixed point of a given function $g: [a, b] \to \mathbb{R}$. A *fixed point* is an $\bar{x}$ such that $g(\bar{x}) = \bar{x}$. Note that any fixed point of $g(x) \equiv f(x) + x$ is also a solution of $f(x) = 0$, that is, if $g(\bar{x}) = f(\bar{x}) + \bar{x} = \bar{x}$, then $f(\bar{x}) = 0$.

Picard Algorithm. Let $g: [a, b] \to \mathbb{R}$ be a "suitable" function. Let $x^0 \in [a, b]$. Then

$$x^{k+1} = g(x^k). \qquad (7.2.2)$$

Example 1. Consider an implicit time discretization of

$$\dot{x}(t) = \frac{x(t)}{1 + x(t)}, \qquad x(0) = 1. \qquad (7.2.3)$$

The solution may be approximated by letting $x^k = x(k\,\Delta t)$ and

$$\frac{x^{k+1} - x^k}{\Delta t} = \frac{x^{k+1}}{1 + x^{k+1}}, \qquad x^0 = 1. \qquad (7.2.4)$$

If $k = 0$, (7.2.4) gives the following nonlinear equation for $x^1 = x$:

$$x = \Delta t \frac{x}{1 + x} + 1 = g(x). \tag{7.2.5}$$

For $\Delta t = 1$ and $x^0 = 1$, we obtain

$$x^0 = 1.0, \qquad x^1 = 1.5,$$
$$x^2 = 1.6, \qquad x^3 = 1.6153,$$
$$x^4 = 1.6176, \qquad x^5 = 1.6180,$$
$$x^6 = 1.6180.$$

Example 2. Consider $f(x) = x^{1/3} - 1$ with root $\bar{x} = 1$. Then $g(x) = f(x) + x = x + x^{1/3} - 1$. Regardless of the initial guess for $\bar{x}$, say $x^0 = 0.5$ or $x^0 = 0.9$ or $x^0 = 1.1$, the iterates x^{k+1} all blow up!

Thus, the Picard iteration method may or may not generate approximations of fixed points. The following theorem gives conditions on $g(x)$ that imply convergence of the Picard algorithm. This theorem may be generalized to much more general maps g (see Ortega [15, Chapter 8]). Since the fixed point $x = g(x)$ is not known, we cannot directly show converges. Therefore, we use the Cauchy criteria, which characterizes convergence of x^k to x as k goes to ∞:

> x^l is *Cauchy* if and only if for all $\epsilon > 0$ there exists an N such that if $k, l \geq N$, then $|x^l - x^k| < \epsilon$.

Note the limit x of the sequence x^l does not appear in the Cauchy criterion.

Theorem 7.2.1. (Contraction-Mapping Theorem) Let $g(x)$ satisfy

(i) $g\colon [a, b] \to [a, b]$. (7.2.6)

(ii) There is a number, r, which is independent of $x, y \in [a, b]$ such that $0 < r < 1$

$$|g(x) - g(y)| \leq r|x - y|. \tag{7.2.7}$$

(g is called a *contraction map* when (7.2.7) holds.)

Then

(a) $x^{k+1} = g(x^k)$, $x^0 \in [a, b]$ converges to $\bar{x} \in [a, b]$ and $\bar{x} = g(\bar{x})$ (i.e., $\bar{x}$ is a fixed point of g).

(b) $\bar{x}$ is the only fixed point.

(c) $|\bar{x} - x^{k+1}| \leq (r^k/(1 - r))|x^1 - x^0|$.

Proof. **(a)** In order to prove x^{k+1} converges, we use the Cauchy criteria, and so we need the estimate

$$|x^l - x^k| < \epsilon. \tag{7.2.8}$$

Note that for $l > k$,

$$x^l - x^k = x^l - x^{l-1} + x^{l-1} - \cdots + x^{k+1} - x^k. \tag{7.2.9}$$

Also

$$\begin{aligned} |x^{k+1} - x^k| &= \left|g(x^k) - g(x^{k-1})\right| && \text{by definition} \\ &\leq r|x^k - x^{k-1}| && \text{by (7.2.7)} \\ &= r\left|g(x^{k-1}) - g(x^{k-2})\right| && \text{by definition} \\ &\leq r^2|x^{k-1} - x^{k-2}| && \text{by (7.2.7)} \\ &\vdots \\ &\leq r^k|x^1 - x^0|. \end{aligned} \tag{7.2.10}$$

By the triangle inequality used on (7.2.9),

$$|x^l - x^k| \leq |x^l - x^{l-1}| + |x^{l-1} - x^{l-2}| + \cdots + |x^{k+1} - x^k|. \tag{7.2.11}$$

By using the inequality (7.2.10) on the terms in (7.2.11),

$$
\begin{aligned}
|x^l - x^k| &\le (r^{k+l-1} + r^{r+l-2} + \cdots + r^k)|x^1 - x^0| \\
&= \left(\sum_{i=k}^{k+l-1} r^i\right)|x^1 - x^0| \\
&\le \left(\sum_{i=k}^{\infty} r^i\right)|x^1 - x^0| \\
&= \frac{r^k}{1-r}|x^1 - x^0| \quad \text{by geometric series.}
\end{aligned}
\tag{7.2.12}
$$

Since $0 < r < 1, r^k/(1-r)$ may be made less than any $\epsilon > 0$ merely by choosing $l(> k)$, $k \ge N$, where N is suitably large. Thus, by the Cauchy criterion, x^{k+1} converges. Since (7.2.6) holds, each $x^{k+1} \in [a, b]$ and so the limit $\bar{x}$ of x^{k+1} is in $[a, b]$. In order to show $\bar{x} = g(\bar{x})$, simply note that (7.2.7) implies $g(x^k)$ converges to $g(\bar{x})$. But, $g(x^k) = x^{k+1}$ and so $g(x^k)$ also converges to $\bar{x}$. Thus $g(\bar{x}) = \bar{x}$.

(b) Let $\bar{\bar{x}}$ be a second fixed point. By (7.2.7) with $x = \bar{x}$ and $y = \bar{\bar{x}}$,

$$
|g(\bar{x}) - g(\bar{\bar{x}})| \le r|\bar{x} - \bar{\bar{x}}| \quad \text{or} \quad |\bar{x} - \bar{\bar{x}}| \le r|\bar{x} - \bar{\bar{x}}|. \tag{7.2.13}
$$

Line (7.2.13) implies $1 \le r$ when $|\bar{x} - \bar{\bar{x}}| \ne 0$. But, $r < 1$ and so $|\bar{x} - \bar{\bar{x}}| = 0$ (i.e., $\bar{x} = \bar{\bar{x}}$).

(c) The inequality is easily established by considering (7.2.12) and taking the limit of both sides as $l \to \infty$. As the right-hand side of (7.2.12) is independent of l and $x^l \to \bar{x}$, we have the error estimate

$$
|\bar{x} - x^k| \le \frac{r^k}{1-r}|x^1 - x^0|. \tag{7.2.14}
$$

This completes the proof of the theorem.

Let us return to the two examples that preceded the contraction-map theorem. Recall the mean-value theorem, which is the usual tool used to verify the contraction-map property (7.2.7).

Mean-Value Theorem. Let $f: [a, b] \to \mathbb{R}$ be continuous on $[a, b]$ and have a derivative on (a, b). Then there is a $c \in [a, b]$ such that

$$f'(c) = \frac{f(b) - f(a)}{b - a}.$$

Example 1. $g(x) = \Delta t[x/(1 + x)] + 1$. In the mean-value theorem let $f(x) = g(x)$, $a = x$, and $b = y$. Then there is a $x < c < y$ such that

$$g(x) - g(y) = g'(c)(x - y) \tag{7.2.15}$$

Thus, (7.2.15) implies

$$|g(x) - g(y)| \leq \max_{a \leq z \leq b} |g'(z)| |x - y|. \tag{7.2.16}$$

If $r = \max_{a \leq z \leq b} |g'(z)| < 1$, then $g(x)$ will be a contraction map. Note, for $[a, b] = [1, 5]$ and $x \in [1, 5]$,

$$0 < g'(x) = \Delta t \frac{1}{(1 + x)^2} \leq \Delta t \frac{1}{4} \tag{7.2.17}$$

So, if $\Delta t < 4$, then g is a contraction map on $[1, 5]$. The first assumption, (7.2.6), holds as $\Delta t < 4$, and $1 \leq x \leq 5$ imply

$$1 < g(x) = \Delta t \frac{x}{1 + x} + 1 < \Delta t + 1 < 5.$$

Example 2. $g(x) = x + x^{1/3} - 1$ for x near $\bar{x} = 1$. $g'(x) = 1 + \frac{1}{3}x^{-2/3}$ and thus $g'(c) > 1$ for all x near $\bar{x} = 1$. Consequently, g cannot be a contraction map and the conclusions of the contraction-map theorem may or may not hold.

We shall also use the contraction-map theorem in the analysis of Newton's method.

Newton Iteration Method. This method is best described by considering Figure 7.2.1. At each step x^{k+1} is obtained by approximating $y = f(x)$ by a straight line. The x^{k+1} value is defined by finding

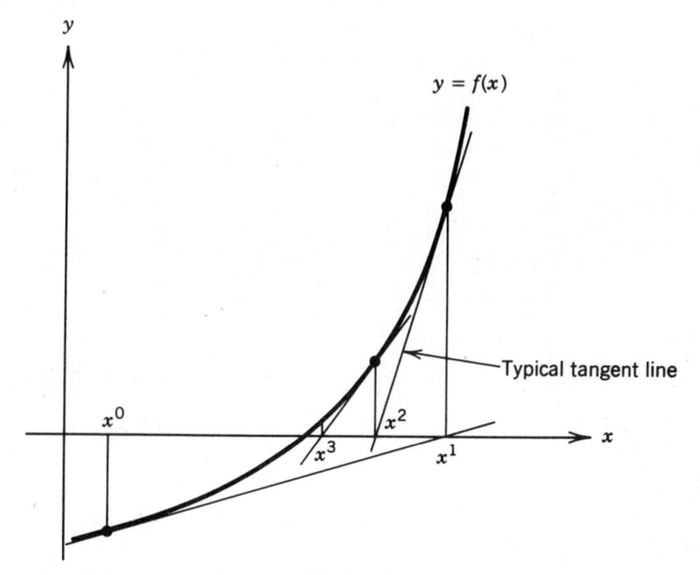

Figure 7.2.1

the intersection of this straight line with the x axis. The equation for a typical tangent line ($k = 1$ in Figure 7.2.1) may be determined since its slope is $f'(x^k)$, and it contains the point $(x^k, f(x^k))$. In particular,

$$y - f(x^k) = f'(x^k)(x - x^k). \tag{7.2.18}$$

x^{k+1} is defined as the x such that $y = 0$, that is,

$$0 - f(x^k) = f'(x^k)(x^{k+1} - x^k). \tag{7.2.19}$$

Line (7.2.19) generates a formula for x^{k+1}.

Newton's Algorithm. Let $f'(x) \neq 0$.

$$x^{k+1} = x^k - \frac{f(x^k)}{f'(x^k)}. \tag{7.2.20}$$

TABLE 7.2.1

x^0	K	x^0	K
10	no conv.	0.1	4
5	no conv.	−0.5	6
4	6	−0.8	8
3	5	−0.9	20
1.8	3	−1.0	no conv.

Example. $f(x) = x^{1/3} - 1$, $f'(x) = \frac{1}{3}x^{-2/3}$, and so (7.2.20) may be rewritten

$$\begin{aligned} x^{k+1} &= x^k - \frac{(x^k)^{1/3} - 1}{(1/3)(x^k)^{-2/3}} \\ &= -2x^k + 3(x^k)^{2/3}. \end{aligned}$$

Table 7.2.1 indicates the number of iterates, K, needed to converge to within 10^{-4} of $\bar{x} = 1$. The x^0 is the initial guess.

Remark. Provided x^0 is suitably close to the root $\bar{x}$, the algorithm converged. This is called the *attractor* property of the algorithm.

In order to examine the properties of Newton's method, we shall need an extended version of the mean-value theorem.

Extended Mean-Value Theorem. Let $f: [a, b] \to \mathbb{R}$, and let $f''(x)$ exist on (a, b). For $x, y \in (a, b)$, there exists a $z \in (x, y)$ such that

$$f(y) = f(x) + f'(x)(y - x) + f''(z)\frac{(x - y)^2}{2}. \quad (7.2.21)$$

Proof. Use the mean-value theorem when $f(x)$ is replaced by

$$F(x) \equiv f(b) - \left(f(x) + f'(x)(b - x) + K\frac{(b - x)^2}{2}\right).$$

Note $F(a) = 0 = F(b)$ when K is defined by the equation

$$f(b) = f(a) + f'(a)(b-a) + K\frac{(b-a)^2}{2}$$

By the mean-value theorem, there is a $c \in (a, b)$ such that $F'(c) = 0$. But, $F'(x) = -(f'(x) + f''(x)(b-x) - f'(x) - K(b-x))$ and, therefore, $f''(c) = K$. Now let $a = x$ and $b = y$ and $c = z$.

In order to illustrate the important properties of Newton's method, we consider a special case, as illustrated in Figure 7.2.1.

Theorem 7.2.2. Let $f: [a, b] \to \mathbb{R}$, $f'(x) \geq m > 0$, and $0 \leq f''(x) \leq M < \infty$. Assume $f(x) \in C^2[a, b]$ and there is a root $\bar{x} \in [a, b]$. For (x^0, x^1) suitably close to the root $\bar{x}$, Newton's algorithm converges to a unique solution $\bar{x} \in (x^0, x^1)$. This is called the *attractor property*. Moreover, the convergence is *quadratic*, that is,

$$|\bar{x} - x^{k+1}| \leq \frac{M}{2m}|\bar{x} - x^k|^2. \qquad (7.2.22)$$

Proof. The convergence is established by using the contraction-map theorem for

$$g(x) = x - \frac{f(x)}{f'(x)}. \qquad (7.2.23)$$

Note that the Picard iterates for this $g(x)$ are just the iterates given by Newton's algorithm. In order to verify the first assumption (7.2.6), examine Figure 7.2.1 where $y = f(x)$ has $f'(x) > 0$ and $f''(x) \geq 0$. One can easily show that for $a = x^0$ and $b = x^1$ that (7.2.6) holds.

In order to verify (7.2.7) we use the mean-value theorem as in (7.2.17).

$$g'(x) = 1 - \frac{f'(x)f'(x) - f''(x)f(x)}{(f'(x))^2}$$

$$= \frac{f''(x)f(x)}{(f'(x))^2}. \qquad (7.2.24)$$

The assumptions of this theorem and (7.2.24) imply

$$|g'(z)| \leq \frac{M}{m^2}|f(z)|. \tag{7.2.25}$$

Since $f(x)$ is continuous at $\bar{x} \in (\bar{x}^0, \bar{x}^1)$ for some $\bar{x}^0, \bar{x}^1 \in (a, b)$, we may choose $(x^0, x^1) \subset (\bar{x}^0, \bar{x}^1)$ such that

$$\max_{x^0 \leq z \leq x^1} |f(z)| = \max_{x^0 \leq z \leq x^1} |f(z) - f(\bar{x})| \leq \frac{m^2}{2M}. \tag{7.2.26}$$

Lines (7.2.25) and (7.2.26) combine to give

$$\max_{x^0 \leq z \leq x^1} |g'(z)| \leq \frac{M}{m^2}\frac{m^2}{2M} = \frac{1}{2} = r. \tag{7.2.27}$$

Thus $g(x)$ is a contraction map on this small interval about $\bar{x}$. The first part of the theorem is proved.

The quadratic convergence follows directly from the extended mean-value theorem applied to $f(x)$ with $x = \bar{x}$ and $y = x^k$ in line (7.2.21):

$$f(\bar{x}) = f(x^k) + f'(x^k)(\bar{x} - x^k) + f''(z)\frac{(\bar{x} - x^k)^2}{2}. \tag{7.2.28}$$

Since $f(\bar{x}) = 0$ and $f(x^k)/f'(x^k) = -x^{k+1} + x^k$, (7.2.28) gives upon division by $f'(x^k)$

$$-x^k + x^{k+1} = -x^k + \bar{x} + \frac{f''(z)}{f'(x^k)}\frac{(\bar{x} - x^k)^2}{2}. \tag{7.2.29}$$

Thus the assumptions of this theorem and (7.2.29) imply

$$\begin{aligned} |\bar{x} - x^{k+1}| &\leq \left(\max_{a \leq x \leq b} |f''(x)| \Big/ \min_{a \leq x \leq b} |f'(x)| \right) \frac{(\bar{x} - x^k)^2}{2} \\ &\leq \frac{M}{2m}|\bar{x} - x^k|^2. \end{aligned} \tag{7.2.30}$$

This completes the proof of the theorem.

Remarks

1. As we shall see, Newton's method extends to N unknowns and N equations. Fortunately, the attractor and quadratic convergence properties usually hold.

2. Quadratic convergence ensures rapid convergence of Newton's algorithm. For example, suppose $M/2m = 1$ and $|\bar{x} - x^k| \leq 0.2$. Then (7.2.22) implies

$$|\bar{x} - x^{k+1}| \leq 1.(0.2)^2 = 0.04,$$

$$|\bar{x} - x^{k+2}| \leq 1.(0.04)^2 = 0.0016,$$

$$|\bar{x} - x^{k+2}| \leq 1.(0.0016)^2 = 0.00000256.$$

3. An algorithm such that for some constant C,

$$|\bar{x} - x^{k+1}| \leq C|\bar{x} - x^k|^1, \tag{7.2.31}$$

is said to *converge linearly*. The Picard algorithm, in general, gives only linear convergence. In general the number r in (7.2.32) measures the *rate of convergence* for a given algorithm.

$$|\bar{x} - x^{k+1}| \leq C|\bar{x} - x^k|^r. \tag{7.2.32}$$

The larger the r, the more rapid the convergence. If

$$\lim_{k \to \infty}(|\bar{x} - x^{k+1}|/|\bar{x} - x^k|) = 0,$$

then the convergence is called *superlinear*.

In some cases the Newton algorithm may be accelerated by using an SOR parameter w. Consider Figure 7.2.2, where $f'(x)$ is large. In this case, because the slope of the tangent line is so large, x^2 changes very little from x^1. The algorithm may be modified by replacing the tangent line by a line (dotted in Figure 7.2.2) that has a smaller slope, namely, $\gamma f'(x^k)$, where $\frac{1}{2} < \gamma \leq 1$. The equation for the new line is

$$y - f(x^k) = \gamma f'(x^k)(x - x^k). \tag{7.2.33}$$

By defining x^{k+1} to be the x intercept of the line (7.2.33), we obtain the following algorithm, where $w = 1/\gamma$.

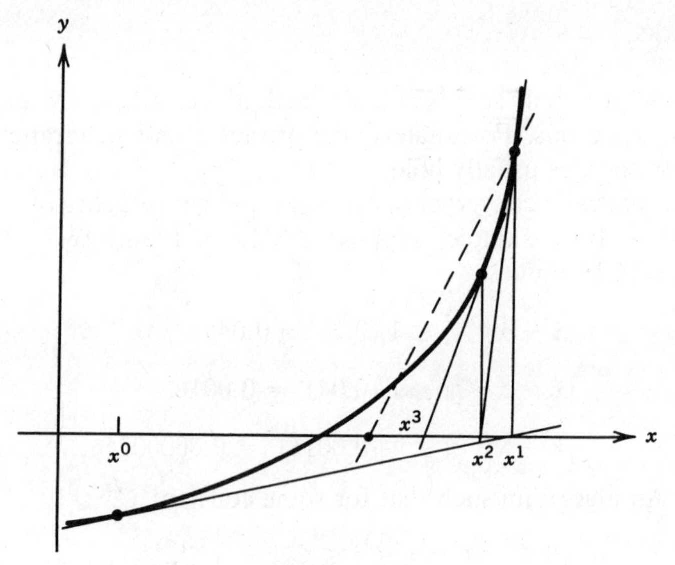

Figure 7.2.2

Newton with SOR Algorithm. Let $1 \le w < 2$.

$$x^{k+1} = x^k - w\frac{f(x^k)}{f'(x^k)}. \tag{7.2.34}$$

7.3 NEWTON'S METHOD FOR *N* UNKNOWNS

First, we describe the two-unknown case. In order to do this, we must examine the concept of a derivative of a function from $F: \mathbb{R}^2 \to \mathbb{R}^2$, that is, $F(x, y) = (f(x, y), g(x, y))^T \in \mathbb{R}^2$. The derivative of a function of one variable has the property, for small Δx,

$$f(x + \Delta x) \simeq f(x) + f'(x)\,\Delta x, \tag{7.3.1}$$

or, in terms of differentials,

$$df = f'(x)\,dx. \tag{7.3.2}$$

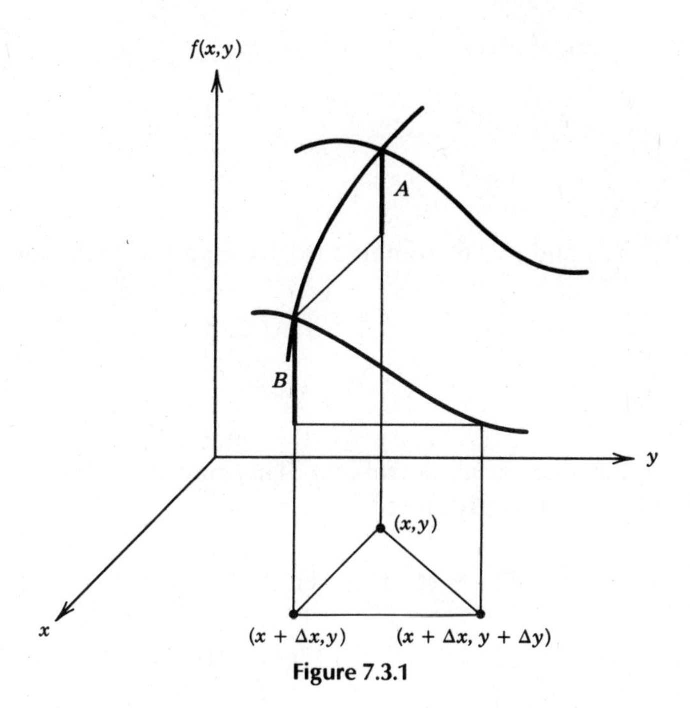

Figure 7.3.1

The approximation property in (7.3.1) is the key property that we shall try to duplicate for $F: \mathbb{R}^2 \to \mathbb{R}^2$.

Consider $f(x, y)$ as given in Figure 7.3.1. We want to estimate $f(x + \Delta x, y + \Delta y) - f(x, y) = \Delta f$. This change in $f(x, y)$ may be broken up into two parts: the change in f due to moving from (x, y) to $(x + \Delta x, y)$, A, and the change in f due to moving from $(x + \Delta x, y)$ to $(x + \Delta x, y + \Delta y)$, B. Because $f(x, y)$ can be viewed as a function of just one variable, we have

$$A \simeq f_x(x, y) \cdot \Delta x, \tag{7.3.3}$$

$$B \simeq f_y(x + \Delta x, y) \cdot \Delta y. \tag{7.3.4}$$

Thus, if f has continuous first-order partial derivatives, (7.3.3) and (7.3.4) imply

$$\Delta f = A + B \cong f_x(x, y)\,\Delta x + f_y(x, y)\,\Delta y. \tag{7.3.5}$$

A similar argument gives

$$\begin{aligned}\Delta g &= g(x+\Delta x, y+\Delta y) - g(x,y) \\ &\simeq g_x(x,y)\,\Delta x + g_y(x,y)\,\Delta y. \end{aligned} \tag{7.3.6}$$

Lines (7.3.5) and (7.3.6) combine to give a vector analog of (7.3.1)

$$\begin{pmatrix} \Delta f \\ \Delta g \end{pmatrix} \simeq \begin{pmatrix} f_x & f_y \\ g_x & g_y \end{pmatrix} \begin{pmatrix} \Delta x \\ \Delta y \end{pmatrix}. \tag{7.3.7}$$

The 2×2 matrix in (7.3.7) serves the same purpose as $f'(x)$ in (7.3.1) for the one-variable problem. This motivates the following definition of the derivative matrix.

Definition. Let $F: \mathbb{R}^2 \to \mathbb{R}^2$ be given by

$$F(x,y) = (f(x,y), g(x,y))^{\mathrm{T}}.$$

$$F'(x,y) = \begin{pmatrix} f_x(x,y) & f_y(x,y) \\ g_x(x,y) & g_y(x,y) \end{pmatrix} \tag{7.3.8}$$

is called the 2×2 *derivative matrix* or *Jacobian* of F.

The two-variable differential analog to (7.3.2) is

$$\begin{pmatrix} df \\ dg \end{pmatrix} = F'(x,y) \begin{pmatrix} dx \\ dy \end{pmatrix} \tag{7.3.9}$$

The two-variable Newton's method is then derived from (7.3.9) by letting $df = 0 - f(x^k, y^k)$, $dg = 0 - g(x^k, y^k)$, $dx = x^{k+1} - x^k$, and $dy = y^{k+1} - y^k$

$$\begin{pmatrix} 0 - f(x^k, y^k) \\ 0 - g(x^k, y^k) \end{pmatrix} = F'(x^k, y^k) \begin{pmatrix} x^{k+1} - x^k \\ y^{k+1} - y^k \end{pmatrix} \tag{7.3.10}$$

Let $(F'(x^k, y^k))^{-1}$ be the inverse matrix of $F'(x^k, y^k)$, and multiply both sides of (7.3.10) by the inverse matrix.

$$\left(F'(x^k, y^k)\right)^{-1}\begin{pmatrix} -f(x^k, y^k) \\ -g(x^k, y^k) \end{pmatrix} = \begin{pmatrix} x^{k+1} - x^k \\ y^{k+1} - y^k \end{pmatrix} \tag{7.3.11}$$

From (7.3.11), we obtain (7.3.12).

Newton's Algorithm for Two Unknowns. Let $(x^0, y^0)^{\mathrm{T}}$ be given. Then

$$\begin{aligned} \begin{pmatrix} x^{k+1} \\ y^{k+1} \end{pmatrix} &= \begin{pmatrix} x^k \\ y^k \end{pmatrix} - \left(F'(x^k, y^k)\right)^{-1}\begin{pmatrix} f(x^k, y^k) \\ g(x^k, y^k) \end{pmatrix} \\ &= \begin{pmatrix} x^k \\ y^k \end{pmatrix} - \frac{1}{\det F'(x^k, y^k)}\begin{pmatrix} g_y(x^k, y^k) & -f_y(x^k, y^k) \\ -g_x(x^k, y^k) & f_x(x^k, y^k) \end{pmatrix} \\ &\quad \times \begin{pmatrix} f(x^k, y^k) \\ g(x^k, y^k) \end{pmatrix} \end{aligned} \tag{7.3.12}$$

Example. Let $f(x, y) = x^2 + y^2 - 1$ and $g(x, y) = x^2 - y^2 + \frac{1}{2}$. Then $f_x = 2x$, $f_y = 2y$, $g_x = 2x$, and $g_y = -2y$.

$$F'(x, y) = \begin{pmatrix} 2x & 2y \\ 2x & -2y \end{pmatrix}.$$

$\det F'(x, y) = -8xy \neq 0$, provided $x \neq 0$ and $y \neq 0$. Algorithm (7.3.12) is easily executed. If $(x^0, y^0) = (1, 3)$, then the algorithm converges to one of the four solutions, $(\pm 1/2, \pm\sqrt{3}/2)$ (see Figure 7.3.2).

Remarks

1. For suitable $F(x, y) = (f(x, y), g(x, y))^{\mathrm{T}}$, the attractor property and the quadratic convergence property of Theorem 7.2.2 may be established. See Ortega [15, Chapter 8] for its generalization to N dimensions.

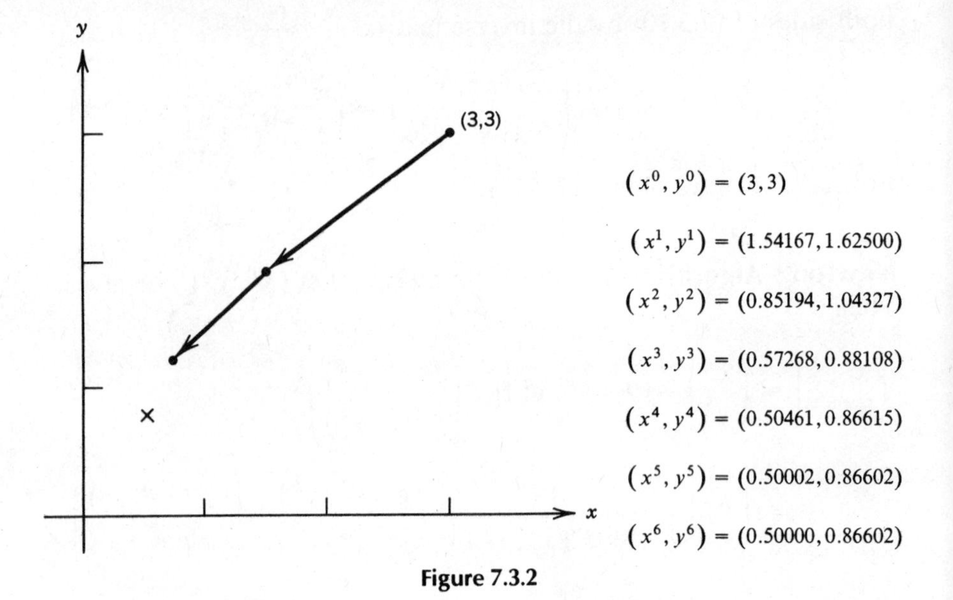

Figure 7.3.2

2. Often $|x^{k+1} - x^k|$ and $|y^{k+1} - y^k|$ might both be small, but $f(x^{k+1}, y^{k+1})$ and $g(x^{k+1}, y^{k+1})$ may not be small! See exercise 7-5. Therefore both the relative error in the variables and the size of the functions should be considered as criteria for convergence.

The extension of Newton's method to N-unknowns is straightforward. We introduce the following notation.

Notation. $F: \mathbb{R}^N \to \mathbb{R}^N$ is given by

$$F(x_1, \ldots, x_N) = (F_1(x_1, \ldots, x_N), \ldots, F_N(x_1, \ldots, x_N))^{\mathrm{T}} \in \mathbb{R}^N.$$

Alternatively, let $x = (x_1, \ldots, x_N)^{\mathrm{T}}$ be a vector, and we write $F(x) = (F_1(x), \ldots, F_N(x))^{\mathrm{T}}$. When $N = 2$, $x_1 = x$, and $x_2 = y$, $F_1(x_1, x_2) = f(x, y)$ and $F_2(x_1, x_2) = g(x, y)$. The analogs of lines (7.3.5) and (7.3.6) may be stated as for $i = 1, \ldots, N$:

$$\Delta F_i \simeq \sum^{N} F_{ix_j} \Delta x_j \tag{7.3.13}$$

where

$$F_{ix_j} = \frac{\partial F_i}{\partial x_j}.$$

Definition. Let $F: \mathbb{R}^N \to \mathbb{R}^N$. The *derivative of F* is an $N \times N$ matrix

$$F'(x) = \left(F_{ix_j}\right), \qquad i, j = 1, \ldots, N. \tag{7.3.14}$$

Newton's Algorithm for N Unknowns. Let $x^0 \in \mathbb{R}^N$.

$$x^{k+1} = x^k - \left(F'(x^k)\right)^{-1} F(x^k). \tag{7.3.15}$$

Remarks.

1. (7.3.15) is a vector equation. It is not, for large N, efficient to compute the inverse of $F'(x^k)$. In practice (7.3.15) may be broken into two parts:

(i) Solve for $x^{k+1/2}$ where $F'(x^k)x^{k+1/2} = F(x^k)$.
(ii) Compute $x^{k+1} = x^k - x^{k+1/2}$.

2. Let one arithmetic operation equal either one ($\pm$ and $\times$) or ($\div$). When Gaussian elimination is used, the number of arithmetic operations to do step (i) is of order N^3. Also, in $F(x^k)$ and $F'(x^k)$ there are $N + N^2$ function evaluations. Thus, if N is large or $F_i(x)$ are complicated, then the execution of Newton's method may be time consuming.

Example 1. Consider the nonlinear algebraic system that evolves from finite difference equations for

$$-(K(u)u')' = f(x), \tag{7.3.16}$$

$$u(0) = 0, \tag{7.3.17}$$

$$u'(1) = 0. \tag{7.3.18}$$

Let $\Delta x = 0.1$, $N = 10$, $(u_i) = (x_i)$ in line (7.3.15), and $u_0 = 0$.

$1 \leq i \leq 9$:

$$F_i(u) \equiv f_i \Delta x + \frac{K(u_{i+1}) + K(u_i)}{2} \frac{u_{i+1} - u_i}{\Delta x} - \frac{K(u_i) + K(u_{i-1})}{2} \frac{u_i - u_{i-1}}{\Delta x}. \tag{7.3.19}$$

$i = 10$:

$$F_{10}(u) \equiv f_{10} \frac{\Delta x}{2} + \left(0 - \frac{K(u_{10}) + K(u_9)}{2} \frac{u_{10} - u_9}{\Delta x}\right). \tag{7.3.20}$$

Note that $F_i(u)$ depends only on u_{i-1}, u_i, and u_{i+1}. Thus, when $j \neq i-1, i, i+1$, $\partial F_i / \partial u_j = F_{iu_j} = 0$. Hence, $F'(x)$ is tridiagonal and we may use the tridiagonal algorithm to compute $(F'(x^k))^{-1} F(x^k)$. For example, when $1 \leq i \leq 9$,

$$b_i = \frac{\partial F_i}{\partial u_i} = 0 + \frac{K'(u_i)}{2} \frac{u_{i+1} - u_i}{\Delta x} + \frac{K(u_{i+1}) + K(u_i)}{2}\left(-\frac{1}{\Delta x}\right) - \frac{K'(u_i)}{2} \frac{u_i - u_{i-1}}{\Delta x} - \frac{K(u_i) + K(u_{i-1})}{2} \frac{1}{\Delta x}, \tag{7.3.21}$$

$$-a_i = \frac{\partial F_i}{\partial u_{i-1}} = 0 - \frac{K'(u_{i-1})}{2} \frac{u_i - u_{i-1}}{\Delta x} - \frac{K(u_i) + K(u_{i-1})}{2}\left(-\frac{1}{\Delta x}\right), \tag{7.3.22}$$

$$-c_i = \frac{\partial F_i}{\partial u_{i+1}} = 0 + \frac{K'(u_{i+1})}{2} \frac{u_{i+1} - u_i}{\Delta x} + \frac{K(u_{i+1}) + K(u_i)}{2} \frac{1}{\Delta x} + 0. \tag{7.3.23}$$

When $i = 1$ or $i = 10$, the formulas for a_i, b_i and c_i are slightly different.

Example 2. Consider $Ax = d$, where $x, d \in \mathbb{R}^N$ and A is an $N \times N$ matrix. If we write $A = (a_{ij})$, then $Ax = d$ becomes

$$\sum_j a_{ij} x_j = d_i \tag{7.3.24}$$

or

$$\sum_j a_{ij} x_j - d_i = 0.$$

So, define

$$F_i(x) = \sum_l a_{il} x_l - d_i.$$

Then

$$F'(x) = \left(\frac{\partial F_i(x)}{\partial x_j} \right) = \left(\sum_l \frac{\partial}{\partial x_j} (a_{il} x_l) - \frac{\partial}{\partial x_j} d_i \right) = (a_{ij}). \tag{7.3.25}$$

Consequently, Newton's method gives the exact solution in the first step, that is,

$$\begin{aligned} x^{0+1} &= x^0 - \left(F'(x^0)\right)^{-1} F(x^0) \\ &= x^0 - A^{-1}(Ax^0 - d) \\ &= x^0 - x^0 + A^{-1}d \\ &= x, \quad \text{the solution of } Ax = d. \end{aligned}$$

Often $F(x)$ has special forms and, consequently, $F'(x)$ may be easily computed. The next proposition summarizes some of the more useful cases.

Proposition 7.3.1. Let $F: \mathbb{R}^N \to \mathbb{R}^N$ and $G: \mathbb{R}^N \to \mathbb{R}^N$.

(a) If $F(x) = Ax - d$, then

$$F'(x) = A. \tag{7.3.26}$$

(b) If $F(x) = B(\phi_i(x_i))$, then

$$F'(x) = \left(b_{ij}\phi_j'(x_j)\right). \tag{7.3.27}$$

(c) If $F(x) = A(x)x = (a_{ij}(x))x$, then

$$F'(x) = \left(\left(\sum_l a_{il,x_j}(x) \cdot x_l\right) + a_{ij}(x)\right). \tag{7.3.28}$$

(d) If $F(x) = C(x)x$ where $C(x) = (\Sigma_l c_{ijl}x_l)$, then

$$F'(x) = \left(\sum_l (c_{ilj} + c_{ijl})x_l\right). \tag{7.3.29}$$

(e) $(F + G)'(x) = F'(x) + G'(x).$ (7.3.30)

Proof. (7.3.26) has already been discussed. For (7.3.27), we let $F_i(x) = \Sigma_l b_{il}\phi_l(x_l)$ and $F'(x) = (\partial F_i/\partial x_j)$.

$$\begin{aligned}\frac{\partial F_i}{\partial x_j} &= \frac{\partial}{\partial x_j}\sum_l b_{il}\phi_l(x_l)\\ &= b_{ij}\frac{\partial \phi_j}{\partial x_j}(x_j) = b_{ij}\phi_j'(x_j).\end{aligned}$$

In (7.3.28) let $F_i(x) = \Sigma_l a_{il}(x)x_l$. Then we have

$$\begin{aligned}\frac{\partial F_i}{\partial x_j} &= \frac{\partial}{\partial x_j}\sum_l a_{il}(x)x_l\\ &= \sum_l\left(\frac{\partial a_{il}(x)}{\partial x_j}x_l + a_{il}(x)\frac{\partial x_l}{\partial x_j}\right)\\ &= \left(\sum_l a_{il,x_j}(x)x_l\right) + a_{ij}(x).\end{aligned}$$

For the proof of (7.3.29), use (7.3.28), where $a_{ij}(x) = \Sigma_k c_{ijk}x_k$:

$$a_{ij,x_j}(x) = \frac{\partial}{\partial x_j}\sum_k c_{ilk}x_k = c_{ilj} \cdot 1.$$

Then (7.3.28) implies

$$\begin{aligned}\frac{\partial F_i}{\partial x_j} &= \sum_l c_{ilj} x_l + \sum_k c_{ijk} x_k \\ &= \sum_l c_{ilj} x_l + \sum_l c_{ijl} x_l \\ &= \sum_l (c_{ilj} + c_{ijl}) x_l.\end{aligned}$$

The proof of (7.3.30) is straightforward.

Application of Proposition 7.3.1. In Chapter 9 we consider Burger's equation (8.3.12). An implicit time discretization coupled with finite differences gives $F(u) = Bu + C(u)u + Au - d$, where $C(u)$ has components of the form in (7.3.29). Since the derivative of the sum equals the sum of the derivatives, (7.3.26) and (7.3.29) yield

$$F'(u) = B + \left(\sum_l (c_{ilj} + c_{ijl}) u_l\right) + A.$$

Modifications of Newton's Method

1. Newton with SOR. Let $1 \le w < 2$.

$$x^{k+1} = x^k - w\left(F'(x^k)\right)^{-1} F(x^k). \tag{7.3.31}$$

2. Approximation of $F'(x) = (\partial F_i / \partial x_j)$. Let

$$\delta F(x^k) \equiv \left(\frac{F_i\left(x^k + e_j h_j\right) - F_i(x^k)}{h_j}\right),$$

where $e_j = (0, \ldots, 0, 1, 0 \ldots, 0)$, 1 is in the jth component.

$$x^{k+1} = x^k - \left(\delta F(x^k)\right)^{-1} F(x^k). \tag{7.3.32}$$

3. $F'(x^k)$ is singular, that is, $(F'(x^k))^{-1}$ does not exist. In some cases for small $\lambda \in \mathbb{R}$, $F'(x^k) + \lambda I$ is nonsingular.

$$x^{k+1} = x^k - \left(F'(x^k) + \lambda I\right)^{-1} F(x^k). \tag{7.3.33}$$

4. m-step methods. For large N, the time-consuming part is the computation of $(F'(x^k))^{-1}F(x^k)$ at each step. We simply avoid putting in a new x^k into $F'(x^k)$ every m steps:

$$x^{k,l} = x^{k,l-1} - \left(F'(x^k)\right)^{-1} F(x^{k,l-1}), \qquad 1 \leq l \leq m, \tag{7.3.34}$$

$$x^{k+1} = x^{k,m}. \tag{7.3.35}$$

If $m = 2$, then

$$\begin{aligned} x^{k,1} &= x^k - \left(F'(x^k)\right)^{-1} F(x^k), \\ x^{k+1} &= x^{k,2} = x^{k,1} - \left(F'(x^k)\right)^{-1} F(x^{k,1}) \\ &= x^k - \left(F'(x^k)\right)^{-1} F(x^k) \\ &\quad - \left(F'(x^k)\right)^{-1} F\left(x^k - \left(F'(x^k)\right)^{-1} F(x^k)\right) \\ &= x^k - \left(F'(x^k)\right)^{-1}\left(F(x^k) + F\left(x^k - \left(F'(x^k)\right)^{-1} F(x^k)\right)\right). \end{aligned} \tag{7.3.36}$$

5. $F'(x^k)^{-1}F(x^k)$ may be approximated by some iterative schemes. Two types of Gauss–Seidel schemes are described in Section 7.4.

6. Quasi-Newton methods. In Section 7.5 we shall discuss one version of this method. In this case $(F(x^k))^{-1}$ is replaced by a matrix H_k. The advantage of this method is that it involves only N^2 order of arithmetic operations and N functions evaluations. This is considerably less than those for Newton's method. For very large N, the quasi-Newton method (Broyden) has the disadvantage of destroying any sparseness (the large number of zeros in a matrix) of $F'(x^k)$. However, as we shall see, it is not necessary to store the matrix H_k.

7.4 GAUSS–SEIDEL VARIATIONS OF NEWTON'S METHOD

We shall give two variations on Newton's method. The first method is called the Gauss–Seidel–Newton–SOR method, and the second is the Newton–SOR method. Both are based on the Gauss–Seidel method, and consequently, $F(x)$ should be such that the Gauss–Seidel method converges for $Ax = d$, where $A = F'(x)$ and x is fixed. In particular, if $A = (a_{ij}) = F'(x)$ is strictly diagonally dominant, that is, $|a_{ii}| > \sum_{j \neq i} |a_{ij}|$ for all i, then these methods may converge. (See Ortega and Rheinboldt [14] for a precise discussion of the assumptions needed so that these schemes converge to the solution.)

First, let us consider the Gauss–Seidel–Newton–SOR method. For two or N unknowns, it is obtained by making two modifications:

1. Replace $F'(x^k)$ by $F'(x^k)$'s diagonal matrix.
2. As soon as x_i^{k+1} is computed use x_i^{k+1} in place of x_i^k.

As this is not Newton's method, one cannot expect the attractor and quadratic convergence properties to hold. However, in some cases this method converges, and as $(F'(x^k))^{-1}F(x^k)$ does not have to be computed, the number of arithmetic operations per iteration is much smaller.

Gauss – Seidel – Newton – SOR Algorithm for Two Unknowns. Let $1 \leq w < 2$. Then

$$x^{k+1} = x^k - w\frac{f(x^k, y^k)}{f_x(x^k, y^k)}, \tag{7.4.1}$$

$$y^{k+1} = y^k - w\frac{g(x^{k+1}, y^k)}{g_y(x^{k+1}, y^k)}. \tag{7.4.2}$$

Remark. Note in (7.4.2), x^{k+1} is used in place of x^k. This updating of the components is the reason for using the Gauss–Seidel description.

Example 1. $f(x, y) = x^2 + y^2 - 1$, and $g(x, y) = x^2 - y^2 + \frac{1}{2}$ with $(x^0, y^0) = (1, 3)$, $w = 1$. One easily computes lines (7.4.1) and (7.4.2):

$$x^{k+1} = x^k - \frac{(x^k)^2 + (y^k)^2 - 1}{2x^k}, \tag{7.4.3}$$

$$y^{k+1} = y^k - \frac{(x^{k+1})^2 - (y^k)^2 + 1/2}{-2y^k}. \tag{7.4.4}$$

One can show that (x^{k+1}, y^{k+1}) diverges. Recall that Newton's method converged. The reason for this is that the functions given by the right-hand sides of (7.4.3) and (7.4.4) fail to have the contractive-map property.

Example 2. $f(x, y) = x + 3\log x - y^2$ and $g(x, y) = 2x^2 - xy - 5x + 1$ with $(x^0, y^0) = (3.4, 2.2)$ and $w = 1$.

$$(x^1, y^1) = (3.4899, 2.2633),$$

$$(x^2, y^2) = (3.4874, 2.2616),$$

$$(x^3, y^3) = (3.4874, 2.2616).$$

Gauss – Seidel – Newton – SOR Algorithm for *N* Unknowns. Let $1 \leq w < 2$. Then

$$x_i^{k+1} = x_i^k - w\frac{F_i\left(x_1^{k+1}, \ldots, x_{i-1}^{k+1}, x_i^k, \ldots, x_N^k\right)}{F_{ix_i}\left(x_1^{k+1}, \ldots, x_{i-1}^{k+1}, x_i^k, \ldots, x_N^k\right)}. \tag{7.4.5}$$

Example. Consider $Ax = d$ with $x, d \in \mathbb{R}^N$ and $A = (a_{ij})$ an $N \times N$ matrix. Then we may rewrite this

$$F_i(x) = \sum_j a_{ij}x_j - d_i = 0. \tag{7.4.6}$$

$F'(x) = A$ and (7.4.5) has the form

$$x_i^{k+1} = x_i^k - w\frac{\sum_{j<i}a_{ij}x_j^{k+1} + a_{ii}x_i^k + \sum_{j>i}a_{ij}x_j^k - d_i}{a_{ii}}$$

$$= (1 - w)x_i^k + w$$

$$\times\left(-\left(\sum_{j<i} a_{ij}x_j^{k+1}\right)\Big/a_{ii} - \left(\sum_{j>i} a_{ij}x_j^k\right)\Big/a_{ii} + d_i/a_{ii}\right)$$

(7.4.7)

Line (7.4.7) is the Gauss–Seidel method with SOR. If $w = 1$, then (7.4.7) is the Gauss–Seidel method for linear equations.

The Newton–SOR method is a little different from the technique described above. Consider solving $F'(x^k)^{-1}F(x^k)$ by the Gauss–Seidel method. Let $F'(x^k) = B - C$, where $B = D(x^k) - L(x^k)$, $C = U(x^k)$, $D(x^k)$ is the diagonal of $F'(x^k)$, and $-L(x^k)$, $-U(x^k)$ are the strictly lower and upper triangular parts of $F'(x^k)$. The Gauss–Seidel algorithm is, in matrix form,

$$x^{k+1,0} = x^k$$

$$x^{k+1,m+1} = B^{-1}Cx^{k+1,m} + B^{-1}F(x^k). \tag{7.4.8}$$

$x^{k+1} = x^{k+1,M}$ when the relative error is satisfied for each component, that is,

$$\frac{|x_i^{k+1,m+1} - x_i^{k+1,m}|}{|x_i^{k+1,m+1}|} < \epsilon \quad \text{for all } i.$$

m refers to the Gauss–Seidel (or inner) iteration, and k refers to the Newton (or outer) iteration. By substituting backward in (7.4.8) we have

$$\begin{aligned} x^{k+1,m+1} &= B^{-1}Cx^{k+1,m} + B^{-1}F(x^k) \\ &= B^{-1}C\left(B^{-1}Cx^{k+1,m-1} + B^{-1}F(x^k)\right) + B^{-1}F(x^k) \\ &= \left(B^{-1}C\right)^{m+1}x^k + \left(\left(B^{-1}C\right)^m + \cdots + I\right)B^{-1}F(x^k). \end{aligned} \tag{7.4.9}$$

In order for the scheme to converge to the solution, we must have $\rho(B^{-1}C) < 1$. For an explanation of $\rho(B^{-1}C)$, see Chapter 6 and Proposition 6.7.1; also, when $F'(x)$ is strictly diagonally dominant, then $\rho(B^{-1}C) < 1$. Then as $m \to \infty$, $(B^{-1}C)^{m+1}x^k \to 0$.

The Newton–SOR method varies from the Gauss–Seidel method in three ways. First, the $(B^{-1}C)^{m+1}x^k$ term will be ignored. Second, the iteration m will be stopped according to the Newton iteration. The reason for this is that the first few Newton-like iterations are going to give a crude estimate of the solution to $F(x) = 0$; therefore, one needs only a rough estimate of $F'(x^k)^{-1}F(x^k)$. The inner iteration m will be stopped when $m = M(k)$, for example, $M(k) = 1$ or k or 2^k. The third modification is to include an SOR parameter, w.

There are two ways of expressing this algorithm. The version in line (7.4.10) is exactly as stated above. The other equivalent version is given in line (7.4.11). Readers may consult Sherman [25] or convince themselves that (7.4.10) and (7.4.11) are equivalent.

Newton–SOR Algorithm. Consider $F(x) = 0$ and use the notation given above.

$$x^{k+1} = x^k - \sum_{m=0}^{M} (B^{-1}C)^m B^{-1}F(x^k), \tag{7.4.10}$$

where $M = M(k) = 1$ or k or 2^k. Alternatively,

$$x^{k+1} = x^{k+1,\,M(k)}, \tag{7.4.11}$$

where $x^{k+1,\,M(k)}$ is given by an inner iteration

$$x^{k+1,0} = x^k,$$

$$x^{k+1,\,m+1} = x^{k+1,\,m} - B^{-1}F_k(x^{k+1,\,m}),$$

where

$$F_k(x) \equiv F(x^k) + F'(x^k)(x - x^k), \qquad 0 \le m \le M(k) - 1.$$

When $B = D - L$, the ith component of $B^{-1}F_k(x^{k+1,\,m})$ in line (7.4.11) is

$$g_i = \left(d_i - \sum_{j<i} a_{ij} g_j\right) \Big/ a_{ii},$$

where $F_k(x^{k+1,\,m}) = (d_i)$, $B^{-1}F_k(x^{k+1,\,m}) = (g_i)$, and the i, j component of B is a_{ij} for $j \le i$. When an SOR parameter w is used in the inner iteration, that is, a Gauss–Seidel–SOR method, then $B = D - wL$. Finally, note if $F(x) = Ax - d$, $w = 1$, and $M(k) = 1$, then the Newton–SOR method is the Gauss–Seidel–SOR method.

7.5 QUASI-NEWTON METHOD (BROYDEN)

As mentioned in the section on Newton's method, the operation count for Newton's method is large compared to the operation count for the quasi-Newton method. The idea of this method is to replace $F'(x^k)^{-1}$ by another matrix H_k, which will require fewer operations to compute.

In order to discover an algorithm for H_k, we first consider finding an analog to $F'(x^k)$, B_k. Suppose B, B_k, is given and we want to find $\overline{B}$, B_{k+1}. Since $\overline{B}$ is to behave like $F'(x^{k+1})$, and $F'(x)$ has the approximation property

$$\Delta F \simeq F'(x)\,\Delta x, \tag{7.5.1}$$

we require

$$y \equiv F(\overline{x}) - F(x) = \overline{B}(\overline{x} - x) = \overline{B}s. \tag{7.5.2}$$

In (7.5.2) $\overline{x} = x^{k+1}$ and $x = x^k$. (7.5.2) alone cannot determine $\overline{B}$. The direction given by the vector $s = \overline{x} - x$ should point toward the desired root of $F(x) = 0$ (see Figure 7.3.2). Therefore, $\overline{B}$ may equal B for all z that are perpendicular to s, that is,

$$\overline{B}z = Bz \quad \text{for all } z^{\mathrm{T}}s = (z, s) = 0. \tag{7.5.3}$$

Conditions (7.5.2) and (7.5.3), which are due to Broyden, imply that $\overline{B}$ is uniquely determined from a given B, y, and s.

Lemma 7.5.1. (Broyden) Let B be a given $N \times N$ matrix and $y, s \in \mathbb{R}^N$. $\overline{B}$ satisfies (7.5.2) and (7.5.3) if and only if

$$\overline{B} = B + \frac{(y - Bs)s^{\mathrm{T}}}{s^{\mathrm{T}}s}. \tag{7.5.4}$$

Line (7.5.4) is called the *quasi-Newton equation*.

Proof. Suppose (7.5.2) and (7.5.3) are satisfied. (7.5.4) will hold if and only if for all $w \in \mathbb{R}^N$

$$\overline{B}w = Bw + \frac{(y - Bs)s^{\mathrm{T}}w}{s^{\mathrm{T}}s}. \tag{7.5.5}$$

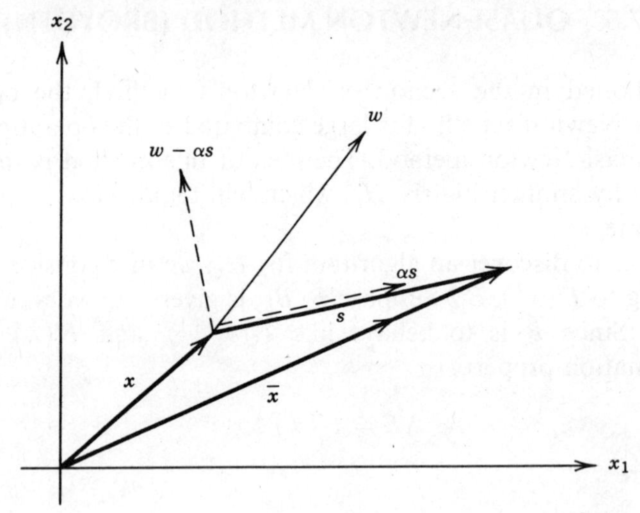

Figure 7.5.1

In order to show (7.5.5), let w be decomposed into a multiple of s and the orthogonal complement of span $\{s\}$. This means, for some $\alpha \in \mathbb{R}$,

$$w = \alpha s + w - \alpha s, \tag{7.5.6}$$

where α is chosen so that $(w - \alpha s, s) = 0$ (see Figure 7.5.1). From $0 = (w - \alpha s, s) = (w, s) - \alpha(s, s)$, we obtain

$$\alpha = \frac{(w, s)}{(s, s)} = \frac{w^{T}s}{s^{T}s} = \frac{s^{T}w}{s^{T}s}. \tag{7.5.7}$$

Apply (7.5.3) with $z = w - \alpha s$

$$\bar{B}w = \bar{B}(w - \alpha s + \alpha s) = \bar{B}(w - \alpha s) + \alpha\bar{B}s = B(w - \alpha s) + \alpha\bar{B}s. \tag{7.5.8}$$

Apply (7.5.2) to line (7.5.8)

$$\bar{B}w = B(w - \alpha s) + \alpha y = Bw + \alpha(y - Bs). \tag{7.5.9}$$

Use (7.5.7) for α in (7.5.9) to obtain

$$\bar{B}w = Bw + \frac{y - Bs}{s^{\mathrm{T}}s}s^{\mathrm{T}}w.$$

Suppose (7.5.5) holds. Line (7.5.2) is easily computed.

$$\bar{B}s = Bs + \frac{y - Bs}{s^{\mathrm{T}}s}s^{\mathrm{T}}s = Bs + (y - Bs) = y$$

Also (7.5.3) is computed by using $(s, z) = s^{\mathrm{T}}z = 0$:

$$\bar{B}z = Bz + \frac{y - Bs}{s^{\mathrm{T}}s}s^{\mathrm{T}}z = Bz + 0.$$

The formula for $\bar{B}$, (7.5.4), may be used to form the pre-Broyden algorithm. Make the following substitutions:

$$\bar{B} \to B_{k+1}, \qquad s \to x^{k+1} - x^k,$$
$$B \to B_k, \qquad y \to F(x^{k+1}) - F(x^k).$$

As in the formulation of Newton's method x^{k+1} is computed from

$$0 - F(x^k) = B_k(x^{k+1} - x^k). \tag{7.5.10}$$

Thus

$$x^{k+1} = x^k - B_k^{-1}F(x^k). \tag{7.5.11}$$

By combining lines (7.5.4) and (7.5.11), we have the following algorithm.

Pre-Broyden Algorithm. Let $F: \mathbb{R}^N \to \mathbb{R}^N$.

1. Select $x^0 \in \mathbb{R}^N$ and B_0, an $N \times N$ matrix, for example, $B_0 = F'(x^0)$.
2. $x^{k+1} = x^k - B_k^{-1}F(x^k)$.
3. Stop or $s_k = x^{k+1} - x^k$.
4. $y_k = F(x^{k+1}) - F(x^k)$.

5. $B_{k+1} = B_k + (y_k - B_k s_k)/(s_k^{\mathrm{T}} s_k) s_k^{\mathrm{T}}$.
6. Increase k by 1 and return to step 2.

Remarks

1. Step 5 requires an order of N^2 arithmetic operations and only N function evaluations.
2. Step 2 still requires the solution of a linear algebraic system.
3. Note from step 5 that B_{k+1} has the form $A + uv^{\mathrm{T}}$, where $A = B_k$, $u = (y_k - B_k s_k)/s_k^{\mathrm{T}} s_k$, and $v = s_k$.

The special form of B_{k+1} suggests that its inverse may also have a particular form. The next lemma is very helpful.

Lemma 7.5.2. (Sherman–Morrison) Let $\sigma \equiv 1 + (v, A^{-1}u) \neq 0$. Then $(A + uv^{\mathrm{T}})^{-1}$ exists and

$$(A + uv^{\mathrm{T}})^{-1} = A^{-1} - \frac{1}{\sigma} A^{-1} uv^{\mathrm{T}} A^{-1}. \tag{7.5.12}$$

Proof. If suffices to show that when (7.5.12) is multiplied from the right by $A + uv^{\mathrm{T}}$ we get the identity matrix. (We shall not prove the existence of an inverse of $A + uv^{\mathrm{T}}$.) Let $\sigma = 1 + (v, A^{-1}u) = 1 + v^{\mathrm{T}} A^{-1} u = 1 + c$.

$$\left(A^{-1} - \frac{1}{\sigma} A^{-1} uv^{\mathrm{T}} A^{-1}\right)(A + uv^{\mathrm{T}})$$

$$= I - \frac{1}{\sigma} A^{-1} uv^{\mathrm{T}} + A^{-1} uv^{\mathrm{T}} - \frac{1}{\sigma} A^{-1} uv^{\mathrm{T}} A^{-1} uv^{\mathrm{T}}$$

$$= I - \frac{1}{1 + c} A^{-1} uv^{\mathrm{T}} + A^{-1} uv^{\mathrm{T}} - \frac{1}{1 + c} A^{-1} ucv^{\mathrm{T}}$$

$$= I + (-A^{-1} uv^{\mathrm{T}} + (1 + c) A^{-1} uv^{\mathrm{T}} - A^{-1} uvc^{\mathrm{T}})/(1 + c)$$

$$= I + 0 = I.$$

Application of Lemma 7.5.2. Let $A = B_k$, $u = (y_k - B_k s_k)/s_k^T s_k$, and $v = s_k$.

$$
\begin{aligned}
B_{k+1}^{-1} &= \left(B_k + \frac{y_k - B_k s_k}{s_k^T s_k} s_k^T \right)^{-1} \\
&= B_k^{-1} - \frac{1}{\sigma} B_k^{-1} \frac{y_k - B_k s_k}{s_k^T s_k} s_k^T B_k^{-1} \\
&= B_k^{-1} - \frac{1}{1 + s_k^T B_k^{-1} \left[(y_k - B_k s_k)/s_k^T s_k \right]} B_k^{-1} \frac{y_k - B_k s_k}{s_k^T s_k} s_k^T B_k^{-1} \\
&= B_k^{-1} - \frac{s_k^T s_k}{s_k^T s_k + s_k^T B_k^{-1} (y_k - B_k s_k)} \frac{B_k^{-1} y_k - s_k}{s_k^T s_k} s_k^T B_k^{-1} \\
&= B_k^{-1} - \frac{B_k^{-1} y_k - s_k}{s_k^T B_k^{-1} y_k} s_k^T B_k^{-1}.
\end{aligned} \tag{7.5.13}
$$

Now define $H_k \equiv B_k^{-1}$ and $H_{k+1} \equiv B_{k+1}^{-1}$. Line (7.5.13) may be rewritten

$$
H_{k+1} = H_k - \frac{H_k y_k - s_k}{s_k^T H_k y_k} s_k^T H_k. \tag{7.5.14}
$$

Consequently, by using (7.5.14), we may write the pre-Broyden algorithm in terms of H_k and H_{k+1}. This has the advantage of avoiding the computation of $B_k^{-1} F(x^k)$. Thus, the following algorithm has N^2-order arithmetic operations and N function evaluations.

Broyden Quasi-Newton Algorithm

1. Select $x^0 \in \mathbb{R}^N$ and an $N \times N$ matrix H_0, for example, $H_0 = (F'(x^0))^{-1}$ or $H_0 = \text{diag}(F'(x^0))$.
2. $x^{k+1} = x^k - H_k F(x^k)$.
3. Stop or $s_k = x^{k+1} - x^k$.
4. $y_k = F(x^{k+1}) - F(x^k)$.
5. $H_{k+1} = H_k + ((s_k - H_k y_k)/(s_k^T H_k y_k)) s_k^T H_k$.
6. Increase k by 1 and return to step 2.

Example. $F\colon \mathbb{R}^2 \to \mathbb{R}^2$ given by

$$F(x_1, x_2) = \begin{pmatrix} (x_1)^2 + (x_2)^2 - 1 \\ (x_1)^2 - (x_2)^2 + 0.5 \end{pmatrix}.$$

Let $x^0 = (x_1^0, x_2^0)^{\mathrm{T}} = (1, 3)^{\mathrm{T}}$ and

$$H_0 = \left(F'(x^0)\right)^{-1} = \begin{pmatrix} 2 & 6 \\ 2 & -6 \end{pmatrix}^{-1} = \begin{pmatrix} 0.25 & 0.25 \\ 0.0833 & -0.0833 \end{pmatrix},$$

$$F(x^0) = \begin{pmatrix} 9.000 \\ -7.500 \end{pmatrix},$$

$$\begin{pmatrix} (x_1)^1 \\ (x_2)^1 \end{pmatrix} = \begin{pmatrix} 1 \\ 3 \end{pmatrix} - \begin{pmatrix} 0.25 & 0.25 \\ 0.0833 & -0.0833 \end{pmatrix} \begin{pmatrix} 9.0 \\ -7.5 \end{pmatrix}$$

$$= \begin{pmatrix} 0.625 \\ 1.625 \end{pmatrix},$$

$$s_0 = \begin{pmatrix} -0.375 \\ -1.375 \end{pmatrix},$$

$$y_0 = \begin{pmatrix} -6.9687 \\ 5.7499 \end{pmatrix},$$

$$H_1 = \begin{pmatrix} 0.25 & 0.25 \\ 0.0833 & -0.0833 \end{pmatrix}$$

$$+ \frac{\begin{pmatrix} -0.375 \\ -1.375 \end{pmatrix} - \begin{pmatrix} 0.25 & 0.25 \\ 0.0833 & -0.0833 \end{pmatrix} \begin{pmatrix} -6.9787 \\ 5.7499 \end{pmatrix}}{(-0.375, -1.375) \begin{pmatrix} 0.25 & 0.25 \\ 0.0833 & -0.0833 \end{pmatrix} \begin{pmatrix} -6.9787 \\ 5.7499 \end{pmatrix}}$$

$$\times (-0.375, -1.375) \begin{pmatrix} 0.25 & 0.25 \\ 0.0833 & -0.0833 \end{pmatrix}$$

$$= \begin{pmatrix} 0.2593 & 0.2491 \\ 0.1258 & -0.0876 \end{pmatrix}$$

$$x^2 = \begin{pmatrix} 0.625 \\ 1.625 \end{pmatrix} - \begin{pmatrix} 0.2593 & 0.2491 \\ 0.1258 & -0.0876 \end{pmatrix} \begin{pmatrix} 2.03125 \\ -1.75000 \end{pmatrix} = \begin{pmatrix} 0.5341 \\ 1.2162 \end{pmatrix}.$$

By continuing the iteration we obtain

$$x^3 = \begin{pmatrix} 0.4932 \\ 0.9615 \end{pmatrix}, \qquad x^4 = \begin{pmatrix} 0.4908 \\ 0.8852 \end{pmatrix}, \ldots, x^{10} = \begin{pmatrix} 0.5000 \\ 0.8660 \end{pmatrix}.$$

Remarks

1. In some cases, as in Newton's method, convergence may be accelerated by inserting an w into step 2, that is,

$$x^{k+1} = x^k - wH_kF(x^k), \qquad 1.0 < w.$$

2. As in Newton's method, we may use an m-step modification. That is, do steps 2–5 for m iterations, then start at step 1 with $x^0 =$ the last iteration.

An interesting application of Broyden's method is the steady-state fluid flow in a cavity. The Navier–Stokes equation must be solved. Usually the number of unknowns, N, will be much larger than the number of iterations, K, needed to each convergence. In Engelman, Strang, and Bathe [10] a clever implementation of the quasi-Newton method is given so that the storage of the $N \times N$ matrix, H_k, is replaced by the storage of $2K$ $N \times 1$ vectors. The derivation is based on the following steps.

Derivation of Quasi-Newton Without Storage of H_k

For $k = 0$: Choose H_0, for example, $H_0 = (F'(x^0))^{-1}$ or $H_0 = (\text{diag}\, F'(x^0))^{-1}$.

$$d^0 = H_0F(x^0),$$

$$x^1 = x^0 - d^0,$$

$$S_0 = x^1 - x^0,$$

$$r_0 = q^0 - d^0 \quad \text{where } q^0 \equiv H_0F(x^1).$$

Store S_0 and r^0.

$$
\begin{aligned}
d^1 = q^1 &= H_1 F(x^1) \\
&= \left(H_0 + \frac{S_0 - H_0 y_0}{S_0^{\mathrm{T}} H_0 y_0} S_0 H_0 \right) F(x^1) \\
&= q^0 + \frac{S_0 - r^0}{S_0^{\mathrm{T}} r^0} S_0 q^0 .
\end{aligned}
$$

For $k = 1$:

$$
\begin{aligned}
x^2 &= x^1 - d^1, \\
S_1 &= x^2 - x^1, \\
r^1 &= q^1 - d^1,
\end{aligned}
$$

where

$$
\begin{aligned}
q^1 &= H_1 F(x^2) \\
&= \left(H_0 + \frac{S_0 - H_0 y_0}{S_0^{\mathrm{T}} H_0 y_0} S_0 H_0 \right) F(x^2) \\
&= q^0 + \frac{S_0 - r^0}{S_0^{\mathrm{T}} r^0} S_0 q^0, \qquad q^0 \equiv H_0 F(x^2).
\end{aligned}
$$

(Note q^0 for $k = 1$ is different from q^0 for $k = 0$.)

Store S_1 and r^1.

$$
\begin{aligned}
d^2 = q^2 &= H^2 F(x^2) \\
&= \left(H_1 + \frac{S_1 - H_1 y_1}{S_1^{\mathrm{T}} H_1 y_1} S_1 H_1 \right) F(x^2) \\
&= q^1 + \frac{S_1 - r^1}{S_1^{\mathrm{T}} r^1} S_1 q^1, \qquad q^1 \equiv H_1 F(x^2).
\end{aligned}
$$

For $k = 2$:

$$x^3 = x^3 - d^2,$$

$$S_2 = x^3 - x^2,$$

$$r^2 = q^2 - d^2,$$

where

$$q^2 = H_2 F(x^3)$$

$$= \left(H_1 + \frac{S_1 - r^1}{S_1^T r^1} S_1 H_1 \right) F(x^3)$$

$$= q^1 + \frac{S_1 - r^1}{S_1^T r^1} S_1 q^1, \qquad q^1 \equiv H_1 F(x^3),$$

$$q^1 = H_1 F(x^3)$$

$$= \left(H_0 + \frac{S_0 - r^0}{S_0^T r^0} S_0 H_0 \right) F(x^3)$$

$$= q^0 + \frac{S_0 - r^0}{S_0^T r^0} S_0 q^0, \qquad q^0 \equiv H_0 F(x^3),$$

(Note q^0 for $k = 2$ is different from q^0 for $k = 0$ or $k = 1$.)
Store S_2 and r^2.

$$d^3 = q^3 = H_3 F(x^3)$$

$$= \left(H_2 + \frac{S_2 - r^2}{S_r^T r^2} S_2 H_2 \right) F(x^3)$$

$$= q^2 + \frac{S_2 - r^2}{S_2^T r^2} S_2 q^2, \qquad q^2 \equiv H_2 F(x^3).$$

For $k = 3$, similar steps are taken (what are they?). For *general* k, this is given by the algorithm below.

Remarks

1. For each quasi-Newton iteration, the $q^k = H_k F(x^{k+1})$ must be computed iteratively, starting with $q^0 = H_0 F(x^{k+1})$ and going up to q^k.
2. In $q^j = H_j F(x^{k+1})$ the x^{k+1} dependence is suppressed.
3. In order to compute q^k, the $2k$ column vectors S_j, r^j for $0 \le j \le k-1$ must have been stored while computing the previous quasi-Newton iterations.

The Broyden Quasi-Newton Algorithm Without Storage of H_k

1. $d^0 = (F'(x^0))^{-1} F(x^0)$ or $d^0 = (\text{diag}(F'(x^0)))^{-1}\, F(x^0)$.
2. $x^{k+1} = x^k - d^k$.
3. Stop, or $s_k = x^{k+1} - x^k$ and store S_k.
4. Let $q^0 \equiv F'(x^0)^{-1} F(x^{k+1})$. Define q^k iteratively by letting $0 \le j \le k-1$.

$$q^{j+1} = q^j + \frac{S_j - r^j}{S_j^{\mathrm{T}} r^j} S_j^{\mathrm{T}} q^j.$$

   Define $r^k \equiv q^k - d^k$ and store r^k.

5. $d^{k+1} \equiv q^{k+1} = q^k + [(S_k - r^k)/S_k^{\mathrm{T}} r^k] S_k^{\mathrm{T}} q^k$.
6. Increase k by 1 and return to step 2.

7.6 CONTINUATION (HOMOTOPY) METHOD

In Sections 7.5 we saw how to modify Newton's method so that the number of computations were reduced. Another problem with Newton's method is that one must make an initial guess, x^0, which must be suitably close to the solution of $F(x) = 0$. One class of techniques is the continuation method which does not, in general, converge quadratically; however, it does tend, under suitable assumptions on $F(x)$, to get close to a root regardless of the choice of x^0. Consequently, the continuation method may be used first, and then second use Newton's method where the last iterate of the continuation method is the first iterate of Newton's method.

Let $F_0, F: \mathbb{R}^N \to \mathbb{R}^N$, where F_0 is "simpler" than F, that is, we know $x^0 \in \mathbb{R}^N$ such that $F_0(x^0) = 0$. In order to describe the continuation method, we must define a homotopy between F_0 and F.

Definition. Let $F_0, F: \mathbb{R}^N \to \mathbb{R}^N$ be given and $F_0(x^0) = 0$, $x^0 \in \mathbb{R}^N$. $H: \mathbb{R}^N \times [0,1] \to \mathbb{R}^N$ is called a *homotopy between* F_0 *and* F if and only if

- **(i)** $H(x,0) = F_0(x)$.
- **(ii)** $H(x,1) = F(x)$.
- **(iii)** H is continuous.

Example. Let $F(x)$ be given and $x^0 \in \mathbb{R}^N$ be any guess for a solution of $F(x) = 0$. Define $F_0: \mathbb{R}^N \to \mathbb{R}^N$ by

$$F_0(x) \equiv F(x) - F(x^0). \tag{7.6.1}$$

Then x^0 is a solution of $F_0(x) = 0$. A simple homotopy between F_0 and F is

$$H(x,t) = (1-t)F_0(x) + tF(x). \tag{7.6.2}$$

For our choice of F_0, (7.6.1), line (7.6.2) becomes

$$\begin{aligned} H(x,t) &= (1-t)\big(F(x) - F(x^0)\big) + tF(x) \\ &= F(x) - (1-t)F(x^0). \end{aligned} \tag{7.6.3}$$

Another homotopy between F_0 and F is

$$\tilde{H}(x,t) = (1-t)F_0(x) + t^2F(x).$$

Idea Behind the Continuation Method. If the solution of $H(x^0,0) = F_0(x^0) = 0$ is known, then for $0 < t$ near zero, we may be able to find the solution of $H(x^1,t_1) = 0$. Then repeat this for $t_2 > t_1$ near t_1 and so on until we have a solution of $F(x^K) = H(x^K,1) = 0$, where

$t_{k+1} - t_k = 1/K$. In practice the $x^1, \ldots, x^K$ are only approximated and, consequently, x^K is an approximation of the solution of $F(x) = 0$. The points $(x^0, x^1, \ldots, x^K)$ are approximations of a *continuation path* $x(t) \in \mathbb{R}^N$ that must satisfy, for $0 \le t \le 1$

$$H(x(t), t) = 0 \tag{7.6.4}$$

Thus, $x(0) = x^0$ is given and $x(1)$ is the solution of $F(x) = 0$.

In order to find $x(t)$ or $x(t_k) = x^k$, $t_k = k/K$, we compute the derivative of (7.6.4) with respect to t. This gives a system of ordinary differential equations for the continuation path $x(t)$. Let $H(x(t), t) = (H_i(x(t), t)) \in \mathbb{R}^N$. Then

$$\begin{aligned} 0 &= \frac{d}{dt} H_i(x(t), t) \quad \text{by (7.6.4)} \\ &= \sum_j H_{ix_j}(x(t), t) \frac{dx_j}{dt} + H_{it}(x(t), t) \cdot 1, \end{aligned}$$

where

$$H_{ix_j}(x(t), t) = \frac{\partial H_i}{\partial x_j}(x(t), t)$$

Or, in matrix notation with $H_x(x(t), t)) = (H_{ix_j}(x(t), t))$ an $N \times N$ matrix and $H_t(x(t), t) = (H_{it}(x(t), t)) \in \mathbb{R}^N$

$$H_t(x(t), t) + H_x(x(t), t)\dot{x}(t) = 0, \qquad \dot{x} = \left(\frac{dx_j}{dt}\right). \tag{7.6.5}$$

Line (7.6.5) is easily used to define a system of ordinary differential equations.

Davidenko's Equation. Let H be a homotopy between F_0 and F.

$$\dot{x} = -(H_x(x(t), t))^{-1} H_t(x(t), t), \tag{7.6.6}$$

$$x(0) = x^0 \in \mathbb{R}^N. \tag{7.6.7}$$

Special Case. Let $F: \mathbb{R}^N \to \mathbb{R}^N$ be given, and F_0 and H be defined by (7.6.1) and (7.6.2), respectively. Then $H(x, t) = -(1 - t)F(x^0) + F(x)$ and $H_t = F(x^0)$ and $H_x = F'(x)$. Davidenko's equation has the form

$$\dot{x} = -(F'(x))^{-1}F(x^0), \tag{7.6.8}$$

$$x(0) = x^0. \tag{7.6.9}$$

Example. Let $N = 1$, $F(x) = x^{1/3} - 1$, and F_0, H as given in the special case. From (7.6.8) and (7.6.9), Davidenko's equation then has the form

$$\dot{x} = -3x^{2/3}\left((x^0)^{1/3} - 1\right), \tag{7.6.10}$$

$$x(0) = x^0. \tag{7.6.11}$$

In this case Davidenko's equation can be explicitly solved by the separation of variables method. Lines (7.6.10) and (7.6.11) imply

$$\int_{x(0)}^{x(t)} x^{-2/3}\,dx = \int_0^t -3\left((x^0)^{1/3} - 1\right),$$

$$3x^{1/3} - 3(x^0)^{1/3} = -3\left((x^0)^{1/3} - 1\right)t,$$

$$x(t) = \left((x^0)^{1/3} - \left((x^0)^{1/3} - 1\right)t\right)^3. \tag{7.6.12}$$

Note that $x(0) = x^0$ and $x(1) = 1 =$ the solution of $F(x) = 0$.

In most cases Davidenko's equation cannot be solved explicitly. We briefly mention some methods that have been used to do this.

Numerical Methods for Davidenko's Equations

1. Euler. In (7.6.8) simply replace $\dot{x}$ by $(x^{k+1} - x^k)/\Delta t$ and x by x^k. Then the equation for the special case given by (7.6.8) and (7.6.9) is, for $h = \Delta t$,

$$x^{k+1} = x^k - h\left(F'(x^k)\right)^{-1}F(x^0) \tag{7.6.13}$$

2. Predictor–Corrector. Again in (7.6.8) replace $\dot{x}$ by $(x^{k+1} - x^k)/\Delta t$, but replace x by x^{k+1}. Then this gives a nonlinear algebraic system to solve

$$x^{k+1} = x^k - h\big(F'(x^{k+1})\big)^{-1}F(x^0). \tag{7.6.14}$$

A simple predictor–corrector method is the Euler–trapezoid technique:

$$x^{k+1,0} = x^k - h\big(F'(x^k)\big)^{-1}F(x^0) \tag{7.6.15}$$

(the predictor given by Euler's method).

$$x^{k+1,m+1} = x^k - h\frac{\big(F'(x^{k+1,m})\big)^{-1}F(x^0) + \big(F'(x^k)\big)^{-1}F(x^0)}{2} \tag{7.6.16}$$

(the corrector given by the trapezoid rule).

The corrector step is repeated until $x^{k+1,m+1} \simeq x^{k+1,m}$. Then $x^{k+1} \equiv x^{k+1,m+1}$ will be an approximate solution of (7.6.14).

The corrector step may be derived from integrating

$$\dot{x} = -\big(F'(x)\big)^{-1}F(x^0) \tag{7.6.17}$$

from t_k to t_{k+1}. Then (7.6.17) becomes

$$x^{k+1} - x^k = -\int_{t_k}^{t_{k+1}}\big(F'(x)\big)^{-1}F(x^0). \tag{7.6.18}$$

If the right-hand side of (7.6.18) is approximated by the trapezoid rule, then for $h = t_{k+1} - t_k$ we obtain (7.6.16). An important observation is that we are using the contraction-map theorem (see Theorem 7.2.1 with $N = 1$) with

$$g(x) \equiv x^k - h\frac{F'(x)^{-1}F(x^0) + \big(F'(x^k)\big)^{-1}F(x^0)}{2}.$$

In order to ensure the iterates $x^{k+1,m+1}$ converge, $|g'(x)|$ must be less than 1 and, hence, $h = \Delta t = 1/K$ should be suitably small.

3. Runge–Kutta. This method gives a higher-order accuracy when used with a variable step size.

Remark. In both the predictor–corrector and the Runge–Kutta methods a number of computations of $(F'(x))^{-1}F(x^0)$ are required. Thus, even though these methods give higher-order accuracy, the additional computing time may be a serious drawback.

Example Using Euler's Method. $N = 1$ and $F(x) = x^{1/3} - 1$ and use the special case given by (7.6.8) and (7.6.9) and, hence, (7.6.13). Line (7.6.13) is

$$x^{k+1} = x^k - h3(x^k)^{2/3}\left((x^0)^{1/3} - 1\right).$$

For different K and x^0, we obtain the results given Table 7.6.1.

Most variations of the continuation method involve a different choice of a homotopy or a different choice of a numerical method for aproximating the solution to Davidenko's equation.

As noted at the beginning of this section one can use a combination of a continuation method and the Newton's method. If we use the special case given by (7.6.8) and (7.6.9) and Euler's method in line (7.6.13), then this combination may be written explicitly.

Continuation – Newton Algorithm. Let $F: \mathbb{R}^N \to \mathbb{R}^N$ be given.

1. Use a continuation method to get close to the root, for example,

$$x^{k+1} = x^k - h\left(F'(x^k)\right)^{-1}F(x^0), \qquad h = 1/K,$$

$$0 \le k \le K - 1. \tag{7.6.19}$$

TABLE 7.6.1

K	x^0	x^K
10	2	0.9815
20	2	0.9908
10	5	0.8806
20	5	0.9416
20	10	0.8643

2. Use Newton's method with starting point x^K

$$x^{k+1} = x^k - \left(F'(x^k)\right)^{-1} F(x^k), \qquad k \geq K. \quad (7.6.20)$$

This method is illustrated in Appendix A.2 where a nonlinear ordinary differential equation is considered.

7.7 NONLINEAR GAUSS–SEIDEL–SOR METHOD

This section generalizes the Gauss–Seidel method from linear problems of the form

$$F(x) = Ax - d = 0 \quad (7.7.1)$$

to certain general nonlinear problems $F(x) = 0$. Recall the Gauss–Seidel–SOR algorithm for (7.7.1) has the form, for $A = (a_{ij})$ and $d = (d_i)$,

$$x_i^{k+1/2} = \left(d_i - \sum_{j<i} a_{ij} x_j^{k+1} - \sum_{j>i} a_{ij} x_j^k\right) \Big/ a_{ii}, \quad (7.7.2)$$

$$x_i^{k+1} = (1 - w) x_i^k + w x_i^{k+1/2}, \qquad 1 \leq w < 2. \quad (7.7.3)$$

Line (7.7.2) when $F_i(x) \equiv \Sigma_j a_{ij} x_j - d_i$, may be written

$$F_i\left(x_1^{k+1}, \ldots, x_{i-1}^{k+1}, x_i^{k+1/2}, x_i^k, \ldots, x_N^k\right) = 0. \quad (7.7.4)$$

For general $F(x) = (F_i(x))$, we may try to duplicate steps (7.7.4) and (7.7.3).

Nonlinear Gauss–Seidel–SOR Algorithm. Let $F: \mathbb{R}^N \to \mathbb{R}^N$.

1. If possible, solve (7.7.4) for $x_i^{k+1/2}$.
2. Compute x_i^{k+1} by (7.7.3).

It may not always be possible to do step 1, and consequently, the method may not be applicable. We give two examples of cases when this method does work. In Chapter 8 we shall illustrate this method for nonlinear heat conduction problems. Also, the Stefan problem generates an algebraic problem similar to the following examples. In particular, in Section 8.6 two more algorithms of this form will be discussed.

Example 1. $F(x) = Ax + (\phi_i(x_i)) - d$, where ϕ_i are nondecreasing and continuous.

$$F_i(x) = \sum_j a_{ij}x_j + \phi_i(x_i) - d_i = 0. \tag{7.7.5}$$

Line (7.7.5) may be written

$$a_{ii}x_i + \phi_i(x_i) = d_i - \sum_{j<i} a_{ij}x_j - \sum_{j>i} a_{ij}x_j. \tag{7.7.6}$$

Define $r_i(z) \equiv a_{ii}z + \phi_i(z)$ and note when $a_{ii} > 0$, $r_i(z)$ is continuous and strictly increasing. Thus, r_i has an inverse, r_i^{-1}, that is, for each $\overline{w} = r_i(z)$ there is a unique z, $r_i^{-1}(\overline{w}) = z$. Consequently, (7.7.6) may be written

$$x_i = r_i^{-1}\left(d_i - \sum_{j<i} a_{ij}x_j - \sum_{j>i} a_{ij}x_j\right) = r_i^{-1}(\overline{w}). \tag{7.7.7}$$

This means that step 1, that is, (7.7.4), can be performed.

Algorithm for $Ax + (\phi_i(x_i)) = d$, ϕ_i Continuous and Nondecreasing

$$x_i^{k+1/2} = r_i^{-1}\left(d_i - \sum_{j<i} a_{ij}x_j^{k+1} - \sum_{j>i} a_{ij}x_j^k\right), \tag{7.7.8}$$

$$x_i^{k+1} = (1-w)x_i^k + wx_i^{k+1/2}, \qquad 1 \le w < 2. \tag{7.7.9}$$

Remark. If A is strictly diagonally dominant, $a_{ii} > 0$ and $a_{ij} \le 0$, $i \ne j$, then (7.7.8), (7.7.9) with $w = 1$ converges to a unique solution of $Ax + (\phi_i(x_i)) = d$ (see Ortega and Rheinboldt [14]).

Illustration of (7.7.8) and (7.7.9). Consider

$$-u'' = -au^2 + d, \qquad a > 0 \tag{7.7.10}$$

$$u(0) = 1 = u(1). \tag{7.7.11}$$

The finite difference method for (7.7.10) gives

$$\frac{2}{\Delta x^2} u_i + au_i^2 = d + \frac{u_{i-1}}{\Delta x^2} + \frac{u_{i+1}}{\Delta x^2}. \tag{7.7.12}$$

(7.7.12) is the analog of (7.7.6), where $a_{ii} = 2/\Delta x^2$, $\phi_i(u_i) = au_i^2$, $u_i = x_i$, $a_{i,i-1} = -1/\Delta x^2$, and $a_{i,i+1} = -1/\Delta x^2$. $\bar{w} = r(z) = (2/\Delta x^2)z + az^2$, and so we can use the quadratic formula to solve

$$\bar{w} = \frac{2}{\Delta x^2} z + az^2 = r(z) \tag{7.7.13}$$

for z in terms of $\bar{w}$. Consequently,

$$z = r^{-1}(\bar{w}) = \frac{1}{a\Delta x^2}\left(-1 + \sqrt{1 + a\bar{w}\Delta x^4}\right) \tag{7.7.14}$$

For $\Delta x = 0.1$, $N = 10$, $u_0 = 1$, $u_{10} = 1$, and $d = 10$, we may explicitly state the algorithm (7.7.8), (7.7.9):

$$u_i^{k+1/2} = r^{-1}\left(d + \frac{u_{i-1}^{k+1}}{\Delta x^2} + \frac{u_{i+1}^{k}}{\Delta x^2}\right), \tag{7.7.15}$$

$$u_i^{k+1} = (1 - w)u_i^k + wu_i^{k+1/2}, \qquad 1 \le w < 2. \tag{7.7.16}$$

In the following calculations the convergence of the algorithm was defined when the relative error was less than 0.001, that is,

$$\left(|u_i^{k+1} - u_i^k| / |u_i^k|\right) < 0.001 \quad \text{for } i = 1, \dots, 9. \tag{7.7.17}$$

The output is presented in graphic form for several different values of a (see Figure 7.7.1). Also, Table 7.7.1 contains a comparison of the number of iterations that were required to reach convergence for different SOR parameters w.

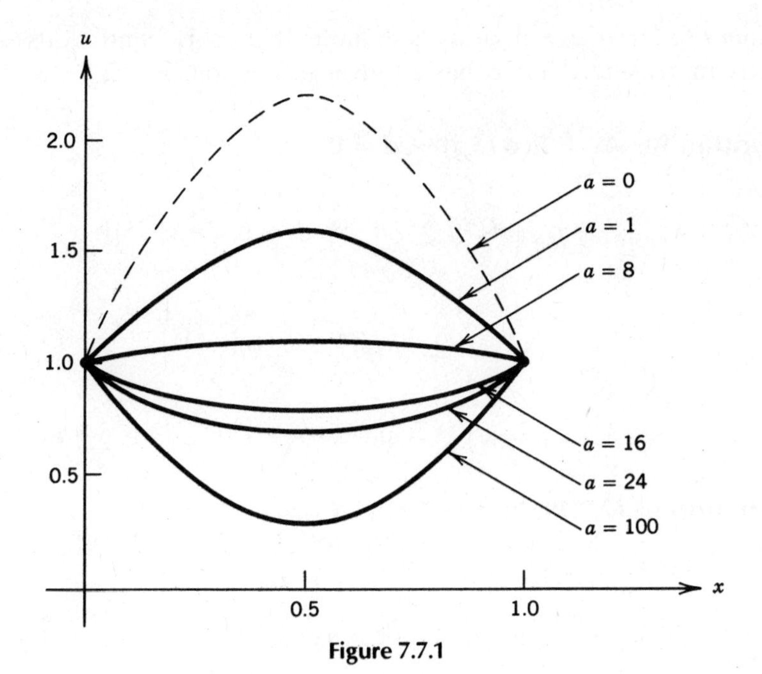

Figure 7.7.1

Example 2. $0 = F(x) = Ax + B(\phi_i(x_i)) - d$, where ϕ_i are nondecreasing and continuous and $b_{ii}, a_{ii} > 0$. This type of system evolves from the finite element method. We can write $F(x) = 0$ as

$$a_{ii}x_i + b_{ii}\phi_i(x_i) = d_i - \sum_{j<i}\left(a_{ij}x_j + b_{ij}\phi_j(x_j)\right) - \sum_{j>i}\left(a_{ij}x_j + b_{ij}\phi_j(x_j)\right). \qquad (7.7.18)$$

TABLE 7.7.1

K	w	a
33	1.0	1
20	1.3	1
13	1.6	1
56	1.9	1

Define $r_i(z) = a_{ii}z + b_{ii}\phi_i(z)$ and note that r_i is continuous and strictly increasing. Thus r_i has an inverse function, $r^{-1}(\overline{w}) = z$.

Algorithm for $Ax + B(\phi_i(x_i)) - d = 0$

$$x_i^{k+1/2} = r_i^{-1}\Big(d_i - \sum_{j<i}\big(a_{ij}x_j^{k+1} + b_{ij}\phi_j\big(x_j^{k+1}\big)\big)$$

$$- \sum_{j>i}\big(a_{ij}x_j^{k} + b_{ij}\phi_j\big(x_j^{k}\big)\big)\Big), \tag{7.7.19}$$

$$x_i^{k+1} = (1-w)x_i^k + wx_i^{k+1/2}, \qquad 1 \le w < 2. \tag{7.7.20}$$

Illustration of (7.7.19) and (7.7.20). Consider

$$-u'' = -au^2 + d, \tag{7.7.21}$$

$$u(0) = 1 = u(1). \tag{7.7.22}$$

We shall use the Galerkin formulation of the finite element method with linear shape functions. The weak formulation of (7.7.21) is for all $\Psi(0) = \Psi(1) = 0$, $\Psi \in C[0,1]$, Ψ' piecewise continuous.

$$\int_0^1 u'\Psi' = -\int_0^1 au^2\Psi + \int_0^1 d\Psi. \tag{7.7.23}$$

Let $u = \Sigma_j u_j \Psi_j$, $u^2 \simeq \Sigma_j u_j^2 \Psi_j$, and $\Psi = \Psi_i$. Then (7.7.23) gives the following equations:

$$\sum_j u_j \int_0^1 \dot{\Psi}_j \dot{\Psi}_i = -\sum_j au_j^2 \int_0^1 \Psi_j \Psi_i + \int_0^1 d\Psi_i. \tag{7.7.24}$$

After evaluating the integrals in (7.7.24), we get

$$-\frac{u_{i-1} - 2u_i + u_{i+1}}{h} = dh - \frac{h}{6}\big(au_{i-1}^2 + 4au_i^2 + au_{i+1}^2\big). \tag{7.7.25}$$

Now divide both sides of (7.7.24) by h and define $r(z) = (2/h^2)z + \frac{2}{3}az^2 = \overline{w}$. Then by the quadratic formula

$$z = r^{-1}(\overline{w}) = \frac{3}{2ah^2}\left(-1 + \sqrt{1 + \frac{2a}{3}wh^4}\right). \qquad (7.7.26)$$

The algorithm (7.7.21), (7.7.22) may be explicitly stated:

$$u_i^{k+1/2} = r^{-1}\left(d + \frac{u_{i-1}^{k+1}}{h^2} - \frac{a\left(u_{i-1}^{k+1}\right)^2}{6} + \frac{u_{i+1}^{k}}{h^2} - \frac{a\left(u_{i+1}^{k}\right)^2}{6}\right), \qquad (7.7.27)$$

$$u_k^{k+1} = (1 - w)u_k^k + wu_i^{k+1/2}, \qquad 1 \le w < 2. \qquad (7.7.28)$$

The computations with (7.7.27) and (7.7.28) give similar results to those of the previous illustration.

In both examples we were able to state explicitly a formula for $r_i^{-1}(\overline{w}) = z$. This may not always be possible. One technique to find z is to use Newton's method where $f(z) \equiv r_i(z) - \overline{w} = 0$. In this case we can use the previous k iterate to start the one-variable Newton method. More precisely, for each fixed i, let

$$z^0 = x_i^k$$

$$z^{n+1} = z^n - \frac{f(z^n)}{f'(z^n)}$$

where

$$f : \mathbb{R} \to \mathbb{R},$$

$$f'(z) = r_i'(z).$$

The n iteration is an inner iteration, that is, it is inside the k loop and is done for each i. Consequently, the stopping criteria for the n iteration should be more restrictive than for the k iteration.

7.8 OBSERVATIONS AND REFERENCES

Given any nonlinear problem there is no general rule that will dictate which method to use. Also, there are many variations on a given method. The reader may choose to experiment with different methods on the same problem. Not all methods will converge to a solution of a given problem. A standard reference is Ortega and Rheinboldt [14], and a more detailed discussion is given about the assumptions on the problem that are needed to give convergence.

Certainly, every reader should be familiar with Newton's method, and some relevant exercises are 7-2, 7-5, and 7-12. An important variation of Newton's method is the quasi-Newton method, and some interesting problems include 7-10 and 7-11.

EXERCISES

7-1 Consider the eigenvalue problem $-u'' = \lambda u$, $u(0) = 0$, and $u'(1) + 2u(1) = 0$. Find an eigenvalue λ for this problem. Use the bisection method, and if possible Picard iteration and Newton methods.

7-2 Consider the initial-value problem $x'(t) = (1 + x(t))^2)^{-1}$ and $x(0) = 1$. Use an implicit time discretization and form a nonlinear algebraic problem. Find $\gamma > 0$ such that when $\Delta t < \gamma$, the Picard iteration method will give a solution. Find γ both by numerical experimentation and by estimating the r in Theorem 7.2.1. Also, try Newton's method for this problem. Compare your answers with the exact solution!

7-3 Consider Theorem 7.2.2. Suppose the condition $0 \le f''(x) \le M$ is replaced by $-M \le f''(x) \le 0$. Can the theorem be proved? What other conditions on $f(x)$ can be used to give the same conclusions?

7-4 Approximate the root of $f(x) = 2 + x - e^{3x}$. Use the Newton with SOR algorithm in (7.2.34) and experiment with different x_0 and w.

7-5 Consider the algebraic system

$$x^2 + xy^3 = 9,$$

$$3x^2y - y^3 = 4.$$

Approximate its solutions by using the Gauss–Seidel–Newton–SOR algorithm for different w and different (x^0, y^0), for example, (1.2, 2.5), (2, 2.5), $(-1.2, -2.5)$, and $(2, -2.5)$. Check your answers by computing $F(x^{k+1}, y^{k+1})$.

7-6 Write a program to solve the problem

$$-(K(x,u)u_x)_x = f(x),$$

$$u(0) = 0,$$

$$u'(1) = 0.$$

Use the finite difference method and Newton's method. In order to test your program, use $K(u) = 1 + u$ and $f(x) = 1.0$ where the exact solution is $u(x) = -1 + \sqrt{2-(x-1)^2}$.

7-7 Consider the problem in exercise 7-6. Find the weak formulation and the use the Galerkin finite element method to form an algebraic system. You may wish to approximate $K(x, u)$ by $K(x,u)_I \equiv \Sigma_j K(x_j, u_j)\Psi_j$. Solve the resulting system by Newton's method.

7-8 Consider the problem in exercise 7-5. Use the Newton with SOR, the finite difference approximation of $F'(x)$, and the two-step modifications of Newton's method. See lines (7.3.31), (7.3.32), and (7.3.36). Compare the methods.

7-9 Consider the problem in exercise 7-5. Use the Gauss–Seidel–Newton–SOR method to approximate its solution. Does this method work for the problem in exercise 7-6?

7-10 Consider the problem in exercise 7-5. Use the quasi-Newton method to approximate its solutions. Compute the matrices H_k.

7-11 Consider the algebraic problem in exercises 7-6 or 7-7. Use the quasi-Newton method to approximate the solutions. Use the version that does not require the storage of H_k.

7-12 Consider the problem in exercise 7-5. Use the continuation method to approximate its solution. Use Euler's method to solve Davidenko's equation.

7-13 Repeat exercise 7-12 but use a predictor–corrector method to solve Davidenko's equation.

7-14 Use a combination of the continuation method and Newton's method to approximate the solution of the algebraic problem in exercises 7-6 and 7-7.

7-15 Consider $-u'' = -bu^{1/2} + 10$, $b > 0$, $u(0) = 1 = u(1)$. Use the finite difference method to form a nonlinear algebraic problem. Approximate its solution by using the nonlinear Gauss–Seidel–SOR method. Experiment with different w.

7-16 Repeat exercise 7-15, but use the Galerkin finite element method to form the system.

7-17 Consider the problem in exercise 7-15. Use the Gauss–Seidel–Newton–SOR method, line (7.4.5), to approximate its solution. Experiment with different SOR parameters w, and find the optimal choice,

7-18 Consider the problem in exercise 7-15. Use the Newton–SOR method as given in line (7.4.11) to approximate its solution. Use $B = D - wL$, where $F'(x^k) = D - L - U$ and $1 \le w < 2$. Experiment with different choices of w, and find the optimal choices.

7-19 Consider $-u'' = a(b^4 - u^4)$, with $a > 0$, $b > 0$, $u(0) = 0$, and $u'(1) = 0$. Use finite differences to form an algebraic system. Use the nonlinear Gauss–Seidel–SOR method to approximate its solution. In order to compute $r_i^{-1}(\overline{w})$, use Newton's method.

8

APPLICATIONS TO NONLINEAR PARTIAL DIFFERENTIAL EQUATIONS

This chapter contains a number of applications of the techniques discussed in Chapter 7 to nonlinear partial differential equations. In Section 8.1 we compare linear and nonlinear aspects of the finite element method. Nonlinear heat-transfer problems are discussed in Section 8.2. Burger's equation, a nonlinear time-dependent problem, is studied in Section 8.3. In Sections 8.4 and 8.5 incompressible viscous fluid flow (Navier–Stokes equations) is studied. The Stefan problem, heat transfer with a phase change, is considered in Section 8.6.

8.1 COMPARISON OF LINEAR AND NONLINEAR FEM PROBLEMS

There are several differences that arise when nonlinear partial differential equations are studied. In the preceding chapters we have discussed two of these. First, one expects to obtain nonlinear algebraic problems. Second, some of the integrals may be more difficult to evaluate. In order to illustrate this, let us suppose the right-hand side of an equation depends on the unknown. Let u be the unknown and $f(u)$ the right-hand side. Then we must compute

$$\int_e f(u^e) N_i^e, \tag{8.1.1}$$

where $u^e = \Sigma_{j=1}^2 u_j^e N_j^e$, N_j^e are the linear shape functions on an interval. One can proceed in a number of ways to compute or approximate (8.1.1).

(i) Direct substitution. Suppose $f(u) = 1 + u^2$. Then (8.1.1) becomes

$$\begin{aligned}\int_e f(u_1^e N_1^e + u_2 N_2^e) N_i^e &= \int_e \left(1 + (u_1^e N_1^e + u_2^e N_2^e)^2\right) N_i^e \\ &= \int_e \left(N_i^e + (u_1^e)^2 N_1^e N_1^e N_i^e + u_1^e u_2^e 2 N_1^e N_2^e N_i^e \right. \\ &\quad \left. + (u_2^e)^2 N_2^e N_2^e N_i^e\right). \end{aligned} \tag{8.1.2}$$

Now use the integral formula

$$\int_e (N_1^e)^m (N_2^e)^n = \frac{m!n!}{(m+n+1)!} h \tag{8.1.3}$$

to evaluate (8.1.2).

(ii) Linear interpolation of $f(u)$. Replace $f(u)$ by its linear interpolation $f_{\text{I}}(u)$ given by (8.1.4)

$$f_{\text{I}}(u) \equiv \sum_{j=1}^{2} f(u_j^e) N_j^e \quad \text{on } e. \tag{8.1.4}$$

In this case (8.1.1) is approximated by

$$\int_e f(u^e)N_i^e \cong \int_e f_I(u)N_i^e$$

$$= \int_e \sum_{j=i}^{2} f(u_j^e)N_jN_i^e$$

$$= \left(1+(u_1^e)^2\right)\int_e N_1^eN_i^e + \left(1+(u_2^e)^2\right)\int_e N_2^eN_i^e. \tag{8.1.5}$$

The integrals in (8.1.5) may also be evaluated by (8.1.3).

A third important difference between linear and nonlinear problems is that admissible energy integrals may or may not exist. When no admissible energy integral exists, we must use Galerkin's formulation of the finite element method. In order to illustrate these points, consider the following sample problem:

$$-(K(u)u_x)_x = f(x), \tag{8.1.6}$$

$$u(0) = 0 = u(1). \tag{8.1.7}$$

It is tempting to define the energy integral of (8.1.6), (8.1.7) as

$$X(u) \equiv \frac{1}{2}\int_0^1 \left(K(u)u_x^2 - 2uf(x)\right). \tag{8.1.8}$$

By the definition of admissible (see Chapter 1), $F(\lambda) \equiv X(u+\lambda\psi)$ and $F'(0) = 0$ must give the weak form of (8.1.6). However, $F'(0) = 0$ gives (8.1.10). Another way to view this is to replace $K(u)$ by $K_I(u)$ and $f(x)$ by $f_I(x)$. We may compute the element matrices given by

(8.1.8) and then the system matrix. This yields

$$h\left[K'(u_i)\left(\frac{u_{i+1}-u_i}{h}\right)^2\frac{1}{4}+K'(u_i)\left(\frac{u_i-u_{i-1}}{h}\right)^2\frac{1}{4}\right]$$

$$-\left[\frac{K(u_i)+K(u_{i+1})}{2}\frac{u_{i+1}-u_i}{h}-\frac{K(u_i)+K(u_{i-1})}{2}\frac{u_i-u_{i-1}}{h}\right]$$

$$=\frac{h}{6}(f_{i-1}+4f_i+f_{i+1}). \tag{8.1.9}$$

If we divide (8.1.9) by h and let $h \to 0$, we get

$$\tfrac{1}{2}K'(u)u_x-(K(u)u_x)_x=f. \tag{8.1.10}$$

Clearly, (8.1.10) may not agree with the original equation (8.1.6)!

Consequently, in all the examples in this chapter, we use Galerkin's method. In order to illustrate this, we consider the problem (8.1.6), (8.1.7). Recall the algebraic system is obtained in two steps: First, find the weak formulation of the problem. Second, replace the test function ψ by ψ_i and u by $\Sigma_j u_j\psi_j$.

In our example the weak formulation is

$$\int_0^1 K(u)u_x\psi_x=\int_0^1 f\psi \tag{8.1.11}$$

for all $\psi \in PC^1[0,1] \equiv \{\psi \in C[0,1] | \psi' \text{ is piecewise continuous on } [0,1] \text{ and } \psi(0)=0=\psi(1)\}$.

If we make the substitutions

$$K(u) \to \sum_l K(u_l)\psi_l,$$

$$f(x) \to \sum_j f_j\psi_j,$$

$$u \to \sum_j u_j\psi_j,$$

$$\psi \to \psi_i,$$

then (8.1.11) yields a nonlinear algebraic system:

$$\int_0^1 \Big(\sum_l K(u_l)\psi_l\Big)\Big(\sum_j u_j \psi_{jx}\Big)\psi_{ix} = \int_0^1 \Big(\sum_j f_j \psi_j\Big)\psi_i \qquad (8.1.12)$$

or

$$\sum_j \Big(\sum_l K(u_l) \int_0^1 \psi_l \psi_{jx} \psi_{ix}\Big) u_j = \sum_j \Big(\int_0^1 \psi_i \psi_j\Big) f_j. \qquad (8.1.13)$$

The integrals in (8.1.12) or (8.1.13) are easily computed by using the definition of ψ_i:

$$\psi_i \equiv \begin{cases} 1, & x = x_i \\ N_1^{e_i} + N_2^{e_{i-1}}, & x \neq x_i. \end{cases} \qquad (8.1.14)$$

Thus, $\psi_i \neq 0$ only on the elements surrounding the ith node, and consequently, (8.1.12) may be rewritten

$$\int_0^1 (K(u_{i-1})\psi_{i-1} + K(u_i)\psi_i + K(u_{i+1})\psi_{i+1})$$

$$\times (u_{i-1}\psi_{i-1,x} + u_i \psi_{ix} + u_{i+1}\psi_{i+1,x})\psi_{i,x}$$

$$= \int_0^1 (f_{i-1}\psi_{i-1} + f_i \psi_i + f_{i+1}\psi_{i+1})\psi_i. \qquad (8.1.15)$$

The integrals in (8.1.15) are easily computed. For example,

$$\int_0^1 \psi_{i+1}\psi_{ix}\psi_{ix} = \int_{e_i} N_2^{e_i} N_{1x}^{e_i} N_{ix}^{e_i}$$

$$= \int_{e_i} N_2^{e_i} \frac{1}{h}\frac{1}{h} = \frac{1}{h^2}\frac{h}{2} = \frac{1}{2h},$$

$$\int_0^1 \psi_{i+1}\psi_{i-1,x}\psi_{ix} = \int_{e_{i-1}} 0 \cdot N_{1x}^{e_{i-1}} N_{2x}^{e_{i-1}} + \int_{e_i} N_2^{e_i} \cdot 0 \cdot N_{1x}^{e_i} = 0.$$

Eventually, (8.1.15) gives

$$-\left[\frac{K(u_{i+1})+K(u_i)}{2}\frac{u_{i+1}-u_i}{h}-\frac{K(u_i)+K(u_{i-1})}{2}\frac{u_i-u_{i-1}}{h}\right]$$

$$=\frac{h}{6}(f_{i-1}+4f_i+f_{i+1}). \tag{8.1.16}$$

With the exception of the right-hand side of (8.1.16), this agrees the finite difference method applied to (8.1.6) and (8.1.7).

8.2 APPLICATION TO NONLINEAR HEAT CONDUCTION

In this section we consider the heat-conduction problem in which the thermal properties ρ, c, and K may depend on space and temperature. Also, we shall assume that there is no phase change, that is, the material is in exactly one of the liquid, solid, or gas states. Heat transfer by convection will be ignored.

When a material has homogeneous thermal properties and its temperature changes over a small range, then the ρ, c, and K can be assumed to be constants. This gives the classical heat equation for a thin rod:

$$\rho c u_t-(Ku_x)_x=f. \tag{8.2.1}$$

If ρ and c depend on u, then the rate of change of the heat energy, given in (8.2.1) by $\rho c u_t=(\rho c u)_t$, becomes more complicated.

The total heat energy of a unit volume is given by $\rho c u$ when ρ and c are constant. If we define the *enthalpy* to be the amount of energy needed to heat a unit volume from absolute zero to a temperature of u, then we may make the following approximation:

$$\text{enthalpy}\simeq\sum_{l=1}^{N}\rho(u_l)c(u_l)\,\Delta u, \tag{8.2.2}$$

where $u_l=l\,\Delta u$ and $(u-0)/N=\Delta u$. This is an approximation because on any small interval (u_l,u_{l+1}), ρ and c are approximated by constants $\rho(u_l)$ and $c(u_l)$, respectively. As there is no phase change, $\rho(u)$ and $c(u)$ are continuous. Thus, as $\Delta u \to 0$, that is, $N \to \infty$, the approximation (8.2.2) converges to an integral. These observations and the definition of the Riemann integral prompt the following quantitative definition of enthalpy.

Definition. Let $\rho(x, u)$ and $c(x, u)$ be a given density and specific heat of a material. We assume they depend on position and temperature, and they are piecewise continuous in (x, u). If there is no change of phase, then we define

$$\begin{aligned} E(x,t) &= \textit{enthalpy} \\ &= \lim_{N\to\infty} \sum_{l=1}^{N} \rho(x,u_l)c(x,u_l)\,\Delta u \\ &= \int_0^{u(x,t)} \rho(x,\bar{u})c(x,\bar{u})\,d\bar{u} = H(x,u(x,t)). \quad (8.2.3) \end{aligned}$$

Example. This example includes phase change. Let the subscripts s and l refer to solid and liquid states. Assume the thermal properties are constants in the two phases. Then the enthalpy $H(u)$ is given by Figure 8.2.1. Note, if $u < u_f$, then $H(u)_t = \rho_s c_s u_t$, or if $u > u_f$, then $H(u)_t = \rho_l c_l u_t$.

The definition of enthalpy enables us to define the enthalpy formulation of the heat equation.

Definition. Let $H(x, u(x, t))$ be the enthalpy of a given material. The *enthalpy formulation of the heat equation* is

$$H(x,u(x,t))_t - \nabla \cdot K(x,u(x,t))\nabla u\,(x,t) = f(x,t,u(x,t)). \quad (8.2.4)$$

We shall examine (8.2.4) for three cases: (1) ρ, c, and K depend only on temperature, and no phase change. (2) ρ, c, and K may depend on space and temperature, and no phase change. (3) In Section 8.6 we consider the phase-change case.

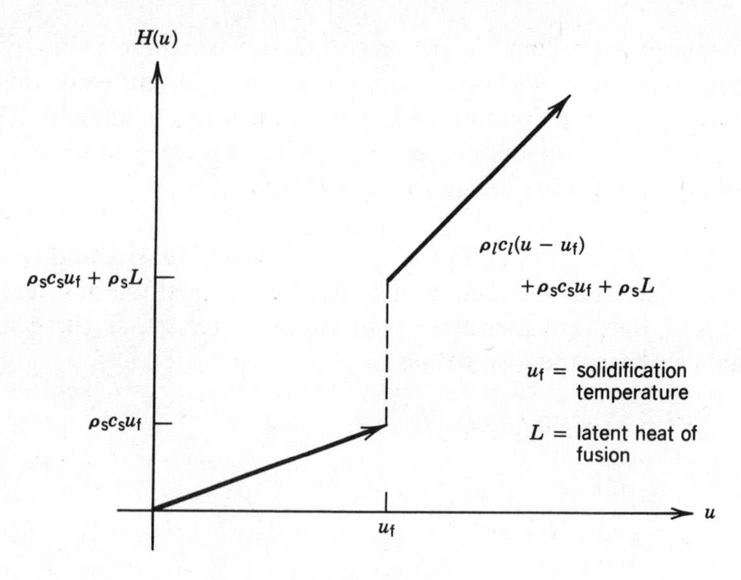

Figure 8.2.1

Problem One. Let $\rho = \rho(u)$, $c = c(u)$, and $K = K(u)$.

$$H(u)_t - \nabla \cdot K(u)\nabla u = f(x,t,u) \qquad \text{on } \Omega \times (0,T), \tag{8.2.5}$$

$$u(x,0) = u_0(x) \qquad \text{on } \Omega, \tag{8.2.6}$$

$$u(x,t) = g_1(x,t) \qquad \text{on } \partial\Omega_1 \times (0,T), \tag{8.2.7}$$

$$\frac{du}{d\nu} + su = g_2(x,t) \qquad \text{on } \partial\Omega_2 \times (0,T), \tag{8.2.8}$$

$$\frac{du}{d\nu} = 0 \qquad \text{on } \partial\Omega_3 \times (0,T), \tag{8.2.9}$$

$$\frac{du}{d\nu} \equiv K(u)\nabla u \cdot \mathbf{n}$$

where $\mathbf{n}$ is the unit outward normal to $\partial\Omega$.

As $K = K(u)$ depends only on the temperature, we may use the Kirchhoff change of dependent variable to simplify (8.2.5).

Kirchhoff Transformation

$$v = F(u) \equiv \int_0^u K(\bar{u})\,d\bar{u}. \tag{8.2.10}$$

Example. Physical properties of H_2O in a solid phase. These values are better estimates than constants:

$$\rho(u) = 920, \qquad u < u_f = 273,$$

$$c(u) = 0.00716u + 0.138,$$

$$K(u) = 0.00224 + 0.00000593(273 - u)^{1.156},$$

$$H(u) = \int_0^u \rho(\bar{u})c(\bar{u})\,d\bar{u}$$

$$= 3.2936u^2 + 126.96u,$$

$$v = F(u) = \int_0^u K(\bar{u})\,d\bar{u}$$

$$= 0.00224u + 0.00000595\left(273^{2.156} - (273 - u)^{2.156}\right)/2.156.$$

This change of variables has the pleasant property that

$$v_{x_i} = \frac{\partial}{\partial x_i} v$$

$$= \frac{\partial}{\partial x_i} F(u)$$

$$= F'(u) \cdot \frac{\partial u}{\partial x_i}$$

$$= K(u) \cdot u_{x_i}. \tag{8.2.11}$$

Line (8.2.11) simply means

$$\Delta v = \nabla \cdot \nabla v = \nabla \cdot K(u) \nabla u. \tag{8.2.12}$$

Thus, we may reformulate problem one in terms of v.

Problem One—v Version

$$H(F^{-1}(v))_t - \Delta v = f(x, t, F^{-1}(v)) \qquad \text{on } \Omega \times (0, T), \tag{8.2.13}$$

$$v(x, 0) = F(u_0) \qquad \text{on } \Omega, \tag{8.2.14}$$

$$v(x, t) = F(g_1) \qquad \text{on } \partial\Omega_1 \times (0, T), \tag{8.2.15}$$

$$\frac{dv}{dn} + sF^{-1}(v) = g_2 \qquad \text{on } \partial\Omega_2 \times (0, T), \tag{8.2.16}$$

$$\frac{dv}{dn} = 0 \qquad \text{on } \partial\Omega_3 \times (0, T), \tag{8.2.17}$$

$$\frac{dv}{dn} \equiv \nabla v \cdot \mathbf{n}.$$

Implicit Method for (8.2.13). Let $v^{k+1} = v(x, (k+1)\Delta t)$ and $v^k = v(x, k\Delta t)$.

$$\frac{H(F^{-1}(v^{k+1})) - H(F^{-1}(v^k))}{\Delta t} - \Delta v^{k+1} = f^{k+1}(x, F^{-1}(v^{k+1})). \tag{8.2.18}$$

Alternatively, when

$$\bar{\phi}^{k+1}(x, v^{k+1}) \equiv \frac{1}{\Delta t} H(F^{-1}(v^{k+1})) - f^{k+1}(x, F^{-1}(v^{k+1})),$$

$$\bar{\Phi}^k(x, v^k) \equiv \frac{1}{\Delta t} H(F^{-1}(v^k)),$$

then (8.2.18) is

$$\bar{\phi}^{k+1}(x, v^{k+1}) - \Delta v^{k+1} = \bar{\Phi}^k(x, v^k). \tag{8.2.19}$$

If we wish to use finite differences on (8.2.19) coupled with conditions (8.2.14)–(8.2.17), we obtain a nonlinear algebraic system, (8.2.20), which is of the form discussed in Section 7.7.

FDM Equation for Problem One—*v* Version. Let $\phi_i^{k+1}(v_i^{k+1}) = \bar{\phi}^{k+1}(x_i, v_i^{k+1})$ plus terms from boundary condition (8.2.16), and $\Phi_i^k(v_i^k) = \bar{\Phi}^k(x_i, v_i^k)$ plus terms from boundary condition (8.2.16). $\phi^{k+1}(v^{k+1}) = (\phi_i^{k+1}(v_i^{k+1})) \in \mathbb{R}^N$ and $\Phi^k(v^k) = (\Phi_i^k(v_i^k)) \in \mathbb{R}^N$, where N is the number of nodes for which v_i^{k+1} are not given:

$$\phi^{k+1}(v^{k+1}) + Av^{k+1} = \Phi^k(v^k) \tag{8.2.20}$$

A is the usual square matrix associated with elliptic boundary-value problems.

If there is no phase change, then H is continuous and hence ϕ^{k+1} is continuous. Also, as $\rho, c, K > 0$, ϕ^{k+1} will be nondecreasing for $f = f(x, t)$. Thus, the nonlinear Gauss–Seidel method of Chapter 7 can be used to approximate the solution of (8.2.20). Equation (8.2.20) has the form whose solution may be approximated by the algorithm (7.7.8) and (7.7.9).

This method requires at each node the solution of an equation $r(z) = w$, where w is given and $r(z) \equiv \phi_i(z) + a_{ii}z$. This solution may be approximated by the one-variable Newton method. In particular, $\phi_i(z)$ has the form $G_i(F^{-1}(z))$, where for interior nodes $G_i(u) = (1/\Delta t)H(u) - f_i(u)$. Therefore, by the inverse function theorem, Newton's method is

$$z^{n+1} = z^n - \frac{r(z^n) - w}{r'(z^n)}, \qquad z^n = F(u^n),$$

$$z^{n+1} = F(u^n) - \frac{(1/\Delta t)H(u^n) - f_i(u^n) + a_{ii}F(u^n) - w}{((1/\Delta t)H'(u^n) - f_i'(u^n))/F'(u^n) + a_{ii}}.$$

Note in this iteration only the given functions, their derivatives, and $u^{n+1} = F^{-1}(z^{n+1})$ are used.

If the finite element method is used, we shall generate an analog of (8.2.20) that is slightly more complicated. As we shall see, the second example of the nonlinear Gauss–Seidel algorithm (7.7.19) and

(7.7.20), can be used to approximate the solution of this nonlinear algebraic system.

In order to formulate the finite element method for problem one —v version, we shall use Galerkin's formulation. Therefore, we must determine the definition of a weak solution of the implicit equation, (8.2.19). This is easily done by multiplying (8.2.19) by a suitable test function and then using Green's theorem.

Weak Formulation of Equation (8.2.19) with (8.2.14)–(8.2.17). Let $\psi \equiv 0$ on $\partial\Omega_1$ and $\psi \in PC^1(\Omega) \equiv \{\psi \in C(\bar{\Omega}) | \psi_x, \psi_y$ are piecewise continuous on $\bar{\Omega}\}$. Then

$$\int_\Omega \bar{\phi}^{k+1}(x, v^{k+1})\psi + \int_\Omega \nabla v^{k+1} \cdot \nabla\psi + \int_{\partial\Omega_2} sF^{-1}(v^{k+1})\psi$$

$$= \int_\Omega \bar{\phi}^k(x, v^k)\psi + \int_{\partial\Omega_2} g_2^{k+1}\psi \tag{8.2.21}$$

In order to obtain the nonlinear algebraic system from (8.2.21), we make the following substitutions:

$$\bar{\phi}^{k+1}(x, v) \to \sum_j \bar{\phi}_j^{k+1}(v_j)\psi_j,$$

$$F^{-1}(v) \to \sum_j F^{-1}(v_j)\psi_j,$$

$$\bar{\Phi}^k(x, v) \to \sum_j \bar{\Phi}_j^k(v_j)\psi_j,$$

$$g_2^k(x) \to \sum_j g_{2j}^k\psi_j,$$

$$v \to \sum_j v_j\psi_j,$$

$$\psi \to \psi_i,$$

where $\psi, \psi_i, \psi_j \in PC^1(\Omega)$. Equation (8.2.21) gives

$$\sum_j \left(\int_\Omega \psi_i \psi_j \right) \bar{\phi}_j^{k+1}\left(v_j^{k+1}\right) + \sum_j \left(\int_\Omega \nabla\psi_i \cdot \nabla\psi_j \right) v_j^{k+1}$$

$$+ \sum_j s \left(\int_{\partial\Omega_2} \psi_i \psi_j \right) F^{-1}\left(v_j^{k+1}\right)$$

$$= \sum_j \left(\int_\Omega \psi_i \psi_j \right) \bar{\Phi}_j^k\left(v_j^k\right) + \sum_j \left(\int_{\partial\Omega_2} \psi_i \psi_j \right) g_{2j}^{k+1}. \quad (8.2.22)$$

We reformulate (8.2.22) in matrix notation.

FEM for Problem One—*v* Version. Let

$$\bar{B} = \left(\int_\Omega \psi_i \psi_j \right), \quad \bar{B}_2 = \left(\int_{\partial\Omega_2} \psi_i \psi_j \right), \quad \bar{A} = \left(\int_\Omega \nabla\psi_i \cdot \nabla\psi_j \right)$$

be $N \times N$ matrices, where N is the total number of nodes. Let B, B_2, A be the matrices associated with $\bar{B}$, $\bar{B}_2$, $\bar{A}$ with the prescribed boundary conditions inserted. Then

$$B\phi^{k+1}\left(v^{k+1}\right) + Av^{k+1} + sB_2F^{-1}\left(v^{k+1}\right) = B\Phi^k\left(v^k\right) + B_2 g_2^{k+1}. \quad (8.2.23)$$

Example. Let there be one space variable and $\partial\Omega_1 = \partial\Omega$, that is, $\partial\Omega_2$ and $\partial\Omega_3$ are empty. Then B_2 is the zero matrix. If there are only three unknowns, then

$$B = \frac{h}{6}\begin{pmatrix} 4 & 1 & 0 \\ 1 & 4 & 1 \\ 0 & 1 & 4 \end{pmatrix} \quad \text{and} \quad A = \frac{1}{h}\begin{pmatrix} 2 & -1 & 0 \\ -1 & 2 & -1 \\ 0 & -1 & 2 \end{pmatrix}.$$

Equation (8.2.23) may be rewritten in component form

$$\frac{h}{6}\left[\phi_{i-1}^{k+1}\left(v_{i-1}^{k+1}\right) + 4\phi_i^{k+1}\left(v_i^{k+1}\right) + \phi_{i+1}^{k+1}\left(v_{i+1}^{k+1}\right)\right]$$

$$+\frac{1}{h}\left[-v_{i-1}^{k+1} + 2v_i^{k+1} - v_{i+1}^{k+1}\right]$$

$$= \frac{h}{6}\left[\Phi_{i-1}^{k}\left(v_{i-1}^{k}\right) + 4\Phi_i^{k}\left(v_i^{k}\right) + \Phi_{i+1}^{k}\left(v_{i+1}^{k}\right)\right]. \quad (8.2.24)$$

The solution of (8.2.24) can be approximated by using the second version, (7.7.19) and (7.7.20), of the nonlinear Gauss–Seidel method.

Problem Two. Let $\rho = \rho(x, u)$, $c = c(x, u)$, and $K = K(x, u)$.

Consider equations (8.2.5)–(8.2.9) with more general thermal properties in $H(x, u) \equiv \int_0^u \rho(x, \bar{u}) c(x, \bar{u})\, d\bar{u}$ and $K = K(x, u)$.

The Kirchhoff change of dependent variable does not simplify (8.2.5), because for $v = F(x, u) = \int_0^u K(x, \bar{u})\, d\bar{u}$,

$$v_{x_i} = F_{x_i} \cdot 1 + F_u \frac{\partial u}{dx_i} = \int_0^u K_{x_i}(x, \bar{u})\, d\bar{u} + K(x, u) u_{x_i}. \quad (8.2.25)$$

The resulting algebraic system will be more complicated than either (8.2.20) or (8.2.23). Consequently, we shall use Newton's method for several variables to approximate the solution to problem two.

In order to find the nonlinear algebraic system for Galerkin's formulation of the finite element method, we need the implicit equation and its weak formulation for problem two.

Implicit Equation for (8.2.5) With $H(x, u)$ and $K(x, u)$

$$\frac{H(x, u^{k+1}) - H(x, u^k)}{\Delta t} - \nabla \cdot K(x, u^{k+1}) \nabla u^{k+1} = f^{k+1}(x, u^{k+1}). \quad (8.2.26)$$

Weak Formulation of Implicit Equation (8.2.26) With (8.2.6)–(8.2.9). Let $\psi \in PC^1(\Omega)$ and $\psi \equiv 0$ on $\partial\Omega_1$. Let

$$\phi^{k+1}(x, u^{k+1}) \equiv \frac{H(x, u^{k+1})}{\Delta t} - f^{k+1}(x, u^{k+1}),$$

$$\Phi^k(x, u^k) \equiv \frac{H(x, u^k)}{\Delta t}.$$

Then

$$\int_\Omega \phi^{k+1}(x, u^{k+1})\psi + \int_\Omega K(x, u^{k+1})\nabla u^{k+1}\nabla\psi + \int_{\partial\Omega_2} su^{k+1}\psi$$
$$= \int_\Omega \Phi^k(x, u^k)\psi + \int_{\partial\Omega_2} g_2^{k+1}\psi. \tag{8.2.27}$$

In order to obtain an algebraic system from (8.2.27), we make the following substitutions:

$$\phi^{k+1}(x, u) \to \sum_j \phi_j^{k+1}(u_j)\psi_j,$$

$$K(x, u) \to \sum_l K_l(u_l)\psi_l,$$

$$\Phi^k(x, u) \to \sum_j \Phi_j^k(u_j)\psi_j,$$

$$g_2(x) \to \sum_j g_{2j}\psi_j,$$

$$u \to \sum_j u_j\psi_j,$$

$$\psi \to \psi_i,$$

where $\psi_i, \psi_j, \psi_l \in PC^1(\Omega)$. The weak formulation (8.2.27) gives

$$\sum_j \left(\int_\Omega \psi_i \psi_j \right) \phi_j^{k+1}(u_j^{k+1}) + \sum_j \sum_l K_l(u_l^{k+1}) \left(\int_\Omega \psi_l \nabla \psi_i \cdot \nabla \Psi_j \right) u_j^{k+1}$$

$$+ \sum_j s \left(\int_{\partial\Omega_2} \psi_j \psi_i \right) u_j^{k+1}$$

$$= \sum_j \left(\int_\Omega \psi_i \psi_j \right) \Phi_j^k(u_j^k) + \sum_j \left(\int_{\partial\Omega_2} \psi_j \psi_i \right) g_{2j}^{k+1}. \tag{8.2.28}$$

In terms of matrix notation we write (8.2.28) as follows.

FEM for Problem Two. Let

$$\bar{B} = \left(\int_\Omega \psi_i \psi_j \right), \quad \bar{B}_2 = \left(\int_{\partial\Omega_2} \psi_i \psi_j \right), \quad \text{and}$$

$$\bar{A}(u^{k+1}) = \left(\sum_l K_l(u_l^{k+1}) \left(\int_\Omega \psi_l \nabla \psi_i \cdot \nabla \psi_j \right) \right)$$

be $N \times N$ matrices. Let B, B_2, A be matrices associated with $\bar{B}$, $\bar{B}_2$, $\bar{A}$ with the prescribed boundary conditions inserted. Then

$$B\phi^{k+1}(u^{k+1}) + sB_2 u^{k+1} + A(u^{k+1})u^{k+1} = B\Phi^k(u^k) + B_2 g_2^{k+1}. \tag{8.2.29}$$

Example. Let there be one space variable and $\partial\Omega = \partial\Omega_1$. Then B_2 is the zero matrix. Suppose there are only three unknowns. The integrals in (8.2.29) may be computed by the usual formulas:

$$B = \frac{h}{6} \begin{pmatrix} 4 & 1 & 0 \\ 1 & 4 & 1 \\ 0 & 1 & 4 \end{pmatrix} \quad \text{and} \quad A(u^{k+1}) = \frac{1}{h} \begin{pmatrix} \alpha + \beta & -\beta & 0 \\ -\alpha & \alpha + \beta & -\beta \\ 0 & -\alpha & \alpha + \beta \end{pmatrix},$$

where $\alpha = (K_{i-1}(u_{i-1}^{k+1}) + K_i(u_i^{k+1}))/2$ and $\beta = (K_{i+1}(u_{i+1}^{k+1}) +$

$K_i(u_i^{k+1}))/2$. Thus, equation (8.2.29) has the form

$$\frac{h}{6}\left[\phi_{i-1}^{k+1}\left(u_{i-1}^{k+1}\right) + 4\phi_i^{k+1}\left(u_i^{k+1}\right) + \phi_{i+1}^{k+1}\left(u_{i+1}^{k+1}\right)\right]$$

$$+\frac{1}{h}\left[\alpha\left(u_i^{k+1} - u_{i-1}^{k+1}\right) - \beta\left(u_{i+1}^{k+1} - u_i^{k+1}\right)\right]$$

$$= \frac{h}{6}\left[\Phi_{i-1}^{k}\left(u_{i-1}^{k}\right) + 4\Phi_i^{k}\left(u_i^{k}\right) + \Phi_{i+1}^{k}\left(u_{i+1}\right)\right]. \quad (8.2.30)$$

As previously mentioned, equation (8.2.29) has nonlinearities in the $\phi(u)$ and $A(u)$ terms, which evolved from $K(x,u)$ and $f(x,u)$. In Section 8.3 we introduce another form of nonlinearity that has the form uu_x.

8.3 BURGER'S EQUATION

This equation evolves from studies of turbulence in fluid flow and of shocks in fluid flow. The techniques of this section will also be used in Sections 8.4 and 8.5, on the Navier–Stokes equations.

Burger's Equation

$$u_t + uu_x - \nu u_{xx} = 0, \quad (8.3.1)$$

$$u(x,0) = f(x), \quad (8.3.2)$$

$$u(0,t) = 0 = u(1,t). \quad (8.3.3)$$

We shall examine two methods for approximating the solution of (8.3.1)–(8.3.3): the explicit finite difference method and the implicit finite element method.

FDM—Explicit—Upwind Difference for uu_x. The upwind difference on uu_x is given by

$$\delta u_i \equiv \begin{cases} \dfrac{u_i - u_{i-1}}{\Delta x}, & u_i \geq 0 \\ \dfrac{u_{i+1} - u_i}{\Delta x}, & u_i < 0. \end{cases} \quad (8.3.4)$$

The explicit finite difference method is given by

$$\frac{u_i^{k+1} - u_i^k}{\Delta t} + u_i^k \,\delta u_i^k - \nu \frac{u_{i+1}^k - 2u_i^k + u_{i-1}^k}{\Delta x^2} = 0. \quad (8.3.5)$$

As the scheme (8.3.4), (8.3.5) is similar to the explicit method for the heat equation, one expects some stability condition on Δt, ν, and Δx. In order to examine the stability for (8.3.4), (8.3.5), we consider a similar equation where uu_x is replaced αu_x, $\alpha \geq 0$. The von Neumann stability analysis is used.

Proposition 8.3.1. Consider $u_t + \alpha u_x - \nu u_{xx} = 0$ with $u(x,0)$, $u(0,t)$, and $u(1,t)$ given. Let the explicit-upwind finite differences be used. If

$$\frac{\Delta t}{\Delta x}|\alpha| + 2\,\Delta t \frac{\nu}{\Delta x^2} < 1,$$

then the algorithm (8.3.5) is stable.

Proof. Let $u_i^{k+1} = A_{k+1}\exp\sqrt{-1}\,Ci\,\Delta x$, $u_i^k = A_k\exp\sqrt{-1}\,Ci\,\Delta x$, and $\theta = C\Delta x$. Substitute these into (8.3.5), where $u_i^k\,\delta u_i^k$ is replaced by $\alpha\,\delta u_i^k$. One easily computes $u_i^{k+1} = Gu_i^k$, where

$$G \equiv 1 - (\bar{d} + 2d)(1 - \cos\theta) + \sqrt{-1}\,\bar{d}\sin\theta,$$

$$\bar{d} = \frac{\Delta t}{\Delta x}|\alpha|, \quad d = \nu\frac{\Delta t}{\Delta x^2}. \quad (8.3.6)$$

Let $x = \cos\theta$ and $1 - x^2 = \sin^2\theta$ with $-1 \leq x \leq 1$. Then (8.3.6) gives

$$\begin{aligned} f(x) &\equiv |G|^2 \\ &= 1 - (\bar{d} + 2d(1-x))^2 + \bar{d}(1 - x^2) \\ &= x^2(4d\bar{d} + 4d^2) + x\big(2(\bar{d} + 2d) - 2(\bar{d} + 2d)^2\big) \\ &\quad + 1 - 2(\bar{d} + 2d) + (\bar{d} + 2d)^2 + \bar{d}^2. \end{aligned} \quad (8.3.7)$$

As $\alpha \geq 0$, (8.3.7) gives $f''(x) = 2(4d\bar{d} + 4d^2) > 0$. Thus, on $[-1, 1]$, $f(x)$ is concave up, and the absolute maximum of $f(x)$ with respect to $[-1, 1]$ must be at the end points. Now, $f(1) = 1$ and $f(-1) = (1 - 2(\bar{d} + 2d))^2$. Since $\bar{d} + 2d < 1$, $1 > 1 - 2(\bar{d} + 2d) > -1$. Thus, $f(-1) \leq 1$ when $\bar{d} + 2d < 1$.

Remark. If $\bar{d} = 0$, that is, $\alpha \equiv 0$, then $2d < 1$ is the usual stability condition on the explicit method for the heat equation. The condition $\bar{d} + 2d < 1$ implies $(\Delta t/\Delta x)|\alpha| < 1$. If $\alpha = \max|u|$, then $(\Delta t/\Delta x)|u| < 1$ and is called the *Courant number* condition. It means that a free particle cannot move more than one grid length in a given time increment Δt.

In order to describe Galerkin's formulation of the finite element method, note that for $\psi \in PC^1(0,1)$ with $\psi(0) = 0 = \psi(1)$

$$uu_x = \frac{1}{2}(u^2)_x,$$

$$\begin{aligned} \int_0^1 uu_x\psi &= \frac{1}{2}\int_0^1 (u^2)_x\psi \\ &= \frac{1}{2}\left[u^2\psi|_0^1 - \int_0^1 u^2\psi_x\right] \\ &= \frac{-1}{2}\int_0^1 u^2\psi_x. \end{aligned} \tag{8.3.8}$$

By forming the implicit equation from (8.3.1) in the usual way, using (8.3.8), and then determining the weak formulation of the implicit equation, we obtain line (8.3.9).

Weak Formulation of Implicit Equation for (8.3.1)–(8.3.3). Let $\psi \in PC^1(0,1)$ and $\psi(0) = 0 = \psi(1)$.

$$\frac{1}{\Delta t}\int_0^1 u^{k+1}\psi - \frac{1}{2}\int_0^1 (u^{k+1})^2\psi_x + \nu\int_0^1 u_x^{k+1}\psi_x = \frac{1}{\Delta t}\int_0^1 u^k\psi. \tag{8.3.9}$$

In order to form the nonlinear algebraic problem, replace u^k by $\Sigma_j u_j^k \psi_j$ and ψ by ψ_i. The only new term in (8.3.9) will have the form

$$\int_0^1 \Big(\sum_j u_j \psi_j\Big)^2 \psi_{ix} = \sum_j \Big[\sum_l u_l \Big(\int_0^1 \psi_l \psi_j \psi_{ix}\Big)\Big] u_j. \tag{8.3.10}$$

The only nonzero terms in (8.3.10) must come from the terms where $l = i,\ i \pm 1$ and $j = i, i \pm 1$.

More precisely, let $C(u) \equiv (\Sigma_l c_{ijl} u_l)$, where $c_{ijl} \equiv -\frac{1}{2}\int \psi_l \psi_j \psi_{ix}$, i is fixed and $c_{ijl} = 0$ if $|j - i| \geq 2$ or $|l - i| \geq 2$ or $|j - l| \geq 2$.

$j = i - 1$:

$$c_{i,i-1,i-1} = -\frac{1}{2}\int \psi_{i-1}\psi_{i-1}\psi_{ix} = -\frac{1}{2}\int_{e_{i-1}} N_1 N_1 N_{2x} = -\frac{1}{6},$$

$$c_{i,i-1,i} = -\frac{1}{2}\int \psi_i \psi_{i-1}\psi_{ix} = -\frac{1}{2}\int_{e_{i-1}} N_2 N_1 N_{2x} = -\frac{1}{12},$$

$$c_{i,i-1,i+1} = 0.$$

$j = i$:

$$c_{i,i,i-1} = -\frac{1}{2}\int \psi_{i-1}\psi_i \psi_{ix} = -\frac{1}{2}\int_{e_{i-1}} N_1 N_2 N_{2x} = -\frac{1}{12},$$

$$c_{i,i,i} = -\frac{1}{2}\int \psi_i \psi_i \psi_{ix} = -\frac{1}{2}\int_{e_{i-1}} N_2 N_2 N_{2x} - \frac{1}{2}\int_{e_i} N_1 N_1 N_{1x}$$

$$= -\frac{1}{6} + \frac{1}{6} = 0,$$

$$c_{i,i,i+1} = -\frac{1}{2}\int \psi_{i+1}\psi_i \psi_{ix} = -\frac{1}{2}\int_{e_i} N_2 N_1 N_{1x} = \frac{1}{12}.$$

$j = i + 1$:

$$c_{i,i+1,i-1} = 0,$$

$$c_{i,i+1,i} = -\frac{1}{2}\int \psi_i \psi_{i+1} \psi_{ix} = -\frac{1}{2}\int_{e_i} N_1 N_2 N_{1x} = \frac{1}{12},$$

$$c_{i,i+1,i+1} = -\frac{1}{2}\int \psi_{i+1} \psi_{i+1} \psi_{ix} = -\frac{1}{2}\int_{e_i} N_2 N_2 N_{1x} = \frac{1}{6}.$$

So, the ith row of $C(u)$ is

$$\langle 0,\ldots,0,\left(-\tfrac{1}{6}u_{i-1} - \tfrac{1}{12}u_i\right),\left(\tfrac{1}{12}u_{i+1} - \tfrac{1}{12}u_{i-1}\right),$$
$$\left(\tfrac{1}{12}u_i + \tfrac{1}{6}u_{i+1}\right),0,\ldots,0\rangle$$

The ith component of $C(u)u$ is

$$\sum_j\left(\sum_l c_{ijl}u_l\right)u_j = \left(-\tfrac{1}{6}u_{i-1} - \tfrac{1}{12}u_i\right)u_{i-1} + \left(\tfrac{1}{12}u_{i+1} - \tfrac{1}{12}u_{i-1}\right)u_i$$
$$+\left(\tfrac{1}{12}u_i + \tfrac{1}{6}u_{i+1}\right)u_{i+1}$$
$$= -\frac{1}{2}\left[\frac{u_{i-1}}{3}u_{i-1} + \frac{1}{3}u_i u_{i-1}\right.$$
$$\left.-\frac{u_i}{3}u_{i+1} - \frac{1}{3}u_{i+1}u_{i+1}\right]$$
$$= -\frac{1}{2}\left(\frac{u_{i-1}}{3}, \frac{u_{i-1} - u_{i+1}}{3}, \frac{-u_{i+1}}{3}\right)\begin{pmatrix} u_{i-1} \\ u_i \\ u_{i+1}\end{pmatrix}.$$

(8.3.11)

Note, $C(u)u = C^*(u)u$, where the ith row of $C^*(u)$ is defined by

$$\langle 0,\ldots,0,\left(-\tfrac{1}{6}u_{i-1}\right),\left(-\tfrac{1}{6}(u_{i-1} - u_{i+1})\right),\left(+\tfrac{1}{6}u_{i+1}\right),0,\ldots,0\rangle.$$

Since the ith row of $C^*(u)$ is less complicated than the ith row of $C(u)$, it is desirable to use $C^*(u)$ in place of $C(u)$.

In matrix notation (8.3.9) gives the following algebraic problem.

FEM for (8.3.1)–(8.3.3). Define the following matrices:

$$B = \frac{1}{\Delta t}\left(\int_0^1 \psi_i \psi_j\right),$$

$$A = \nu\left(\int_0^1 \psi_{ix}\psi_{jx}\right),$$

$$C^*(u) = \text{as given below line 8.3.11},$$

$$Bu^{k+1} + C^*(u^{k+1})u^{k+1} + Au^{k+1} = Bu^k. \qquad (8.3.12)$$

Newton's Method for Approximating the Solution of (8.3.12). Let $F(u^{k+1}) \equiv Bu^{k+1} + C^*(u^{k+1})u^{k+1} + Au^{k+1} - Bu^k$. Let $k \leftrightarrow$ time step, and $m \leftrightarrow$ Newton iteration.

$$u^{k+1,0} \equiv \text{the solution of the linear problem which is formed from (8.3.12) by replacing } C^*(u^{k+1}) \text{ by } C^*(u^k), \qquad (8.3.13)$$

$$u^{k+1,m+1} \equiv u^{k+1,m} - \left(F'(u^{k+1,m})\right)^{-1} F(u^{k+1,m}). \qquad (8.3.14)$$

By Proposition 7.3.1, $F'(u)$ has the special form

$$F'(u) = B + \left(\sum_l \left(c^*_{ilj} + c^*_{ijl}\right)u_l\right) + A$$

where $C^*(u) = (\Sigma_l c^*_{ijl} u_l)$.

8.4 INCOMPRESSIBLE VISCOUS FLUID FLOW—EXPLICIT METHOD

In this section we discuss the finite difference and finite element methods with an explicit time discretization applied to the Navier–Stokes equations. Section 8.5 contains a discussion of the implicit method.

Navier–Stokes Equations. Let (u, v) be normalized velocity, P normalized pressure, and $\mu = 1/\text{Re}$, $\text{Re} = u_\infty L_\infty/\nu$, the Reynolds number with u_∞ the free stream speed, L_∞ the characteristic length, and ν the kinematic viscosity. Let $\partial\Omega_{\text{w}}$ be the wall portion of $\partial\Omega$, and $\mathbf{n}$ be the outward normal to $\partial\Omega_{\text{w}}$.

$$u_t + uu_x + vu_y + P_x - \mu\,\Delta u = f \quad \text{on } \Omega \times (0, T), \tag{8.4.1}$$

$$v_t + uv_x + vv_y + P_y - \mu\,\Delta v = g \quad \text{on } \Omega \times (0, T), \tag{8.4.2}$$

$$u_x + v_y = 0 \quad \text{on } \Omega \times (0, T), \tag{8.4.3}$$

$$u(x, y, 0) = u_0(x, y),$$

$$v(x, y, 0) = v_0(x, y) \quad \text{on } \Omega \tag{8.4.4}$$

$$(u, v) \cdot \mathbf{n} = 0 \qquad \text{on } \partial\Omega_{\text{w}}, \tag{8.4.5}$$

$$(u, v) = (u_1, v_1) \quad \text{on } \partial\Omega_1 = \partial\Omega \setminus \partial\Omega_{\text{w}},\ t > 0. \tag{8.4.6}$$

As in Burger's equation we use upwind finite differences on the uu_x, vu_y, uv_x, and vv_y terms:

$$\delta_x^+ u \equiv \frac{u_{i+1,j} - u_{ij}}{\Delta x}, \tag{8.4.7}$$

$$\delta_x^- u \equiv \frac{u_{ij} - u_{i-1,j}}{\Delta x}, \tag{8.4.8}$$

$$\delta_x u \equiv \begin{cases} \delta_x^- u, & u_{ij} \geq 0 \\ \delta_x^+ u, & u_{ij} < 0. \end{cases} \tag{8.4.9}$$

The finite differences with respect to y are defined as in (8.4.7)–(8.4.9). If it were not for P_x and P_y in the momentum equations, (8.4.1) and

(8.4.2), the explicit method could be used just as in Burger's equation. In the method that we describe, we first assume $P \equiv 0$ and solve (8.4.1) and (8.4.2) for what we call the *auxiliary velocity*, $(u^{\text{aux}}, v^{\text{aux}})$. The auxiliary velocity will not, in general, satisfy the conservation-of-mass equation, (8.4.3). Therefore, in the second step we find a suitable pressure term so that the adjusted velocity will satisfy all equations (8.4.1)–(8.4.6). The third step makes this adjustment and proceeds to the next time step. The nature of the adjusting pressure term needs careful study. However, we now just state the method.

FDM—Explicit—Upwind Differences for (8.4.1)–(8.4.6)

Step I: In (8.4.1) and (8.4.2) set $P \equiv 0$, use upwind finite differences, and solve for $(u^{\text{aux}}, v^{\text{aux}})$, where on $\Omega \cup \partial\Omega_{\text{w}}$

$$\frac{u^{\text{aux}} - u^k}{\Delta t} = F^k, \tag{8.4.10}$$

with

$$F^k \equiv f_1^k - \left[u^k\delta_x u^k + v^k\delta_y u^k - \mu\left(\delta_x^+\delta_x^- + \delta_y^+\delta_y^-\right)u^k\right],$$

and

$$\frac{v^{\text{aux}} - v^k}{\Delta t} = G^k, \tag{8.4.11}$$

with

$$G^k \equiv f_2^k - \left[u^k\delta_x v^k + v^k\delta_y v^k - \mu\left(\delta_x^+\delta_x^- + \delta_y^+\delta_y^-\right)v^k\right].$$

On $\partial\Omega_1$ let $(u^{\text{aux}}, v^{\text{aux}}) = (u_1, v_1)$.

Step II: Solve for the pressure P, where

$$-\Delta t\left(\delta_x^+\delta_x^- + \delta_y^+\delta_y^-\right)P = -\delta_x^+ u^{\text{aux}} - \delta_y^+ v^{\text{aux}} \quad \text{on } \Omega, \tag{8.4.12}$$

$$\Delta t\left(\delta_x^- P, \delta_y^- P\right)\cdot\mathbf{n} = \left(u^{\text{aux}}, v^{\text{aux}}\right)\cdot\mathbf{n} \quad \text{on } \partial\Omega. \tag{8.4.13}$$

Step III: Adjust the auxiliary velocity by letting on $\Omega \cup \partial\Omega_w$

$$u^{k+1} = u^{\text{aux}} - \Delta t\, \delta_x^- P, \tag{8.4.14}$$

$$v^{k+1} = v^{\text{aux}} - \Delta t\, \delta_y^- P. \tag{8.4.15}$$

Proof of (u^{k+1}, v^{k+1}, P) Approximates Equations (8.4.1)–(8.4.6). Subtract $\delta_x^- P$ and $\delta_y^- P$ from (8.4.10) and (8.4.11), respectively. Then (8.4.10), (8.4.14), and (8.4.11), (8.4.15) imply

$$\frac{u^{k+1} - u^k}{\Delta t} = \frac{u^{\text{aux}} - u^k}{\Delta t} - \delta_x^- P = F^k - \delta_x^- P, \tag{8.4.16}$$

$$\frac{v^{k+1} - v^k}{\Delta t} = \frac{v^{\text{aux}} - v^k}{\Delta t} - \delta_x^- P = G^k - \delta_x^- P. \tag{8.4.17}$$

Equations (8.4.16) and (8.4.17) are approximations of (8.4.1) and (8.4.2). The approximation of the left-hand side of (8.4.3) is

$$\begin{aligned} \delta_x^+ u^{k+1} + \delta_y^+ v^{k+1} &= \delta_x^+ \left(u^{\text{aux}} - \Delta t \delta_x^- P\right) + \delta_y^+ \left(v^{\text{aux}} - \Delta t \delta_y^- P\right) \\ &= \delta_x^+ u^{\text{aux}} + \delta_y^+ v^{\text{aux}} - \Delta t \left(\delta_x^+ \delta_x^- + \delta_y^+ \delta_y^-\right) P. \end{aligned} \tag{8.4.18}$$

By the way P is determined, that is, (8.4.12) must hold, the right-hand side of (8.4.18) is zero. Therefore, the approximation of (8.4.3) holds. In a similar way the condition on P in (8.4.13) implies that the approximation of (8.4.5) holds.

Remarks

1. The auxiliary velocity may be determined by other explicit or implicit methods.
2. The pressure may be computed implicitly along with the velocity. As we shall see, this gives a larger algebraic system to solve.
3. If P satisfies (8.4.12) and (8.4.13), then P + constant will also satisfy these conditions. In fact, the matrix associated with

(8.4.12), (8.4.13) has rank one less than full rank. Consequently, any two solutions of (8.4.12), (8.4.13) will differ by a constant.

The nonuniqueness of a solution to (8.4.12), (8.4.13) causes many numerical difficulties. In order to examine this problem more closely, we shall first study the continuum version of (8.4.12), (8.4.13), and then second propose one method for numerically solving (8.4.12), (8.4.13).

Continuum Version of (8.4.12)–(8.4.13) With $\partial\Omega_1 = \phi$

$$-\Delta t\,\Delta P = -u_x^{\text{aux}} - v_y^{\text{aux}} \quad \text{on } \Omega, \tag{8.4.19}$$

$$\frac{dP}{d\nu} \equiv \Delta t\,(P_x, P_y)\cdot\mathbf{n}$$

$$= (u^{\text{aux}}, v^{\text{aux}})\cdot\mathbf{n} \quad \text{on } \partial\Omega. \tag{8.4.20}$$

An important feature of a problem of the form (8.4.19), (8.4.20) is the *consistency condition for $K\Delta u = f$, $K\,du/dn = g$*. This condition, which is stated in (8.4.21), requires f and g to be related in a specific way. Consequently, one must examine the right-hand sides of (8.4.19) and (8.4.20). Use Green's identity, (2.2.6), with $u = u$ and $\psi \equiv 1$, to obtain

$$\int_\Omega K\Delta u = \int_\Omega f = \int_{\partial\Omega} g\,ds = \int_{\partial\Omega} K\frac{du}{dn}\,d\sigma. \tag{8.4.21}$$

In order to show (8.4.19), (8.4.20) satisfy the consistency condition (8.4.21), use Green's theorem, (2.2.5), with $\overline{Q} = u^{\text{aux}}$ and $\overline{P} = -v^{\text{aux}}$. Then (8.4.19) and (8.4.20) imply

$$\int_\Omega \Delta t\,\Delta P = \int_\Omega \nabla\cdot(u^{\text{aux}}, v^{\text{aux}})$$

$$= \int_{\partial\Omega} -v^{\text{aux}}\,dx + u^{\text{aux}}\,dy$$

$$= \int_{\partial\Omega} (u^{\text{aux}}, v^{\text{aux}})\cdot\mathbf{n}\,d\sigma$$

$$= \int_{\partial\Omega} \Delta t\frac{dP}{dn}\,d\sigma$$

There are two important observations to make from this. First, as in the discrete problem (8.4.19), (8.4.20) does not have a unique solution, and any two solutions will differ by a constant. Second, if the consistency condition did not hold, then there would be no solution. In the discrete case this is also true. In fact, round-off error ensures that the discrete analog of the consistency condition never holds! In order to examine this problem more closely, let us consider the following one-variable example.

One-Variable Example of (8.4.19), (8.4.20) and (8.4.12), (8.4.13)

$$\Delta t\, P''(x) = f(x), \tag{8.4.22}$$

$$-\Delta t\, P'(0) = \overline{P}_0, \tag{8.4.23}$$

$$\Delta t\, P'(1) = \overline{P}_1. \tag{8.4.24}$$

The consistency condition has the form, given by integrating (8.4.22) and then using (8.4.23) and (8.4.24),

$$\begin{aligned} \int_0^1 f(x) &= \int_0^1 \Delta t\, P''(x)\, dx \\ &= \Delta t\, P'(x)\big|_0^1 \\ &= \overline{P}_1 + \overline{P}_0. \end{aligned} \tag{8.4.25}$$

Thus, $f(x)$, $\overline{P}_0$, and $\overline{P}_1$ must be related by (8.4.25).

The discrete version of (8.4.22)–(8.4.24) may be formed by finite differences. When $(0, 1)$ is broken into two intervals $(0, \frac{1}{2})$ and $(\frac{1}{2}, 1)$, there will be three unknowns, P_1, P_2, and P_3. The resulting algebraic system is, for $\Delta t = 1$,

$$\begin{pmatrix} 2 & -2 & 0 \\ -1 & 2 & -1 \\ 0 & -2 & 2 \end{pmatrix} \begin{pmatrix} P_1 \\ P_2 \\ P_3 \end{pmatrix} = \left(\frac{1}{2}\right)^2 \begin{pmatrix} f_1 - 4\overline{P}_0 \\ f_2 \\ f_3 - 4\overline{P}_1 \end{pmatrix}. \tag{8.4.26}$$

Note the following:

(i) The matrix in (8.4.26) has zero determinant.

(ii) The matrix has a 2×2 submatrix, which has a nonzero determinant, that is, the 3×3 matrix has rank 2.

(iii) By fixing any one of the components of $(P_1, P_2, P_3)^T$, we can uniquely determine the others.

The discrete version of the consistency condition is that $\frac{1}{4}(f_1 - 4\bar{P}_0, f_2, f_3 - 4\bar{P}_1)^T$ should be in the range of the 3×3 matrix. Use row operations on (8.4.26) until the bottom row is zero. This gives the *discrete consistency condition*, for this particular problem, as

$$0 = \left(\frac{1}{2}\right)^2 \left(\frac{f_1 - 4\bar{P}_0}{2} + f_2 + \frac{f_3 - 4\bar{P}_1}{2}\right)$$

or

$$0 = \frac{1}{2}\frac{f_1 + f_2}{2} + \frac{1}{2}\frac{f_2 + f_3}{2} - (\bar{P}_1 + \bar{P}_0). \qquad (8.4.27)$$

Note that (8.4.27) is an approximation, when the trapezoid rule is used, of the consistency condition of the continuum problem, (8.4.25).

The important feature of this example is that it suggests the following procedure for approximating the pressure.

A Method for Solving for P in (8.4.12) and (8.4.13)

1. Fix P at exactly one boundary node, (x^0, y^0).
2. Solve the reduced version of system (8.4.12), (8.4.13), where $\partial\tilde{\Omega} = \partial\Omega \setminus \{(x^0, y^0\}$, that is, the new $\tilde{\Omega} \equiv \Omega \setminus \{(x_0, y_0)\}$.

As in Burger's equation there is a stability condition on Δt, Δx, Δy, and μ. The von Neumann stability analysis can be used when $|u| \le \alpha$ and $|v| \le \beta$.

Stability Condition

$$\frac{\Delta t}{\Delta x}|u| + \frac{\Delta t}{\Delta y}|v| + 2\mu\left(\frac{\Delta t}{\Delta x^2} + \frac{\Delta t}{\Delta y^2}\right) < 1. \qquad (8.4.28)$$

If (8.4.28) is violated, then (u^{k+1}, v^{k+1}) may blow up as $k \to \infty$. Even for implicit methods, the condition given by the *Courant number*

$$\frac{\Delta t}{\Delta x}|u| + \frac{\Delta t}{\Delta y}|v| < 1 \qquad (8.4.29)$$

must hold.

In order to discuss the finite element version of the explicit method, we must have a weak formulation of (8.4.1)–(8.4.6). Because of the integration (by Green's theorem), we must require a stronger condition on (u, v) than is given by (8.4.5). Line (8.4.5) states that the velocity normal to the wall must be zero. We shall simply replace (8.4.5) by imposing a condition of zero velocity at the wall:

$$(u, v) = (0, 0) \quad \text{on } \partial\Omega_{\mathrm{w}},\ t > 0 \qquad (8.4.30)$$

This is often called the *no-slip* boundary condition because it implies that both the tangential velocity and normal velocity at $\partial\Omega_{\mathrm{w}}$ must be zero.

Another useful fact that we shall need for the statement of the weak formulation of (8.4.1)–(8.4.4), (8.4.6), (8.4.30) is the conservative form of the *advection* terms $uu_x + vu_y$ and $uv_x + vv_y$. Let $u_x + v_y = 0$ hold (hence, the term *conservative*). Then we easily compute

$$\begin{aligned}
(u^2)_x + (uv)_y &= 2uu_x + u_y v + uv_y \\
&= uu_x + u_y v + uu_x + uv_y \\
&= uu_x + vu_y + u(u_x + v_y) \\
&= uu_x + vu_y. \qquad (8.4.31)
\end{aligned}$$

Also,

$$(uv)_x + (v^2)_y = uv_x + vv_y. \tag{8.4.32}$$

In order to derive the weak formulation of (8.4.1)–(8.4.4), (8.4.6), (8.4.30), we let $\psi \in PC^1(\Omega)$ and, as $(u, v) = (0, 0)$ or $(u, v) = (u_1, v_1)$ on $\partial\Omega$, $\psi \equiv 0$ on $\partial\Omega$. The only new terms will come from the advection terms, where we use (8.4.31) and (8.4.32):

$$\begin{aligned}\int_\Omega (uu_x + vu_y)\psi &= \int_\Omega ((u^2)_x + (uv)_y)\psi \\ &= \int_\Omega (\psi u^2)_x + (\psi uv)_y - \int_\Omega \psi_x u^2 + \psi_y uv\end{aligned} \tag{8.4.33}$$

Now, in (8.4.33) use Green's theorem, (2.2.5), with $\overline{P} = -\psi uv$ and $\overline{Q} = \psi u^2$:

$$\begin{aligned}\int_\Omega (uu_x + vu_y)\psi &= \int_{\partial\Omega} -\psi uv\,dx + \psi u^2\,dy - \int_\Omega \psi_x u^2 + \psi_y uv \\ &= \int_{\partial\Omega} (\psi u^2, \psi uv)\cdot \mathbf{n} - \int_\Omega \left(\psi_x u^2 + \psi_y uv\right).\end{aligned}$$

Since $\psi \equiv 0$ on $\partial\Omega$,

$$\int_\Omega (uu_x + vu_y)\psi = -\int_\Omega \left(\psi_x u^2 + \psi_y uv\right). \tag{8.4.34}$$

A similar analysis of (8.4.32) gives

$$\int_\Omega (uv_x + vv_y)\psi = -\int_\Omega \left(\psi_x uv + \psi_y v^2\right). \tag{8.4.35}$$

The finite element method for (8.4.1)–(8.4.4), (8.4.6), (8.4.30) follows the three-step procedure described in (8.4.10)–(8.4.15). The algebraic equations are formed from the weak formulation by means of Galerkin's formulation of the finite element method. The weak formulation of (8.4.1) and (8.4.2) for $P \equiv 0$ is easily determined by using (8.4.34) and (8.4.35).

Weak Formulation of (8.4.1) and (8.4.2) With $P \equiv 0$. Let $\psi \in PC^1(\Omega)$ and $\psi \equiv 0$ on $\partial\Omega$.

$$\int_\Omega u_t \psi - \int_\Omega \left(\psi_x u^2 + \psi_y uv\right) + \mu \int_\Omega \left(u_x \psi_x + u_y \psi_y\right) = \int_\Omega f\psi, \tag{8.4.36}$$

$$\int_\Omega v_t \psi - \int_\Omega \left(\psi_x uv + \psi_y v^2\right) + \mu \int_\Omega \left(v_x \psi_x + v_y \psi_y\right) = \int_\Omega g\psi. \tag{8.4.37}$$

Note (8.4.36) and (8.4.37), as now stated, depend on time. In fact, we want u_t and v_t to be approximated by $(u^{\text{aux}} - u^k)/\Delta t$ and $(v^{\text{aux}} - v^k)/\Delta t$, respectively, and all other u and v to be u^k and v^k, respectively. The algebraic system is formed as usual by the following substitutions:

$$u \to \sum_j u_j \psi_j,$$

$$v \to \sum_j v_j \psi_j,$$

$$\psi \to \psi_i.$$

The new matrices of interest come from the advection terms. For example, $\int_\Omega \psi_x u^2$ gives

$$\int_\Omega \psi_{ix}\left(\sum_j u_j \psi_j\right)^2 = \sum_j \sum_l \left(\int_\Omega \psi_{ix} \psi_j \psi_l\right) u_l u_j. \tag{8.4.38}$$

(8.4.38) can be written in matrix form

$$\left(\int_\Omega \psi_{ix}\left(\sum_j u_j\psi_j\right)^2\right) = C_x(u)u, \tag{8.4.39}$$

where $C_x(u)$ is a square matrix given by

$$C_x(u) \equiv \left(\sum_l \left(\int_\Omega \psi_{ix}\psi_j\psi_l\right)u_l\right). \tag{8.4.40}$$

By considering the other advection terms in (8.4.36) and (8.4.37), we find it useful to define

$$C_y(v) \equiv \left(\sum_l \left(\int_\Omega \psi_{iy}\psi_i\psi_l\right)v_l\right). \tag{8.4.41}$$

Finally, by using (8.4.40) and (8.4.41) we can state the finite element method for our problem.

FEM—Explicit for (8.4.1)–(8.4.4), (8.4.6), and (8.4.30)

Step I: Set $P \equiv 0$ and solve for $(u^{\text{aux}}, v^{\text{aux}})$, where on $\Omega \cup \partial\Omega_{\text{w}}$

$$B\frac{u^{\text{aux}} - u^k}{\Delta t} - \left(C_x(u^k)u^k + C_y(v^k)u^k\right) + Au^k = Bf, \tag{8.4.42}$$

$$B\frac{v^{\text{aux}} - v^k}{\Delta t} - \left(C_x(u^k)v^k + C_y(v^k)v^k\right) + Av^k = Bg. \tag{8.4.43}$$

$C_x(u^k)$ and $C_x(v^k)$ are defined in (8.4.40) and (8.4.41).

$$B \equiv \left(\int_\Omega \psi_i\psi_j\right) \quad \text{and} \quad A \equiv \left(\mu\int_\Omega\left(\psi_{ix}\psi_{jx} + \psi_{iy}\psi_{jy}\right)\right).$$

On $\partial\Omega_1$ let $(u^{\text{aux}}, v^{\text{aux}}) = (u_1, v_1)$.

Step II: Solve for the pressure P where

$$\tilde{A}P = B(-\delta_x^+ u^{\text{aux}} - \delta_y^+ v^{\text{aux}}) + B_2((u^{\text{aux}}, v^{\text{aux}}) \cdot \mathbf{n}) \tag{8.4.44}$$

$$B \equiv \left(\int_\Omega \psi_i \psi_j\right),$$

$$\tilde{A} \equiv \left(\Delta t \int_\Omega (\psi_{ix}\psi_{jx} + \psi_{iy}\psi_{jy})\right),$$

$$B_2 \equiv \left(\int_{\partial\Omega} \psi_i \psi_j\right).$$

Step III: Adjust the auxiliary velocity by letting on $\Omega \cup \partial\Omega_w$

$$u^{k+1} = u^{\text{aux}} - \Delta t\, \delta_x^- P, \tag{8.4.45}$$

$$v^{k+1} = v^{\text{aux}} - \Delta t\, \delta_y^- P, \tag{8.4.46}$$

Remarks

1. In order to perform step I, we must solve for Bu^{aux} and Bv^{aux} given. This can be avoided by replacing B by the diagonal matrix $\text{diag}(\int_\Omega \psi_i)$.
2. In steps II and III we have used the finite difference notation for approximation of the partial derivatives. If we do not have a rectangular grid, then we must adjust the notation in (8.4.44)–(8.4.46) to something meaningful, for example, use the results of Proposition 2.1.1 where linear shape functions are used.

8.5 INCOMPRESSIBLE VISCOUS FLUID FLOW—IMPLICIT METHOD

We shall describe a fully implicit finite element method for approximating the solution of (8.4.1)–(8.4.4), (8.4.6), and (8.4.30). In Section 8.4 the velocity and the pressure were decoupled. The momentum equations were solved explicitly for the auxiliary velocity.

Then the conservation-of-mass equation was used to couple the velocity and pressure functions in the appropriate way. By a fully implicit method we mean solving for the (u, v, P) implicitly at each time step.

As usual, we need the weak formulation of this problem. The implicit form of (8.4.1)–(8.4.3) with the advection terms given by (8.4.31) and (8.4.32) is

Navier–Stokes—Implicit

$$\frac{u^{k+1}-u^k}{\Delta t}+\left((u^{k+1})^2\right)_x+(u^{k+1}v^{k+1})_y-\mu\,\Delta u^{k+1}+P_x^{k+1}=f^{k+1}, \tag{8.5.1}$$

$$\frac{v^{k+1}-v^k}{\Delta t}+(u^{k+1}v^{k+1})_x+\left((v^{k+1})^2\right)_y-\mu\,\Delta v^{k+1}+P_y^{k+1}=g^{k+1}, \tag{8.5.2}$$

$$u_x^{k+1}+v_y^{k+1}=0. \tag{8.5.3}$$

By multiplying (8.5.1)–(8.5.3) by ψ and using (8.4.34) and (8.4.35) we obtain a weak formulation of (8.5.1)–(8.5.3). Also, $\int P_x\psi = -\int P\psi_x$ and $\int P_y\psi = -\int P\psi_y$.

Weak Formulation of Equations (8.5.1)–(8.5.3). Let $\psi \in PC^1(\Omega)$ and $\psi = 0$ on $\partial\Omega$.

$$\int_\Omega \frac{u^{k+1}-u^k}{\Delta t}\psi-\int_\Omega\left((u^{k+1})^2\psi_x+u^{k+1}v^{k+1}\psi_y\right)$$
$$+\mu\int_\Omega\left(u_x^{k+1}\psi_x+u_y^{k+1}\psi_y\right)-\int_\Omega P\psi_x=\int_\Omega f^{k+1}\psi, \tag{8.5.4}$$

$$\int_\Omega \frac{v^{k+1}-v^k}{\Delta t}\psi-\int_\Omega\left(u^{k+1}v^{k+1}\psi_x+(v^{k+1})^2\psi_y\right)$$
$$+\mu\int_\Omega\left(v_x^{k+1}\psi_x+v_y^{k+1}\psi_y\right)-\int_\Omega P\psi_y=\int_\Omega g^{k+1}\psi, \tag{8.5.5}$$

$$-\int_\Omega\left(u_x^{k+1}+v_y^{k+1}\right)\psi=0. \tag{8.5.6}$$

In order to obtain an algebraic system from the weak equations (8.5.4)–(8.5.6), we make the following substitutions:

$$u \to \sum_j u_j \psi_j,$$

$$v \to \sum_j v_j \psi_j,$$

$$P \to \sum_j P_j \bar{\psi}_j,$$

$$\psi \to \psi_i, \quad \text{for equations (8.5.4) and (8.5.5)}$$

$$\psi \to \bar{\psi}_i, \quad \text{for equation (8.5.6).}$$

Note the shape functions ψ_j that are used to approximate the velocity may be different from those $\bar{\psi}_j$ that are used to approximate the pressure. For example, good results have been obtained when ψ_j are quadratic and $\bar{\psi}_j$ are linear. This is necessary because in (8.5.4)–(8.5.6) the approximation of the first-order derivatives of u and v and the approximation of P should have the same order, linear. The only new matrices that evolve from the integrals are $\int_\Omega P\psi_x$, $\int_\Omega P\psi_y$, $\int_\Omega u_x\bar{\psi}$, and $\int_\Omega v_y\bar{\psi}$. For example, $\int_\Omega P\psi_x$ gives

$$\int_\Omega \left(\sum_j P_j \bar{\psi}_j \right) \psi_{ix} = \sum_j \left(\int_\Omega \bar{\psi}_j \psi_{ix} \right) P_j. \tag{8.5.7}$$

(8.5.7) can be written in matrix form as $D_x P$, where

$$D_x \equiv \left(\int_\Omega \bar{\psi}_j \psi_{ix} \right), \quad D_y = \left(\int_\Omega \bar{\psi}_j \psi_{iy} \right), \quad D_x^{\mathrm{T}} = \left(\int \psi_{jx} \bar{\psi}_i \right),$$

$$D_y^{\mathrm{T}} = \left(\int \psi_{jy} \bar{\psi}_i \right). \tag{8.5.8}$$

Other matrices of interest are

$$B \equiv \left(\frac{1}{\Delta t} \int_\Omega \psi_i \psi_j \right), \qquad A \equiv \mu \left(\int_\Omega \psi_{ix} \psi_{jx} + \psi_{iy} \psi_{jy} \right), \tag{8.5.9}$$

$$C_x(u) \equiv \left(\sum_l \left(\int_\Omega \psi_{ix} \psi_j \psi_l \right) u_l \right), \qquad C_y(v) \equiv \left(\sum_l \left(\int_\Omega \psi_{iy} \psi_j \psi_l \right) v_l \right), \tag{8.5.10}$$

$$\mathscr{B}(u, v) \equiv B - C_x(u) - C_y(v) + A. \tag{8.5.11}$$

The algebraic system (8.5.14) has an unknown $(\bar{u}^{k+1}, \bar{v}^{k+1}, P^{k+1})$ in $\mathbb{R}^{NN} \times \mathbb{R}^{NN} \times \mathbb{R}^N$, where NN is N minus the number of prescribed nodes. Since the pressure may vary by a constant, we assume $dP/dn = 0$ on $\partial\Omega$ and, consequently, P has N unknowns.

FEM—Fully Implicit for (8.4.1)–(8.4.4), (8.4.6), and (8.4.30). Let the matrices be given as in (8.5.8)–(8.5.11). Define $d_1, d_2 \in \mathbb{R}^N$:

$$d_1 \equiv \Delta t\, B f^{k+1} + B u^k, \tag{8.5.12}$$

$$d_2 \equiv \Delta t\, B g^{k+1} + B v^k. \tag{8.5.13}$$

Define $\tilde{d}_1, \tilde{d}_2 \in \mathbb{R}^{NN}$ by omitting the rows of d_1, d_2 that correspond to the prescribed values of (u_j, v_j). Define $\bar{d}_1, \bar{d}_2 \in \mathbb{R}^{NN}$, $\bar{d}_3 \in \mathbb{R}^N$ from $\tilde{d}_1, \tilde{d}_2$ and the prescribed values of (u_j, v_j). Also, let $\overline{\mathscr{B}}(\bar{u}, \bar{v}) \in \mathbb{R}^{NN} \times \mathbb{R}^{NN}$, $\overline{D}_x, \overline{D}_y \in \mathbb{R}^{NN} \times \mathbb{R}^N$ and $\overline{D}_x^{\mathrm{T}}, \overline{D}_y^{\mathrm{T}} \in \mathbb{R}^N \times \mathbb{R}^{NN}$ be the corresponding matrices.

$$\begin{pmatrix} \overline{\mathscr{B}}(\bar{u}^{k+1}, \bar{v}^{k+1}) & 0 & -\overline{D}_x \\ 0 & \overline{\mathscr{B}}(\bar{u}^{k+1}, \bar{v}^{k+1}) & -\overline{D}_y \\ -\overline{D}_x^T & -\overline{D}_y^T & 0 \end{pmatrix} \begin{pmatrix} \bar{u}^{k+1} \\ \bar{v}^{k+1} \\ P^{k+1} \end{pmatrix} = \begin{pmatrix} \bar{d}_1 \\ \bar{d}_2 \\ \bar{d}_3 \end{pmatrix}. \tag{8.5.14}$$

Remark. Suppose we wish to approximate the solution of (8.5.14) by Newton's method. Because the components in the matrix are linear combinations of the components in $(\bar{u}^{k+1}, \bar{v}^{k+1}, P^{k+1})^{\mathrm{T}}$, we can use, as in Burger's equation, Proposition 7.3.1. This will simplify the computation of $F'(\bar{u}, \bar{v}, P)$, where $F: \mathbb{R}^M \to \mathbb{R}^M$, $M = 2 \cdot NN + N$ and is

$$F(\bar{u}, \bar{v}, P) = \begin{bmatrix} \overline{\mathscr{B}}(\bar{u}, \bar{v})\bar{u} - \overline{D}_x P - \bar{d}_1 \\ \overline{\mathscr{B}}(\bar{u}, \bar{v})\bar{v} - \overline{D}_y P - \bar{d}_2 \\ -\overline{D}_x^{\mathrm{T}}\bar{u} - \overline{D}_y^{\mathrm{T}}\bar{v} - \bar{d}_3 \end{bmatrix} \in \mathbb{R}^M. \tag{8.5.15}$$

8.6 STEFAN PROBLEM

The classical Stefan problem refers to heat conduction in melting ice. Any heat-conduction problem that involves a change of phase is often called a Stefan problem. Let us briefly describe two problems.

Solidification of Water. Suppose the sides of a thin bar of water are insulated, and the left end is at a fixed temperature $\bar{u}_0$ below freezing. If the bar is initially at $\bar{u}_1 > u_{\mathrm{f}}$, the freezing temperature, then a *sharp interface* is formed which separates the solid and liquid regions. See Figure 8.6.1 where $s(t)$ is the position of the sharp interface.

Welding of a Pure Substance. Suppose two plates are to be welded together by sending an electric current between two probes (see Figure 8.6.2). The resistance of the metal plates will cause the plates to heat up. The largest temperature will be at the point of contact of the plates, that is, $x = l$. In this case a sharp interface will not form. Instead a *mushy region* $\{x | u(x, t) = u_{\mathrm{f}}\}$ will form once the temperature reaches u_{f} (see Figure 8.6.3). An explanation of why this mushy region evolves will be given after the classical formulation of the Stefan problem is discussed.

It is important to note if the metal is not pure (e.g., if it is an alloy), then the solidification temperature u_{f} will depend on the concentration of the impurity. The concentration of the impurity depends on the temperature. That is, the heat conduction in an alloy must be described by a coupled system of partial differential equa-

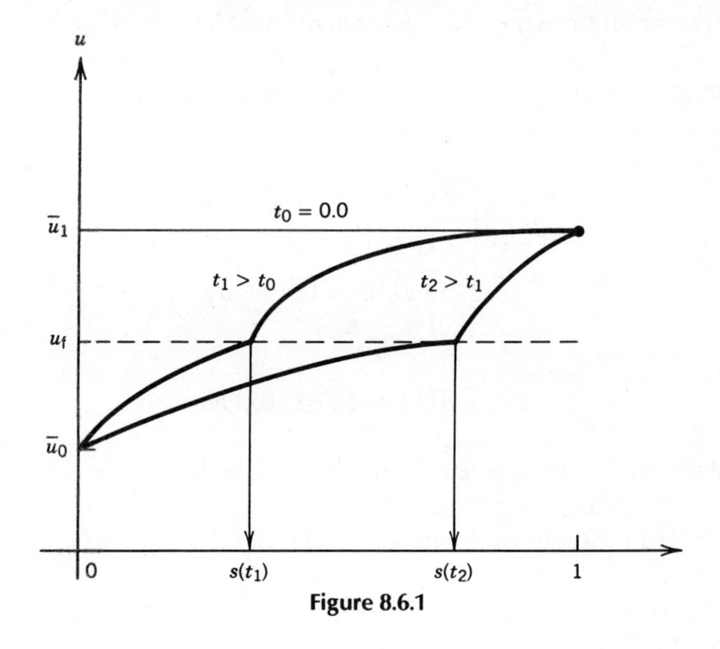

Figure 8.6.1

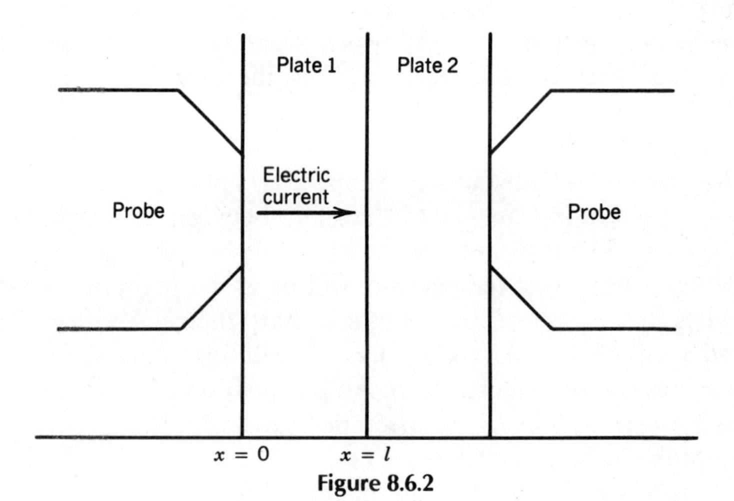

Figure 8.6.2

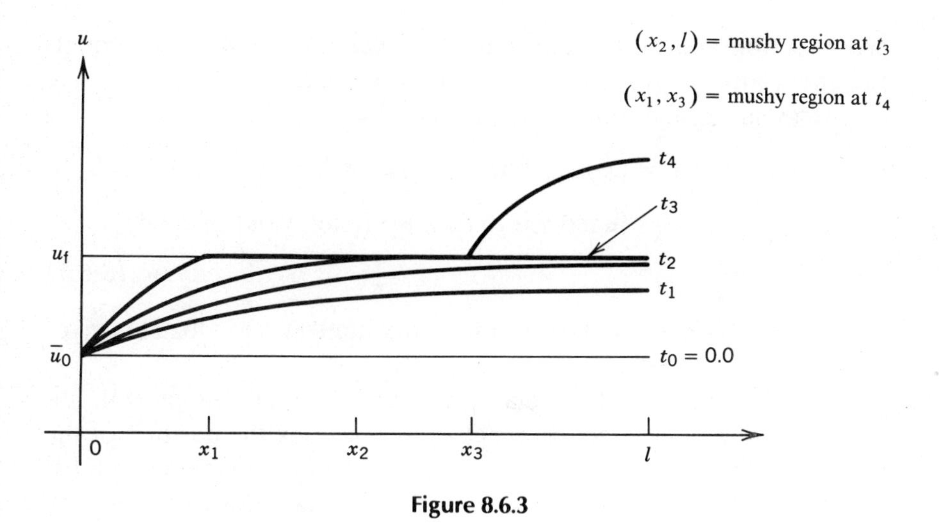

Figure 8.6.3

tions—one for the temperature and one for the concentration of the impurity.

Let us consider the one-space variable problem with Dirichlet boundary conditions.

Classical Formulation With Sharp Interface

$$\rho_i c_i u_{it} - (k_i u_{ix})_x = 0, \qquad i = 1, 2, \tag{8.6.1}$$

$$u_1(0, t) = \bar{u}_0 < u_f \quad (1 \to \text{solid}), \tag{8.6.2}$$

$$u_2(l, t) = \bar{u}_l > u_f \quad (2 \to \text{liquid}), \tag{8.6.3}$$

$$u_i(s(t), t) = u_f, \qquad i = 1, 2, \tag{8.6.4}$$

$$k_2 u_{2x} - k_1 u_{1x} = -\rho_1 L\dot{s}, \tag{8.6.5}$$

$$s(0) = 0, \qquad u_2(x, 0) = \psi(x) > u_f. \tag{8.6.6}$$

Except for the condition given by (8.6.5), these conditions follow in a straightforward manner. Line (8.6.5) is a consequence of the Fourier heat law being applied at $x = s(t)$, the solid–liquid interface. As $s(t)$ is increasing, $\Delta s = s(t + \Delta t) - s(t) > 0$. If A is the cross-section area of the thin bar of ice water, then $\rho_1 L \Delta s A$ equals the energy that is required to melt the volume $\Delta s A$. The only source of energy is through diffusion, and therefore, by the heat law.

$$\begin{aligned} -\rho_1 L \Delta s A &\simeq (k_2 u_{2x} A \Delta t) - (k_1 u_{1x} A \Delta t) \\ &= (\text{energy into } \Delta s A) - (\text{energy out of } \Delta s A) \end{aligned} \tag{8.6.7}$$

Now, divide (8.6.7) by $A \Delta t$ and take the limit as $\Delta t \to 0$. This gives (8.6.5).

Analytic solutions of the classical formulation are, in general, not easy to find. Some numerical methods try to track the position of the interface. The heat balance at $x = s(t)$, (8.6.5), can be formulated for more than one space variable. However, the numerical methods for tracking the interface become more complicated. Indeed, in more than one space variable the interface is now a surface. This surface may have a very complicated shape, for example, consider a melting ice "cube." Moreover, the classical formulation simply fails to describe heat conduction in problems in which the interface is not sharp. Before we discuss the enthalpy formulation of the Stefan problem, we shall justify the existence of mushy regions.

The Evolution of a Mushy Region in the Welding Problem. Consider the welding problem as previously described. At $x = l$, the temperature will be the highest, and so $u_x(l, t) = 0$. As time increases, $u(l, t)$ eventually reaches u_f; $t = t_2$ indicates, in Figure 8.6.3, this time. The only sources of heat are diffusion and heating due to the electric current. Since the temperature $u(l, t)$ is already at the highest possible, there is no gain in energy by diffusion. If F is the energy per unit (volume times time), which is a result of the electric current, then the amount of energy produced by the current in the volume $A \Delta x$ in time Δt is

$$\text{energy} \simeq F A \Delta x \Delta t. \tag{8.6.8}$$

The volume $A \Delta x$ refers to a volume near $x = l$. Since the energy needed to melt $A \Delta x$ is $\rho_1 L A \Delta x$, we may have for "small" Δt

$$\rho_1 L A \Delta x > F A \Delta x \Delta t. \tag{8.6.9}$$

Thus, (8.6.9) indicates that it may take several time iterations k until

$$\rho_1 LA\,\Delta x \simeq FA\,\Delta x\,\Delta t\,k. \tag{8.6.10}$$

When (8.6.10) holds, the temperature of the volume $A\Delta x$ may rise above u_f. Consequently, a mushy region has formed.

Recall the enthalpy formulation of the heat equation given in line (8.2.4). Now we assume the thermal properties are constant in each phase. For simplicity we refer to (8.6.11), or (8.6.13), as the *preenthalpy formulation of the Stefan problem.*

$$H(u)_t - (K(u)u_x)_x = f, \tag{8.6.11}$$

where

$$H(u) \equiv \begin{cases} \rho_1 c_1 u, & u < u_f \\ \rho_2 c_2 (u - u_f) + \rho_1 L + \rho_1 c_1 u_f, & u > u_f, \end{cases}$$

$$K(u) \equiv \begin{cases} k_1, & u < u_f \\ k_2, & u > u_f. \end{cases}$$

Usually $\rho_2 c_2 > \rho_1 c_1$ and $k_1 > k_2$. Note both $H(u)$ and $K(u)$ are discontinuous at $u = u_f$. (8.6.11) can be simplified by using the Kirchhoff change of variables:

$$v \equiv F(u) = \int_0^u K(\bar{u})\, d\bar{u} = \begin{cases} k_1 u, & u < u_f \\ k_2(u - u_f) + k_1 u_f, & u > u_f. \end{cases} \tag{8.6.12}$$

Thus, (8.6.11) becomes

$$H(F^{-1}(v))_t - v_{xx} = f, \tag{8.6.13}$$

where

$$u = F^{-1}(v) \equiv \begin{cases} \dfrac{v}{k_1}, & v < v_f = k_1 u_f \\ \dfrac{v + (k_2 - k_1)u_f}{k_2}, & v > v_f = k_1 u_f, \end{cases}$$

$$H(F^{-1}(v)) \equiv \begin{cases} \dfrac{\rho_1 c_1}{k_1} v, & v < v_f \\ \dfrac{\rho_2 c_2}{k_2} v + \left(\dfrac{\rho_1 c_1}{k_1} - \dfrac{\rho_2 c_2}{k_2} \right) v_f + \rho_1 L, & v > v_f. \end{cases}$$

It is important to note $H(F^{-1}(v))/\Delta t$ is also discontinuous in v. Thus, when an implicit time discretization of (8.6.13) is used, we encounter problems. When finite differences are used, we obtain

$$\frac{H(F^{-1}(v_i^{k+1})) - H(F^{-1}(v_i^k))}{\Delta t} - \frac{v_{i+1}^{k+1} - 2v_i^{k+1} + v_{i-1}^{k+1}}{\Delta x^2} = f^{k+1} \tag{8.6.14}$$

(8.6.14) generates a semilinear algebraic problem of the form $\phi(v^{k+1}) + Av^{k+1} = \eta$, which was discussed in Section 7.7. The nonlinear Gauss–Seidel algorithm was used provided the $\phi_i(v)$ were nondecreasing and continuous. In (8.6.14) $\phi_i(v) = H(f^{-1}(v))/\Delta t$ is nondecreasing and *not* continuous!

Another problem with the model in (8.6.14) is indicated by the welding problem and the evolution of a mushy region. If we wish to develop a numerical model of the mushy region, we must realize that the energy in the volume $A\Delta x$ at $t = t_2$ is $\rho_1 c_1 u_f \cdot A\Delta x$. As time increases, the energy increases in a continuous manner until it equals $\rho_1 c_1 u_f A\Delta x + \rho_1 LA\Delta x$. At this time the temperature may get larger than u_f. In fact, we are interested in the energy in a given volume and not its temperature. In the temperature description of the enthalpy, given by (8.6.11), the continuous transition from $\rho_1 c_1 u_f A\,\Delta x$ to $\rho_1 c_1 u_f A\,\Delta x + \rho_1 LA\,\Delta x$ is destroyed.

The enthalpy formulation has the enthalpy as the primary unknown, and the temperature is to be computed from the enthalpy. In

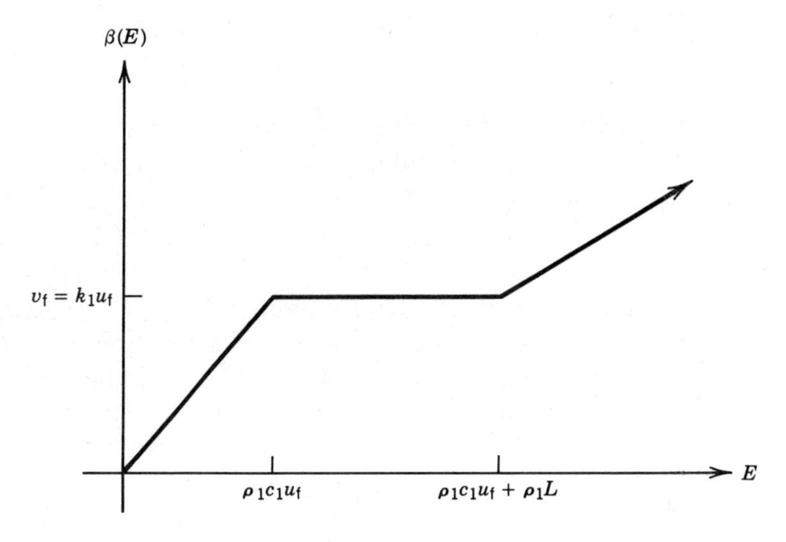

Figure 8.6.4

the preenthalpy model, (8.6.11) or (8.6.13), the temperature was the essential unknown and the "enthalpy" was computed by $H(u)$. In order to compute the temperature from the enthalpy, we use the inverse of $E = H(F^{-1}(v))$.

Definition. Let $H(u)$ be given by (8.6.11) and call it the *enthalpy function*. Let $\beta(E) = v$ be the "inverse" of $E = H(F^{-1}(v))$, (see Figure 8.6.4). Let $E_l = \rho_1 c_1 u_f + \rho_1 L$.

$$\beta(E) \equiv \begin{cases} E\dfrac{k_1}{\rho_1 c_1}, & E < \rho_1 c_1 u_{\mathrm{f}} \\ v_{\mathrm{f}} = k_1 u_{\mathrm{f}}, & \rho_1 c_1 u_{\mathrm{f}} \le E \le E_l \\ \left(E - \left(\dfrac{\rho_1 c_1}{k_1} - \dfrac{\rho_2 c_2}{k_2}\right) k_1 u_{\mathrm{f}} - \rho_1 L\right)\dfrac{k_2}{\rho_2 c_2}, & E_l < E. \end{cases}$$

(8.6.15)

Remarks

1. It is important to note that $\beta(E)$ is continuous.
2. The function $E(x,t)$ = enthalpy will be defined as a solution to a nonlinear partial differential equation—see (8.6.16)–(8.6.19).

Enthalpy Formulation of the Stefan Problem. Let $\beta(E)$ be defined from (8.6.15), or and an equivalent function defined from the thermal properties which may not be constants. $E(x,t)$ = the enthalpy is defined as the solution of (8.6.16)–(8.6.19). The temperature is computed from E by $F^{-1}(\beta(E))$.

$$E_t - \beta(E)_{xx} = f(x,t), \qquad f(x,t) \text{ bounded}, \tag{8.6.16}$$

$$E(x,0) = H(\psi_0), \qquad \psi_0 > u_f, \tag{8.6.17}$$

$$\beta(E(0,t)) = F(\bar{u}_0), \qquad \bar{u}_0 < u_f, \tag{8.6.18}$$

$$\beta(E(l,t)) = F(\bar{u}_l), \qquad \bar{u}_l > u_f. \tag{8.6.19}$$

Since $\beta(E)_{xx}$ may not exist in a classical sense, we need to define a weak solution of problem (8.6.16)–(8.6.19). Since this is a time-dependent problem, our test functions ψ must depend on x and t. In order to ensure that any weak solution is unique, we must assume that $\psi(0,t) = \psi(l,t) = \psi(x,T) = 0$. Details of this can be found in White [30]. By multiplying (8.6.16) by $\psi(x,t)$ and integrating, we obtain the weak equation in the following definition.

Definition. Consider the enthalpy formulation of the Stefan problems given in lines (8.6.16)–(8.6.19). Let $\psi \in PC^1((0,l) \times (0,T))$ and $\psi(0,t) = \psi(l,t) = \psi(x,T) = 0$. $E \in L_2((0,l) \times (0,T))$ is called a *weak solution* if and only if

(i) $\beta(E(x,t)), \beta(E(x,t))_x \in L_2((0,l) \times (0,T))$,
(ii) $\beta(E(0,t)) = F(\bar{u}_0)$ and $\beta(E(l,t)) = F(\bar{u}_l)$,

(iii) For all test functions $\psi(x,t)$ the weak equation holds:

$$-\int_0^T\int_0^l E\psi_t + \int_0^T\int_0^l \beta(E)_x\psi_x$$

$$= \int_0^l E(x,0)\psi(x,0) + \int_0^T\int_0^l \psi f. \qquad (8.6.20)$$

If (8.6.19) is replaced by

$$\beta(E(l,0))_x = 0, \qquad (8.6.21)$$

then for f a positive constant, (8.6.16)–(8.6.18), (8.6.21) will give a mushy region. If $f = 0$, then this may not be true. When the boundary condition (8.6.19) is used, the following propositions give interesting related results.

Proposition 8.6.1. If E is a solution of (8.6.16) and the interface is sharp, that is, $\{x \in [0,l] | u(x,t) = u_f\} = \{s(t)\}$, then (8.6.5) holds.

Proof. Since the interface is sharp, we have $E(s(t)^-, t) = \rho_1 c_1 u_f$. Thus

$$\frac{d}{dt}E(s(t)^- t) = 0 = E_x\frac{ds}{dt} + E_t \cdot 1. \qquad (8.6.22)$$

(8.6.22) and (8.6.16) imply

$$-E_x \cdot \dot{s}(t) = \beta(E)_{xx} + f. \qquad (8.6.23)$$

Integrate (8.6.23) from $x = s^-$ to $x = s + \Delta s$:

$$\int_{s^-}^{s+\Delta s} - E_x \cdot \dot{s}(t) = \int_{s^-}^{s+\Delta s}(\beta(E)_{xx} + f),$$

$$-\dot{s}(t)\int_{s^-}^{s+\Delta s} E_x \cong \int_{s^-}^{s+\Delta s}\beta(E)_{xx} + \int_{s^-}^{s+\Delta s} f,$$

$$-\dot{s}(E|_{s+\Delta s} - E|_{s^-}) \cong \beta(E)_x|_{s+\Delta s} - \beta(E)_x|_{s^-} + \int_{s^-}^{s+\Delta s} f. \qquad (8.6.24)$$

Since f is bounded, $\int_{s^-}^{s+\Delta s} f \to 0$ as $\Delta s \to 0$. Also, $E|_{s+\Delta s} \cong \rho_1 c_1 u_f + \rho_1 L$ and $E|_{s^-} \cong \rho_1 c_1 u_f$. Therefore, the left-hand side goes to $-\dot{s}\rho_1 L$ as $\Delta s \to 0$. Also, $\beta(E)_x|_{s+\Delta s} \simeq k_2 u_{2x}$ and $\beta(E)_x|_{s^-} = k_1 u_{1x}$. Thus, as $\Delta s \to 0$ (8.6.24) approximates (8.6.5).

Proposition 8.6.2. Consider the classical formulation, (8.6.1)–(8.6.6), and the enthalpy formulation, (8.6.16)–(8.6.19), of the Stefan problem. If $f \equiv 0$, then the interface must be sharp, and consequently, the two formulations are equivalent.

Proof. Consider the enthalpy formulation and suppose the interface is not sharp. Then for $x \in (s_1(t), s_2(t))$, $\beta(E) = k_1 u_f$. Without loss of generality, assume at $x = s_1(t)$ $E = \rho_1 c_1 u_f$, and at $x = s_2(t)$ $E = \rho_1 c_1 u_f + \rho_1 L$. Then $E_t = \beta(E)_{xx} = 0$ for $x \in (s_1(t), s_2(t))$. So, either $E = \rho_1 c_1 u_f$ or $E = \rho_1 c_1 u_f + \rho_1 L$ for $x \in (s_1(t), s_2(t))$. E must satisfy the classical heat equation on either $(0, s_2(t))$ or $(s_1(t)), l)$. Then it is analytic on either $(0, s_2(t))$ or $(s_1(t), l)$, and consequently, cannot be a constant on any interval. In either case $s_1(t) = s_2(t)$, that is, the interface is sharp. By Proposition 8.6.1, line (8.6.5) holds. The other lines in the classical formulation easily follow from the definition of $\beta(E)$. Also, it is easy to show that any classical solution is an enthalpy solution in the weak sense as given by line (8.6.20).

Remark. The enthalpy formulation of the Stefan problem is quite general in that it describes heat conduction without phase change, the classical Stefan problem with a sharp interface, and the Stefan problem with a mushy region.

In order to approximate a solution of (8.6.16)–(8.6.19), we must solve nonlinear algebraic systems of the form $bE + A\beta(E) = \eta$ for the FDM or $BE + A\beta(E) = \eta$ for the FEM. Before we state the nonlinear Gauss–Seidel algorithms for doing this we derive these systems.

FDM, Implicit Time Discretization of (8.6.16)–(8.6.19)

$$\frac{E_i^{k+1} - E_i^k}{\Delta t} - \frac{1}{\Delta x}\left(\frac{\beta(E_{i+1}^{k+1}) - \beta(E_i^{k+1})}{\Delta x} - \frac{\beta(E_i^{k+1}) - \beta(E_{i-1}^{k+1})}{\Delta x}\right) = f_i^{k+1}, \quad (8.6.25)$$

$$E_i^0 = H(\psi_0), \quad (8.6.26)$$

$$\beta(E_0^k) = F(\bar{u}_0), \qquad k = 0, 1, \ldots, \quad (8.6.27)$$

$$\beta(E_N^k) = F(\bar{u}_l), \qquad \Delta x = l/N. \quad (8.6.28)$$

The matrix form of (8.6.25)–(8.6.28) is

$$BE + A\beta(E) = \eta, \tag{8.6.29}$$

where

$$B = \frac{1}{\Delta t} I, \qquad E = \left(E_i^{k+1}\right) \in \mathbb{R}^{N-1},$$

$$\beta(E) = \left(\beta\left(E_i^{k+1}\right)\right) \in \mathbb{R}^{N-1}, \qquad \eta = BE^k + f^{k+1},$$

and A is $(N-1) \times (N-1)$ matrix. For $N = 4$, A has the form

$$A = \frac{1}{\Delta x^2} \begin{pmatrix} 2 & -1 & 0 \\ -1 & 2 & -1 \\ 0 & -1 & 2 \end{pmatrix}.$$

The implicit method has been used by Shamsundar and Sparrow [24]. There is no stability condition. The explicit method is discussed in Atthey [1], and there is the usual stability constraint.

Algorithm for $bE + A\beta(E) = \eta$. Let $A = (a_{ij})$, $a_{ii} > 0$, $b > 0$. $R_i(z) \equiv bz + a_{ii}\beta(z)$ is continuous and strictly increasing. Let $R_i^{-1}(\overline{w})$ denote its continuous inverse function. Note $bE + A\beta(E) = \eta$ can be written

$$R_i(E_i) = \eta_i - \sum_{j<i} a_{ij}\beta(E_j) - \sum_{j>i} a_{ij}\beta(E_j).$$

The algorithm with an SOR parameter is

$$E_i^{k+1/2} = R_i^{-1}\left(\eta_i - \sum_{j<i} a_{ij}\beta\left(E_j^{k+1}\right) - \sum_{j>i} a_{ij}\beta\left(E_j^k\right)\right), \tag{8.6.30}$$

$$E_i^{k+1} = \left(1 - WW\left(E_i^k\right)\right)E_i^k + WW\left(E_i^k\right)E_i^{k+1/2}, \tag{8.6.31}$$

where $WW(E)$ is given by Figure 8.6.5.

Remark. The SOR parameter is used only in the nonliquid region. If $WW(E) \equiv w$, then the algorithm fails to converge. If $A \equiv (a_{ij})$, $a_{ij} \leq 0$ for $i \neq j$, $\eta \geq 0$, and $a_{ii} \geq \Sigma_{j \neq i}|a_{ij}| + \delta$, $\delta > 0$, then the algorithm converges to a unique solution of $bE + A\beta(E) = \eta$. A less constrictive condition is that A should be an M matrix.

Example. Let $\rho_i = c_i = k_i = 1.0$, $i = 1, 2$, and $u_{\mathrm{f}} = 1.0$, $\psi_0 = 1.5$, $\bar{u}_0 = 0.1$, $\bar{u}_l = 1.5$, $f = 0.0$.

$$\beta(u) = \begin{cases} u, & u < 1 \\ 1, & 1 \leq u \leq 1 + L \\ u - L, & 1 + L < u, \end{cases}$$

$$\bar{w} = R(z) = \frac{1}{\Delta t} z + \frac{2}{\Delta x^2} \beta(z)$$

$$= \begin{cases} \dfrac{1}{\Delta t} z + \dfrac{2}{\Delta x^2} z = az, & z < 1, \quad a \equiv \dfrac{1}{\Delta t} + \dfrac{2}{\Delta x^2} \\ \dfrac{1}{\Delta t} z + \dfrac{2}{\Delta x^2} \cdot 1, & 1 \leq z \leq 1 + L \\ az - \dfrac{2L}{\Delta x^2}, & 1 + L < z, \end{cases}$$

$$R^{-1}(\bar{w}) = \begin{cases} \bar{w}/a, & \bar{w} < a \\ \left(\bar{w} - \dfrac{2}{\Delta x^2}\right) \Delta t, & a \leq \bar{w} \leq a + L \\ \left(\bar{w} + \dfrac{2L}{\Delta x^2}\right) / a, & a + L < \bar{w}. \end{cases}$$

Let $\bar{E}_i$ be the previous time and E_i^{k+1} the $(k + 1)$th iterate of the algorithm for the present time at the ith node. $N = 20$ and $l = 1$. $E_i^0 = H(1.5) = 1.5 + L$, $E_0^{k+1} = H(0.1) = 0.1$, $E_{10}^{k+1} = H(1.5) = 1.5 + L$. Line (8.6.30) may be rewritten

$$E_i^{k+1/2} = R^{-1}\left(\frac{\bar{E}_i}{\Delta t} + \frac{1}{\Delta x^2} \left(\beta\left(E_{i-1}^{k+1}\right) + \beta\left(E_{i+1}^{k}\right) \right) \right). \tag{8.6.32}$$

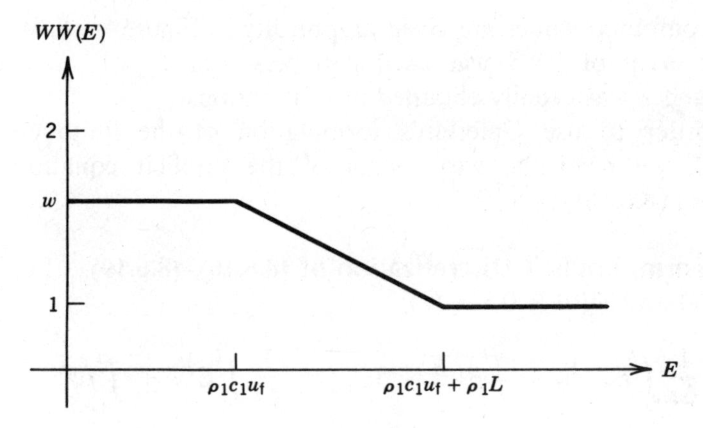

Figure 8.6.5

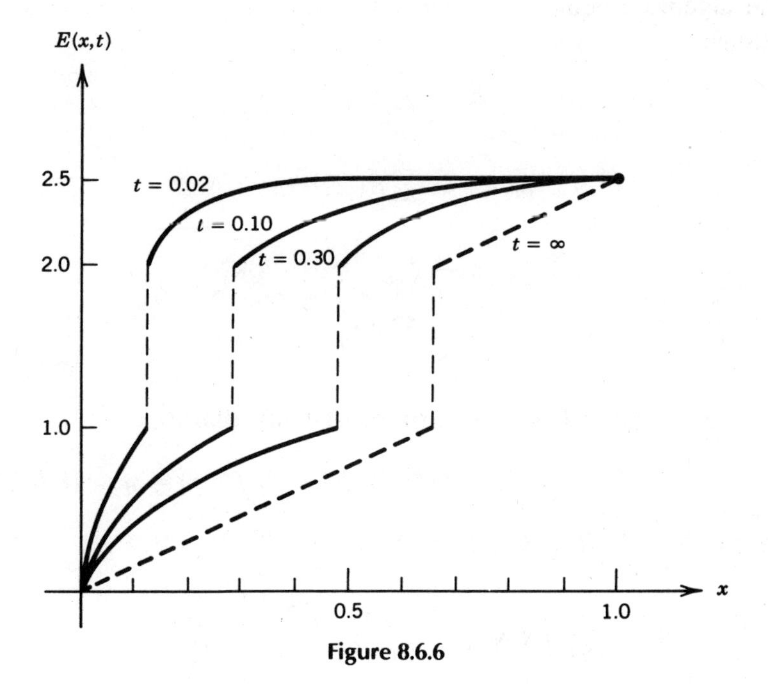

Figure 8.6.6

Some computed values are given graphically in Figure 8.6.6. When a relative error of 10^{-3} was used with $w = 1.0$, $L = 1$, $\Delta t = 0.01$, convergence was usually obtained in 15 iterations.

In order to use Galerkin's formulation of the finite element method, we need the weak form of the implicit equation that describes (8.6.16)–(8.6.19).

Weak Form, Implicit Discretization of (8.6.16)–(8.6.19). Let $\psi \in PC^1(0, l)$ and $\psi(0) = 0 = \psi(l)$.

$$\frac{1}{\Delta t}\int_0^l E^{k+1}\psi + \int_0^l \beta(E^{k+1})_x\psi_x = \frac{1}{\Delta t}\int_0^l E^k\psi + \int_0^l f\psi. \tag{8.6.33}$$

The algebraic equation is derived by making the following substitutions:

$$E \to \sum_j E_j\psi_j,$$

$$\beta(E) \to \sum_j \beta(E_j)\psi_j,$$

$$\psi \to \psi_i,$$

$$f \to \sum_j f_j\psi_j.$$

FEM, Implicit Time Discretization of (8.6.16)–(8.6.19)

$$BE^{k+1} + A\beta(E^{k+1}) = B(E^k + \Delta t f^{k+1}) = \eta \tag{8.6.34}$$

where for $N = 4$, B and A have the forms

$$B = \left(\frac{1}{\Delta t}\int_0^l \psi_j\psi_i\right) = \frac{\Delta x}{\Delta t \cdot 6}\cdot\begin{pmatrix} 4 & 1 & 0 \\ 1 & 4 & 1 \\ 0 & 1 & 4 \end{pmatrix},$$

$$A = \left(\int_0^l \psi_{jx}\psi_{ix}\right) = \frac{1}{\Delta x}\cdot\begin{pmatrix} 2 & -1 & 0 \\ -1 & 2 & -1 \\ 0 & -1 & 2 \end{pmatrix}.$$

Algorithm for $BE + A\beta(E) = \eta$. Let $A = (a_{ij})$, $a_{ii} > 0$, $B = (B_{ij})$, and $b_{ii} > 0$. $R_i(z) \equiv b_{ii}z + a_{ii}\beta(z)$. Let $R_i^{-1}(\overline{w})$ be its inverse function. Note $BE + A\beta(E) = \eta$ can be rewritten

$$R_i(E_i) = \eta_i - \sum_{j<i}\left(a_{ij}\beta_j(E_j) + b_{ij}E_j\right) - \sum_{j>i}\left(a_{ij}\beta(E_j) + b_{ij}E_j\right).$$

The algorithm with an SOR parameter is

$$E_i^{k+1/2} = R_i^{-1}\left(\eta_i - \sum_{j<i}\left(a_{ij}\beta_j\left(E_j^{k+1}\right) + b_{ij}E_j^{k+1}\right)\right.$$
$$\left. - \sum_{j>i}\left(a_{ij}\beta_j\left(E_j^k\right) + b_{ij}E_j^k\right)\right) \tag{8.6.35}$$

$$E_i^{k+1} = \left(1 - WW\left(E_i^k\right)\right)E_i^k + WW\left(E_i^k\right)E_i^{k+1/2}, \tag{8.6.36}$$

where $WW(E)$ is given in Figure 8.6.5.

Example. Let us consider the previous example, but in this case use the FEM.

$$R(z) = \frac{2\Delta x}{3\Delta t}z + \frac{2}{\Delta x}\beta(z),$$

and for $a \equiv 2/\Delta x$, $b \equiv 2\Delta x/3\Delta t$,

$$R^{-1}(\overline{w}) \equiv \begin{cases} \overline{w}/(a+b), & \overline{w} < a+b \\ (\overline{w} - a)/b, & a+b \le \overline{w} \le a+b+Lb \\ (\overline{w} + aL)/(a+b), & a+b+Lb < \overline{w}. \end{cases}$$

Line (8.6.35) may be rewritten

$$E_i^{k+1/2} = R^{-1}\left(\frac{\Delta x}{\Delta t\cdot 6}\left(\overline{E}_{i+1} + 4\overline{E}_i + \overline{E}_{i-1}\right)\right.$$
$$+ \frac{1}{\Delta x}\left(\beta\left(E_{i-1}^{k+1}\right) + \beta\left(E_{i+1}^k\right)\right)$$
$$\left. - \frac{\Delta x}{\Delta t\cdot 6}\left(E_{i-1}^{k+1} + E_{i+1}^k\right)\right). \tag{8.6.37}$$

The results of these computations agree with those in the previous example.

8.7 OBSERVATIONS AND REFERENCES

The particular problems in this chapter are meant to serve as an introduction to the difficulties in applying the finite element method to nonlinear problems. Often the programming becomes very time consuming. For example, the reader should consult Appendixes A.2 and A.3. Students would be wise to choose carefully the type of problems they want or need to program.

The interested reader may consult the papers in Gallagher et al. [12, Chapter 5], where numerous applications of the FEM to fluid flow are discussed. Further study of the Stefan problem may be found in Shamsundar and Sparrow [24], and White [30].

EXERCISES

8-1 Verify (8.1.16).

8-2 Consider problem (8.2.13)–(8.2.17). Describe the Crank–Nicolson method for both finite differences and finite elements.

8-3 Repeat exercise 8-2 for problem two in Section 8.2.

8-4 Consider a nonlinear heat-conduction problem that is interesting to you. If possible, approximate the resulting nonlinear problem by both the nonlinear Gauss–Seidel method and Newton's method.

8-5 Verify line (8.3.6).

8-6 Consider Burger's equation (8.3.1)–(8.3.3) with

$$f(x) = \frac{2\pi\left(\frac{1}{4}\sin \pi x + \sin 2\pi x\right)}{1 + \frac{1}{4}\cos \pi x + \frac{1}{2}\cos 2\pi x} \quad \text{and} \quad \nu = 1.0.$$

Show

$$u(x,t) = \frac{2\pi\left(\frac{1}{4}\exp(-\pi^2 t)\sin \pi x + \exp(-4\pi^2 t)\sin 2\pi x\right)}{1 + \frac{1}{4}\exp(-\pi^2 t)\cos \pi x + \frac{1}{2}\exp(-4\pi^2 t)\cos 2\pi x}$$

is a solution.

8-7 Use the explicit finite difference method to solve (8.3.1)–(8.3.3) with the data in exercise 8-6. Observe stability.

8-8 Use the implicit finite element method to solve (8.3.1)–(8.3.3) with the data in exercise 8-6. Use Newton's method as indicated in (8.3.13), (8.3.14) to solve the resulting nonlinear equations. Observe stability.

8-9 Consider the Stefan problem given by lines (8.6.16)–(8.6.19) where $\rho_i = c_i = k_i = 1.0$, $i = 1, 2$, $u_f = 1.0$, $\psi_0 = 2.0$, $\bar{u}_0 = 0.1$, $\bar{u}_l = 2.0$ and $f \neq 0.0$.

(a) Let $f = 2.0$ and use an implicit time discretization, the FEM, and algorithm (8.6.35), (8.6.36) to approximate the solution.

(b) Let $f = 2.0$ and change the boundary condition (8.6.19) to $\beta(E(l, t))_x = 0.0$. Apply the same method as in part (a). Can you find a mushy region? What happens when f changes?

9

VARIATIONAL INEQUALITIES

Variational inequalities (VI) may be considered as a generalization of certain differential equations. They often evolve from problems that have discontinuous nonlinear forcing terms and free boundaries. In this chapter we shall consider the obstacle, the porous medium, and the one-phase Stefan problems. As we shall see, the inequalities may be derived from an energy formulation of these problems. The essential difference between these problems and the standard differential equations is that the minimization of the energy integral is over a convex set of functions where this set is not linear. Consequently, the weak formulations are no longer equalities!

Section 9.1 contains a discussion of an ideal string with an obstacle against it. This will serve as motivation for the general definition of an elliptic variational inequality (EVI), which is presented in Section 9.2. The discretization of an EVI and a Gauss–Seidel type of minimization algorithm will be given in Section 9.3. Section 9.4 contains a two-space-variable example that describes fluid flow in a porous medium. Section 9.5 contains an introduction to parabolic variational inequalities (PVI), and an application is given to the one-phase Stefan problem.

9.1 OBSTACLE PROBLEM ON A STRING—A MOTIVATING EXAMPLE

In Chapters 1 and 5 we discussed some differential equations that have an equivalent energy formulation. For example, a steady-state ideal string problem is

$$-(Tu_x)_x = f \quad \text{and} \quad u(0) = 0 = u(1), \tag{9.1.1}$$

where u is the small deflection of the string, T the tension, and f the pressure. This has an equivalent energy formulation

$$X(u) = \min X(v), \tag{9.1.2}$$

where the minimization is over the linear set $v \in H_0^1(0, 1)$, and

$$X(v) \equiv \frac{1}{2}\int_0^1 (Tu_x^2 - 2uf),$$

is the potential energy of the string. Moreover, (9.1.1) and (9.1.2) are equivalent to the variational *equality*, which we called the weak formulation, for all $\psi \in H_0^1(0, 1)$:

$$a(u, \psi) = (f, \psi),$$

where

$$a(u, \psi) \equiv \int_0^1 Tu_x\psi_x,$$

$$(f, \psi) \equiv \int_0^1 f\psi. \tag{9.1.3}$$

If a convex obstacle is pressed against the string, then there will be free boundaries $x = s$ and $x = S$ separating the string and the obstacle (see Figure 9.1.1). By demanding continuity of u and u_x at $x = s$ and $x = S$, we may find the exact solution. The two-space-variable problem is more complicated and requires a numerical solution. In order to investigate a possible numerical method, we shall restrict the discussion to the following string problem.

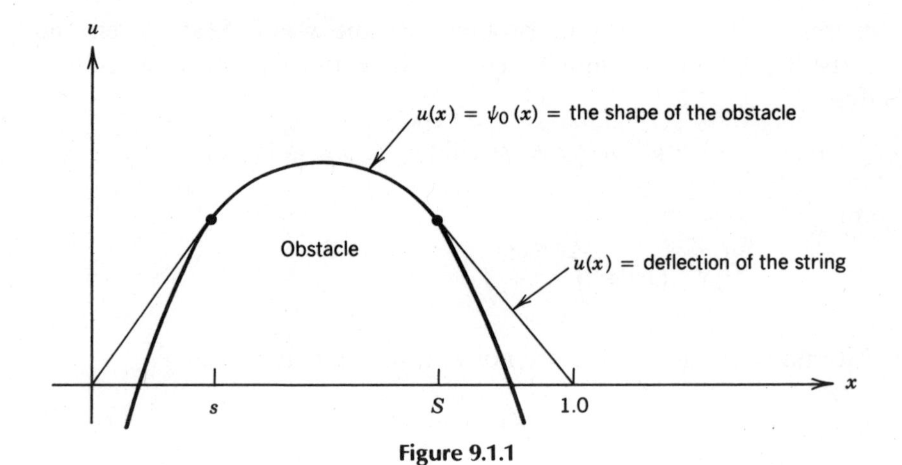

Figure 9.1.1

Classical Formulation of the Obstacle on a String. Let $\psi_0(x) \in C^2[0,1]$ represent the obstacle and assume $\psi_{0xx} \leq 0$.

$$u = \psi_0 \quad \text{on } [s, S], \tag{9.1.4}$$

$$-u_{xx} = 0 \quad \text{on } [0,1] \setminus [s, S], \tag{9.1.5}$$

$$-u_{xx} \geq 0 \quad \text{on } [0,1], \tag{9.1.6}$$

$$u \geq \psi_0 \quad \text{on } [0,1], \tag{9.1.7}$$

$$u, u_x \text{ are continuous at } x = s \text{ and } S, \tag{9.1.8}$$

$$u(0) = 0 = u(1). \tag{9.1.9}$$

Note that conditions (9.1.4) and (9.1.5) may be written $-u_{xx} = F(u)$, where $F(u)$ is a discontinuous nonincreasing function of the solution

$$F(u) = \begin{cases} -\psi_{0xx}, & u = \psi_0 \\ 0.0, & u > \psi_0. \end{cases}$$

A related property is that these problems may be viewed as energy-minimization problems over a certain convex set of functions. For example, in the preceding problem assume s and S are given and consider $X^{\sim}(u^{\sim}) = \min X^{\sim}(v^{\sim})$, where the minimization is over the set

$$K^{\sim} \equiv \{v \in H_0^1(0,1) | v = \psi_0 \text{ on } [s, S]\}$$

and

$$X^{\sim}(v) \equiv \int_0^s v_x^2 + \int_S^1 v_x^2.$$

Alternatively, if s and S are not known, we may consider

$$X(u) = \min X(v) \tag{9.1.10}$$

where the minimization is over the set

$$K \equiv \{v \in H_0^1(0,1) | v \geq \psi_0\}$$

and

$$X(v) \equiv \frac{1}{2}\int_0^1 v_x^2.$$

One might expect the solution of the classical problem to be the solution of the minimization problem (9.1.10). Indeed, this is true. In order to prove this, we shall need to note that K is convex and closed in $H_0^1(0,1)$. K being *convex* means that whenever $u, v \in K$, then for all $\lambda \in [0,1]$, $\lambda u + (1-\lambda)v \in K$. K is *closed* in $H_0^1(0,1)$ means that whenever $u^n \in K$ and u^n converges to $u \in H_0^1$, then $u \in K$. See exercise 5-14 in Chapter 5.

Theorem 9.1.1. Consider the model problem (9.1.4)–(9.1.9) and let K be as in (9.1.10).

(a) $u \in K$ satisfies (9.1.10) (u is called an *energy* solution) if and only if for all $\psi \in K$, $\int_0^1 u_x(\psi_x - u_x) \geq 0$ (u is called a *variational inequality* solution).

(b) A classical solution of (9.1.4)–(9.1.9) is a variational inequality solution.

(c) The variational inequality solution is unique.

Proof. **(a)** Let (9.1.10) hold. Since K is convex, for all $\lambda \in [0,1]$, $(1-\lambda)u + \lambda\psi = u + \lambda(\psi - u) \in K$. Choose $v = u + \lambda(\psi - u)$ in (9.1.10):

$$
\begin{aligned}
X(u) &\le X(u + \lambda(\psi - u)) \\
&= \frac{1}{2}\int_0^1 \left(u_x^2 + \lambda 2u_x(\psi - u)_x + \lambda^2(\psi - u)_x^2\right) \\
&= X(u) + \lambda\int_0^1 u_x(\psi_x - u_x) + \lambda^2 X(\psi - u). \qquad (9.1.11)
\end{aligned}
$$

So, by subtracting $X(u)$ from both sides of (9.1.11) we obtain

$$
0 \le \lambda\left[\int_0^1 u_x(\psi_x - u_x) + \lambda X(\psi - u)\right]. \qquad (9.1.12)
$$

If $\int_0^1 u_x(\psi_x - u_x)$ were negative, then for some small $\lambda > 0$ we would contradict (9.1.12). Therefore, u must satisfy the variational inequality.

Let the variational inequality hold. Then (9.1.12) holds and so (9.1.11) holds for all $\psi \in K$ and $\lambda \in [0,1]$. Choose $\lambda = 1$ and note $u + \lambda(\psi - u) = \psi$. Then (9.1.11) gives $X(u) \le X(\psi)$ for any $\psi \in K$. This shows that (9.1.10) holds.

(b) Let u satisfy (9.1.4)–(9.1.9) and consider

$$
\begin{aligned}
\int_0^1 u_x(\psi - u)_x &= \int_0^s u_x(\psi - u)_x \\
&\quad + \int_s^S u_x(\psi - u)_x + \int_S^1 u_x(\psi - u)_x \\
&= u_x(\psi - u)|_0^s - \int_0^s u_{xx}(\psi - u) \\
&\quad + u_x(\psi - u)|_s^S - \int_s^S u_{xx}(\psi - u) \\
&\quad + u_x(\psi - u)|_S^1 - \int_S^1 u_{xx}(\psi - u). \qquad (9.1.13)
\end{aligned}
$$

Since $u = \psi = 0$ at $x = 0$ and 1, the evaluations at $x = 0$ and 1 are zero. Also, by the continuity of ψ, u and u_x at $x = s$ and S, the evaluations at $x = s$ and S are zero. On $[0, s]$ and $[s, S]$, $u_{xx} = 0$, and therefore (9.1.13) gives

$$\int_0^1 u_x(\psi - u)_x = -\int_s^S u_{xx}(\psi - u)$$

$$= -\int_s^S \psi_{0xx}(\psi - \psi_0) \quad \text{by (9.1.4)}.$$

Since $\psi \in K$, $\psi \geq \psi_0$. By the assumption on ψ_0, $-\psi_{0xx} \geq 0$. Thus, u satisfies the variational inequality.

(c) Let u and v satisfy the variational inequality

$$\int u_x(\psi - u)_x \geq 0, \tag{9.1.14}$$

$$\int v_x(\psi - v)_x \geq 0. \tag{9.1.15}$$

In (9.1.14) choose $\psi = v$ and in (9.1.15) choose $\psi = u$. Then add the two inequalities to obtain

$$\int u_x(v - u)_x + \int v_x(u - v)_x \geq 0. \tag{9.1.16}$$

Since the left-hand side of (9.1.16) is $-\int (u - v)_x^2$, $(u - v)_x = 0$ and so $u - v = \text{const}$. Because $(u - v)(0) = 0$, the constant is zero.

The variational inequality in Theorem 9.1.1 may be written $a(u, \psi - u) \geq 0$ for all $\psi \in K$ and $a(u, v) \equiv \int u_x v_x$. This should be contrasted with the variational equality in (9.1.3) for the problem $-(Tu_x)_x = f$. If $T = 1$ and $f = 0$, then (9.1.3) is $a(u, \psi) = 0$. If K is the linear set $H_0^1(0, 1)$, then the variational inequality will imply the variational equality. In order to establish this, simply choose $\psi = \psi^{\sim} + u$ and $\psi = -\psi^{\sim} + u$. This gives $a(u, \psi^{\sim}) = 0$ for all choices of $\psi^{\sim} \in H_0^1(0, 1)$.

In Section 9.2 we introduce the general definition of an elliptic variational inequality. A numerical example is given in Section 9.3, and it describes an obstacle on a string with a nonzero forcing term. The reader may find it interesting to look at this example where the exact solution is given.

9.2 ELLIPTIC VARIATIONAL INEQUALITIES

The results of this section will be stated in terms of a Hilbert space H. The reader may choose to let H equal $H^1(0,1)$ or $H_0^1(0,1)$. A *Hilbert* space is defined to be a vector space with an inner product, $(u, v)_H$, and with the Cauchy property. The latter means that convergence of a sequence u^n to u in H may be established by showing that for all $\epsilon > 0$, there is an N such that if $n, m > N$, then

$$\|u^n - u^m\| \equiv (u^n - u^m, u^n - u^m)_H^{1/2} < \epsilon.$$

This criterion is useful because it does not require the proposed limit u. This has been used in the existence proofs of Theorem 5.7.1 and Theorem 7.2.1.

The EVI will be formulated in terms of the following:

$$a\colon H \times H \to \mathbb{R} \quad \text{with } a(u,v) \in \mathbb{R},$$

$$f\colon H \to \mathbb{R} \quad \text{with } (f,u) \in \mathbb{R},$$

$$K \subset H.$$

We shall need the following assumption on $a(u, v)$, (f, u), and K. For the special case in Chapter 5, these conditions were studied.

Assumptions

1. a is *symmetric*, that is, $a(u, v) = a(v, u)$.
2. a is *bilinear*, that is, for all α and β in $\mathbb{R}$ $a(u, \alpha v + \beta w) = \alpha a(u, v) + \beta a(u, w)$ and similarly in the first variable.
3. a is *coercive*, that is, there is a constant c_1 such that

$$c_1\|u\|^2 \le a(u,u).$$

4. a is *continuous*, that is, there is a constant c_2 such that

$$|a(u,v)| \leq c_2\|u\|\,\|v\|.$$

5. f is *linear*, that is, for all α and β in $\mathbb{R}$

$$(f, \alpha u + \beta v) = \alpha(f,u) + \beta(f,v).$$

6. f is *continuous*, that is, there is a constant c_3 such that

$$|(f,u)| \leq c_3\|u\|.$$

7. K is *convex* in H, that is, for all λ in $[0,1]$ and $u, v \in K$, $\lambda u + (1-\lambda)v \in K$.
8. K is *closed* in H, that is, when $u^n \in K$ and u^n converges to u in H, then $u \in K$.

Example. As results of Chapter 5, the following satisfy the assumptions 1–8:

$$H = H_0^1(0,1),$$

$$K = \{u \in H | u \geq \psi_0 \in C^2[0,1]\},$$

$$a(u,v) = \int_0^1 u_x v_x,$$

$$(f,u) = \int gu,$$

where $g \in L_2(0,1)$.

Definition. Let H be a Hilbert space, and let $a(u,v)$, (f,u), and K be as above. An *elliptic variational inequality* is

$$a(u, \psi - u) \geq (f, \psi - u). \tag{9.2.1}$$

The solution is a $u \in K$ such that (9.2.1) holds for all $\psi \in K$.

The following result was proved by Stampacchia [27]. Since the proof is similar to the discussions in Chapter 5 and in Section 9.1, we give only a sketch of the proof.

Theorem 9.2.1. Let H be a Hilbert space and $a(u, v)$, (f, v), and K be as above. If assumptions 1–8 hold, then the following are true:

(a) The EVI is equivalent to the minimization problem

$$X(u) = \min X(v) \tag{9.2.2}$$

where the $v \in K$ and $X(v) \equiv \frac{1}{2}(a(v, v) - 2(f, v))$.

(b) The solution of the EVI is unique.

(c) The solution of the EVI exists.

Sketch of the Proof. The proofs of (a) and (b) are similar to those given in the proof of Theorem 9.1.1; the only additional complication comes from the nonzero (f, u) term. The existence proof follows the same pattern as the existence result in Theorem 5.7.1. In Theorem 5.7.1 the K equaled H and was linear; the interested reader should repeat the proof with K being convex and closed in H.

Remark. If K is linear, then replace ψ with $\psi + u$ and $\psi - u$ and conclude $a(u, \psi) = (f, \psi)$. Thus, the EVI may be considered as a generalization of certain elliptic partial differential equations.

9.3 THE DISCRETE PROBLEM AND AN ALGORITHM

A numerical solution of the EVI in (9.2.1) may be obtained from its equivalent minimization problem (9.2.2). In order to illustrate this, we restrict H and K to

$$H = H_0^1(0, 1),$$

$$K = \{v \in H | v \geq \psi_0\}$$

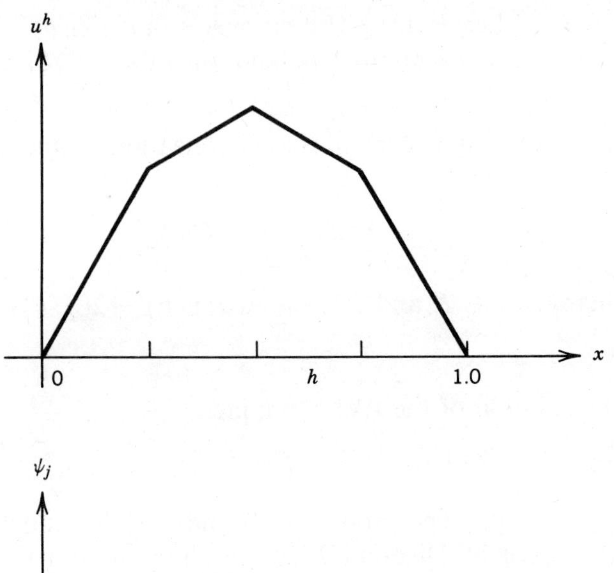

Figure 9.3.1

and approximate H and K by

$$S^h = \left\{ u^h \in H | u^h = \sum_{j=1}^{N-1} u_j \psi_j, \quad \psi_j \text{ are linear shape functions} \right\},$$

$$K^h = \left\{ v^h \in S^h | v_j \geq \psi_{0j} = \psi_0(jh) \text{ where } h = 1/N \right\}.$$

The functions u^h and ψ_j are illustrated in Figure 9.3.1.

The discrete analog of problem (9.2.2) is

$$X(u^h) = \min X(v^h), \tag{9.3.1}$$

where the minimization is over K^h. $X(v^h)$ may be written in matrix form as given below, where all summations are from 1 to $N-1$.

$$\begin{aligned}
X(v^h) &= \tfrac{1}{2}a(v^h, v^h) - (f, v^h) \\
&= \tfrac{1}{2}a\Big(\sum_j v_j\psi_j, \sum_i v_i\psi_i\Big) - \Big(f, \sum_i v_i\psi_i\Big) \\
&= \tfrac{1}{2}\sum_i v_i \sum_j a(\psi_i, \psi_j)v_j - \sum_i v_i(f, \psi_i) \\
&= \tfrac{1}{2}v^{\mathrm{T}}Av - v^{\mathrm{T}}d,
\end{aligned}$$

where $A = (a(\psi_i, \psi_j))$ is an $(N-1)\times(N-1)$ matrix, $v = (v_i) \in \mathbb{R}^{N-1}$, $d = ((f, \psi_i)) \in \mathbb{R}^{N-1}$. Thus, (9.3.1) may be written

$$X(u^h) = \min_{v_i \geq \psi_{0j}} \big(\tfrac{1}{2}v^{\mathrm{T}}Av - v^{\mathrm{T}}d\big). \tag{9.3.2}$$

Since $a(u, v)$ is symmetric, the matrix A is symmetric. As $a(u, v)$ is coercive, A is positive definite. Recall that for such matrices the linear algebraic problem $Au = d$ is equivalent to the unconstrainted minimization problem

$$X(u) = \min_{v \in \mathbb{R}^{N-1}} \big(\tfrac{1}{2}v^{\mathrm{T}}Av - v^{\mathrm{T}}d\big). \tag{9.3.3}$$

Moreover, the solution of $Au = d$ may be approximated by the Gauss–Seidel–SOR algorithm

$$u_i^{n+1/2} = \Big(d_i - \sum_{j<i} a_{ij}u_j^{n+1} - \sum_{j>i} a_{ij}u_j^n\Big)\Big/a_{ii},$$

$$u_i^{n+1} = (1-\omega)u_i^n + \omega u_i^{n+1/2}, \quad \text{where } 1 \leq \omega < 2 \text{ and } A = (a_{ij}). \tag{9.3.4}$$

The only difference between the minimization problems (9.3.2) and (9.3.3) is that (9.3.2) requires $v_j \geq \psi_{0j}$. This suggests the following modification of algorithm (9.3.4) for problem (9.3.2).

Algorithm for Problem (9.3.2). Let $u_i^{n+1/2}$ be as in (9.3.4). Then

$$u_i^{n+1} = \max\left(\psi_{0i}, (1-\omega)u_i^n + \omega u_i^{n+1/2}\right). \tag{9.3.5}$$

The proof of the following theorem may be found in Céa [7].

Theorem 9.3.1. Consider the minimization problem (9.3.2) and assume A is symmetric and positive definite. If $1 \leq \omega < 2$, then the algorithm (9.3.5) converges to the unique solution of (9.3.2).

Example. Consider the classical obstacle-on-a-string problem (9.1.4)–(9.1.9). Assume there is a pressure function g, so that (9.1.5) and (9.1.6) are replaced by

$$-u_{xx} = g \quad \text{on } [0,1]\setminus[s,S], \tag{9.1.5a}$$

$$-u_{xx} \geq g \quad \text{on } [0,1]. \tag{9.1.6a}$$

Lines (9.1.4) and (9.1.6a) will hold if $-\psi_{0xx} \geq g$. One can show that a classical solution will satisfy the EVI (9.2.1), where H, K, $a(u,v)$, and (f,u) are as in the example preceding (9.2.1). In this case algorithm (9.3.5) has the form

$$u_i^{n+1/2} = \left(D_i + u_{i-1}^{n+1} + u_{i+1}^n\right)/2,$$

$$u_i^{n+1} = \max\left(\psi_{0i}, (1-\omega)u_i^n + \omega u_i^{n+1/2}\right),$$

where

$$D_i = (h^2/6)(g_{i+1} + 4g_i + g_{i-1}) \cong h(f, \psi_i),$$

$$\psi_i = \text{linear shape function},$$

$$i = 1, \ldots, N-1,$$

$$h = 1/N,$$

$$1 \leq \omega < 2.$$

In the numerical experiments we let $g = -16.0$ and

$$\psi_0(x) = -8\left(x - \tfrac{1}{2}\right)^2 + B,$$

where $\psi_0(0) = -2 + B < 0$. When $B = 1.0$, the exact solution is symmetric about $x = \frac{1}{2}$ and is given by

$$u(x) = \begin{cases} 8x^2, & 0 \le x \le s = \frac{1}{4} \\ \psi_0(x), & s < x \le \frac{1}{2}. \end{cases}$$

Convergence is defined by the relative error at each node being less than 0.0001. The initial guess is $u(x) = 0.001$. When $N = 20$ and $\omega = 1.3$, convergence is reached in 12 iterations of the preceding algorithm. If B is decreased to $B = 0.5$, then s increases to $(3/2)^{1/2}/4$, and the convergence is obtained in 16 iterations (see exercises 9-4 and 9-5).

9.4 FLUID FLOW IN A POROUS MEDIUM

Fluid flow through a porous medium can be modeled by the diffusion equation. In this case Darcey's law is the analog of the Fourier's law for heat conduction. In this section we restrict the discussion to steady-state problems in which the porous medium is either saturated with moisture or completely dry. Thus, as in the Stefan problem, there is a free boundary or interface that separates the saturated and the dry portions of the medium. We mention two examples.

Earthen Dam. It is hoped that in an earthen dam we have a steady-state interface that separates the saturated and dry regions (see Figure 9.4.1).

Cylindrical Water Filter. Consider a water filter that purifies water by letting it flow from the exterior to the interior of a cylindrical shell. The cross section is indicated in Figure 9.4.2.

Problems of this nature can be described by EVIs. In order to develop this, one must use a change of dependent variable given by the Baiocchi function. This will give an elliptic partial differential equation with a discontinuous nonlinear forcing function. Before defining the Baiocchi function, we shall state the classical form and then the weak form of the water filter problem. We closely follow the

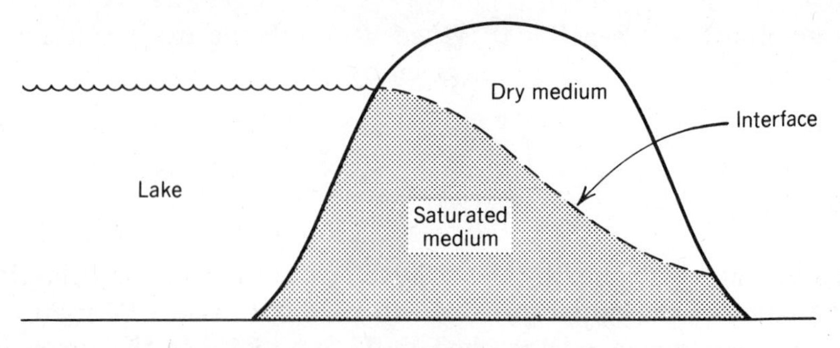

Figure 9.4.1

discussion given in Cryer and Fetter [8]. Appendix A.4 contains a computer code which further describes the following problem.

Classical Formulation of the Water-Filter Problem. Let a cross section of a cylindrical water filter be labeled as in Figure 9.4.3. We have used the following notations:

x	radial variable
y	vertical variable
D	cross-section
Ω	saturated medium
Γ_0	interface
$u =$	$y + P/(\rho g)$ hydraulic head
P	pressure
ρ	density
g	gravitation constant
k	porosity/(kinematic viscosity)

In a porous medium, Darcey's law states that the velocity of the fluid is proportional to the gradient of the pressure. Since the hydraulic head equals $y + P/\rho g$, we may write the velocity in terms of the gradient of the hydraulic head. The steady-state assumption, Darcey's law, and the mass continuity equation give equation (9.4.1). This and the boundary conditions form the classical model of this problem.

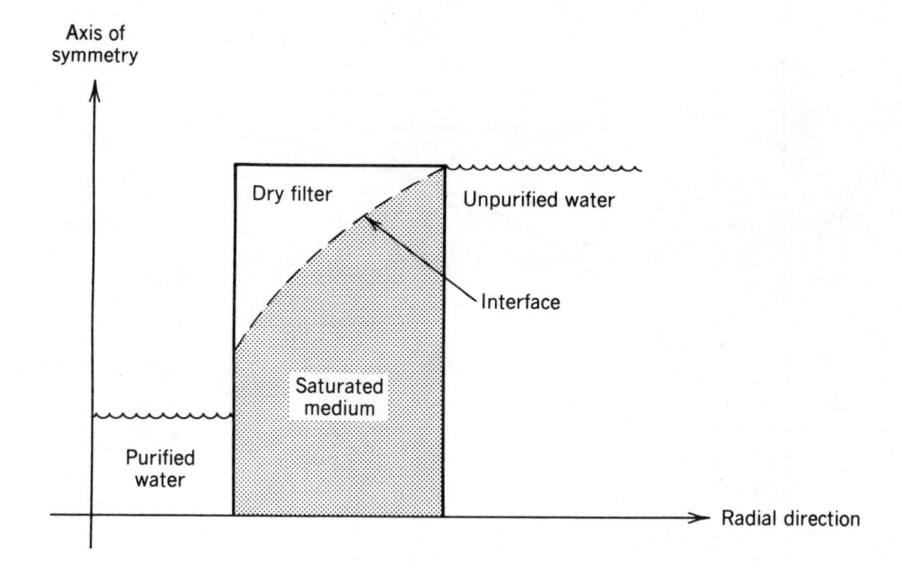

Figure 9.4.2

Classical Formulation of Fluid Flow in a Porous Medium

$$(kxu_x)_x + (kxu_y)_y = 0 \qquad \text{on } \Omega, \tag{9.4.1}$$

$$u = 0 \qquad \text{on } D \setminus \Omega, \tag{9.4.2}$$

$$u = \psi_0(x) \qquad \text{on } \Gamma_0, \tag{9.4.3}$$

$$(d/d\nu)u = 0 \qquad \text{on } \Gamma_0, \tag{9.4.4}$$

$$(d/d\nu)u \equiv (kxu_x, kxu_y) \cdot \mathbf{n},$$

where $\mathbf{n}$ = unit outward normal,

$$u = H \qquad \text{on } \Gamma_1, \tag{9.4.5}$$

$$u = h_w \qquad \text{on } \Gamma_2, \tag{9.4.6}$$

$$u = y \qquad \text{on } \Gamma_3', \tag{9.4.7}$$

$$(d/d\nu)u = 0 \qquad \text{on } \Gamma_4. \tag{9.4.8}$$

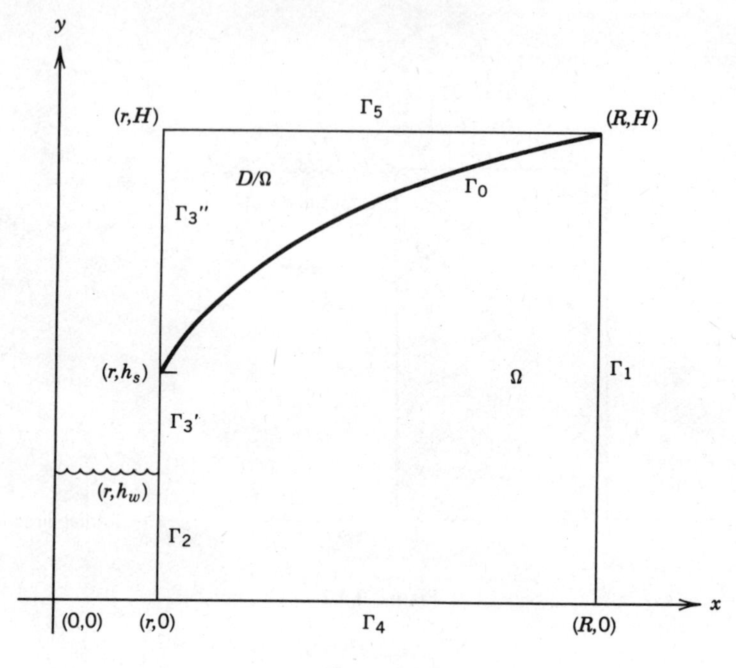

Figure 9.4.3

In order to define the weak version, we define the following space of functions:

$$H = H^1(D) \equiv \{u \in L_2(D) | u_x, u_y \in L_2(D)\}.$$

The derivatives u_x and u_y are viewed as weak derivatives, which are defined as in Chapter 5. This space of functions has an inner product

$$(u, v)_H \equiv \int_D (uv + u_x v_x + u_y v_y).$$

As shown in Friedman [11], these functions have the following properties:

1. H is a Hilbert space.
2. If C is a smooth curve in D, then the restriction or trace of u to C is defined.
3. Green's theorem holds for these functions.

We may derive the weak form of (9.4.1)–(9.4.8) by multiplying (9.4.1) by ψ, where $\psi = 0$ on $\Gamma_1 \cup \Gamma_2 \cup \Gamma_3'$, and then using Green's theorem.

Definition. Consider the classical problem (9.4.1)–(9.4.8). $u \in H^1(D)$ is a *weak solution* if and only if for all $\psi \in H^1(D)$ with $\psi = 0$ on $\Gamma_1 \cup \Gamma_2 \cup \Gamma_3'$

$$\int_\Omega (kxu_x\psi_x + kxu_y\psi_y) = 0, \tag{9.4.9}$$

and lines (9.4.2), (9.4.3) and (9.4.5)–(9.4.7) hold.

The Baiocchi function is defined in such a way as to give the properties of Proposition 9.4.1. Note u is discontinuous on Γ_0, but the Baiocchi function w is continuous on Γ_0.

Definition. Let u be a weak solution of (9.4.1)–(9.4.8). The *Baiocchi function* is

$$w \equiv \begin{cases} \displaystyle\int_y^{\psi_0(x)} (u(x,t) - t)\,dt & \text{on } \Omega \\ 0 & \text{on } D \setminus \Omega. \end{cases}$$

Proposition 9.4.1. Let u be a weak solution of (9.4.1)–(9.4.8), and let w be the Biaocchi function defined from u. Then w has the following properties:

(a) $P = \rho g(u - y) \geq 0$ and so $w \geq 0$.

(b) Let $Lw \equiv -((kxw_x)_x + (kxw_y)_y)$, and then

$$Lw = \begin{cases} -kx & \text{on } \Omega \\ 0 & \text{on } D \setminus \Omega. \end{cases}$$

(c) $w = w_x = w_y = (d/d\nu)w = 0$ on Γ_0.

(d) $w(x, H) = 0$ on Γ_5.

(e) $w(R, y) = H^2/2 - Hy + y^2/2$ on Γ_1.

(f) $w(r, y) = h_w^2/2 - h_w y + y^2/2$ on Γ_2.

(g) $w(r, y) = 0$ on $\Gamma_3' \cup \Gamma_3''$.

(h) $w(x,0) = w(r,0) + (w(R,0) - w(r,0))F(x)/F(R)$ on Γ_4, where $F(x) = \ln(x/r)$.

Proof. **(a).** Let Lw be as in part (b) and note

$$\begin{aligned} L(u-y) &= Lu = 0 && \text{on } \Omega, \\ u - y &\geq 0 && \text{on } \partial\Omega \setminus \Gamma_4, \\ (d/d\nu)(u-y) &= -kx(u-y)_y = kx \geq 0 && \text{on } \Gamma_4. \end{aligned}$$

Therefore, by the maximum principle (see Protter and Weinberger [16]) we have $u - y \geq 0$ on Ω, and so $w \geq 0$ on D.

(b). The following are a result of the definition of the Baiocchi function.

$$w = \int_y^{\psi_0(x)} (u(x,t) - t)\, dt$$

$$w_x = (u(x,\psi_0(x)) - \psi_0(x))\psi_{0x}(x) + \int_y^{\psi_0(x)} u_x(x,t)\, dt$$

$$= \int_y^{\psi_0(x)} u_x(x,t)\, dt \quad \text{by (9.4.3)}, \tag{9.4.10}$$

$$w_y = -u(x,y) + y, \tag{9.4.11}$$

$$\begin{aligned} Lw &= -(kxw_x)_x - (kxw_y)_y \\ &= -\left(kx\int_y^{\psi_0(x)} u_x(x,t)\, dt\right)_x - (kx(-u(x,y) + y))_y \\ &= -kx\left(u_x(x,\psi_0(x))\psi_{0x}(x) + \int_y^{\psi_0(x)} u_{xx}(x,t)\, dt\right) \\ &\quad -k\int_y^{\psi_0(x)} u_x(x,t)\, dt + kxu_y(x,y) - kx \\ &= -\int_y^{\psi_0(x)} \left((kxu_x)_x + (kxu_y)_y\right) dt \\ &\quad + kxu_y(x,\psi_0(x)) - kxu_x(x,\psi_0(x))\psi_{0x}(x) - kx. \end{aligned} \tag{9.4.12}$$

Since (9.4.9), or equivalently (9.4.1), holds, the first term on the right-hand side of (9.4.12) is zero. The sum of the next two terms is $((\psi_0')^2 + 1)^{1/2}((d/d\nu)u)$ on Γ_0, and by (9.4.4) it is also zero. Thus, $Lw = -kx$ on Ω.

(c)–(h). The proof of (c) follows from the definition of w and lines (9.4.10) and (9.4.11). The proofs of (d)–(g) are just substitutions into the definition of w. The formula in (h) for $w(x,0)$ is derived from the solution of the ordinary differential equation $(kxw_x)_x = 0$.

The equation in part (b) of Proposition 9.4.1 may be viewed as an elliptic partial differential equation with a discontinuous nonlinear forcing term

$$Lw = F(w) = \begin{cases} -kx, & w > 0 \\ 0, & w \leq 0. \end{cases}$$

This suggests that this problem may be reformulated as an EVI. In order to do this, note that the boundary conditions for w as a function on D depend only on the physical parameters (see parts (e)–(h) of Proposition 9.4.1). Let $w^{\sim}$ be a function in $H^1(D)$ such that on the boundary of D it satisfies these conditions. For this problem, we choose the following:

$$K \equiv \{\psi \in H^1(D) | \psi = w^{\sim} \text{ on } \partial D, \psi \geq 0\},$$

$$a(w, \psi) \equiv \int_D kx(w_x\psi_x + w_y\psi_y),$$

$$(f, \psi) \equiv \int_D -kx\psi.$$

Proposition 9.4.2. Let u be a weak solution of (9.4.1)–(9.4.8), and w be the Baiocchi function defined from u. Then $w \in K$ and for all $\psi \in K$, w satisfies the EVI

$$a(w, \psi - w) - (f, \psi - w) \geq 0. \tag{9.4.13}$$

Proof.

$$a(w, \psi - w) - (f, \psi - w) = \int_D kx\big(w_x(\psi - w)_x + w_y(w - \psi)_y\big)$$

$$+ \int_D kx(\psi - w)$$

$$\geq \int_\Omega kx\big(w_x(w - \psi)_x + w_y(w - \psi)_y\big)$$

$$+ \int_\Omega kx(\psi - w),$$

because $w = 0$ on $D \setminus \Omega$ and $\psi \geq 0$ on D. Then, by Green's theorem, $\psi - w = 0$ on $\partial\Omega \setminus \Gamma_0$, and $(d/d\nu)w = 0$ on Γ_0,

$$a(w, \psi - w) - (f, \psi - w) \geq \int_\Omega (Lw + kx)(\psi - w).$$

Then, by part (b) of Proposition 9.4.1,

$$a(w, \psi - w) - (f, \psi - w) \geq \int_\Omega 0 \cdot (\psi - w) = 0.$$

9.5 PARABOLIC VARIATIONAL INEQUALITIES

In order to motivate the definition of a parabolic variational inequality (PVI), we consider a one-phase Stefan problem. A classical version with a sharp solid–liquid interface and one space variable has the following form.

Classical One-Phase Stefan Problem. Let $u(x, t)$ equal the temperature of the liquid phase and assume the physical properties are constant:

ρ density

c specific heat

K thermal conductivity

L latent heat of fusion.

Let $x = s(t)$ be the location of the solid–liquid interface, with $x > s(t)$ being the liquid region and $x \le s(t)$ the solid region, where $u(x, t) = 0$ is the solidifcation temperature. Then

$$\rho c u_t - (K u_x)_x = 0, \qquad x > s(t) \tag{9.5.1}$$

$$u(x,t) = 0, \qquad x \le s(t) \tag{9.5.2}$$

$$u(x,0) = 0, \qquad 0 < x \le 1 \tag{9.5.3}$$

$$u(0,t) = 0 \quad \text{and} \quad u(1,t) = g(t) > 0, \qquad t > 0 \tag{9.5.4}$$

$$K u_x(s(t),t) = -\rho L s'(t), \qquad x = s(t). \tag{9.5.5}$$

As in the porous-medium problem, we use a change of dependent variable. Let $l(x)$ equal the time such that the mass at x changes phase. Thus, $l(s(t)) = t$ and $l_x(s(t))s'(t) = 1$. The *Duvant function*

$$w \equiv \begin{cases} \int_{l(x)}^{t} u(x,\tau)\, d\tau, & t > l(x) \\ 0, & t \le l(x). \end{cases}$$

Then we easily compute

$$w_t(x,t) = u(x,t)$$

$$w_x(x,t) = -u(x,l(x))l_x(x) + \int_{l(x)}^{t} u_x(x,\tau)\,d\tau$$

$$= \int_{l(x)}^{t} u_x(x,\tau)\,d\tau$$

$$(Kw_x)_x = -Ku_x(x,l(x))l_x(x) + \int_{l(x)}^{t} (Ku_x)_x\,d\tau$$

$$= \rho L + \int_{l(x)}^{t} \rho c u_t\,d\tau, \quad \text{by (9.5.), (9.5.5)}$$

$$= \rho L + \rho c(u(x,t) - u(x,l(x)))$$

$$= \rho L + \rho c w_t, \qquad x > s(t).$$

Thus,

$$\rho c w_t - (Kw_x)_x = \begin{cases} -\rho L, & x > s(t) \\ 0, & x \le s(t). \end{cases}$$

This may be written as a time-dependent problem with a discontinuous nonlinear forcing term

$$\rho c w_t - (Kw_x)_x = F(w) \equiv \begin{cases} -\rho L, & w > 0 \\ 0, & w \le 0, \end{cases} \tag{9.5.6}$$

$$w(x,0) = 0 \tag{9.5.7}$$

$$w(0,t) = 0 \quad \text{and} \quad w(1,t) = G(t) \equiv \int_0^t g(\tau)(d\tau). \tag{9.5.8}$$

Line (9.5.6) indicates that we may be able to view this as a time-dependent variational inequality. In order to study this, we shall consider functions of (x,t) such that when t is fixed these functions will be in $H^1(0,1)$ and such that all integrals exist. Since we want the solution to satisfy (9.5.8), we define $K(t)$ as

$$K(t) \equiv \{\psi(x,t) \in H^1(0,1) | \psi(0,t) = 0, \psi(1,t) = G(t), \psi(x,t) \geq 0\}.$$

Let $\psi \in K(t)$, multiply (9.5.6) by $\psi - w$, integrate from $x = 0$ to $x = 1$ and use $w = 0$ on $[0, s]$.

$$\begin{aligned}\int_s^1 - \rho L(\psi - w) &= \rho c \int_0^1 w_t(\psi - w) - \int_s^1 (K w_x)_x (\psi - w) \\ &= \rho c \int_0^1 w_t(\psi - w) - K w_x(\psi - w)|_s^1 \\ &\quad + \int_s^1 K w_x (\psi - w)_x. \end{aligned} \tag{9.5.9}$$

Since $(\psi - w) = 0$ for $x = 0$ and $x = 1$ and since $K w_x$ is continuous at $x = s(t)$, the evaluations in (9.5.9) sum to zero. Since $w = 0$ on $[0,s]$ and $\psi \geq 0$ on $[0,s]$, (9.5.9) gives

$$\rho c \int_0^1 w_t(\psi - w) + \int_0^1 K w_x (\psi - w)_x \geq \int_0^1 - \rho L(\psi - w). \tag{9.5.10}$$

By dividing by ρc and using the notation

$$(u, v) \equiv \int_0^1 uv,$$

$$a(u, v) \equiv \int_0^1 (K/\rho c) u_x v_x,$$

we have $w \in K(t)$, and for all $\psi \in K(t)$,

$$(w_t, \psi - w) + a(w, \psi - w) \geq (-L/c, \psi - w). \quad (9.5.11)$$

Definition. Let $K(t) \subset H$ be such that for each t, $K(t)$ is closed and convex in the Hilbert space H. Let a: $H \times H \to \mathbb{R}$ be symmetric, bilinear, continuous, and coercive. Let f: $H \to \mathbb{R}$ be linear and continuous. Let $w(x, t)$ be such that $w_t \in H$. (The precise conditions on $w(x, t)$ are stated in Elliott and Ockendon [9] and in Céa [7].) $w \in K(t)$ is called a *solution of the PVI* if and only if for almost all $t \in (0, T)$ and for all $\psi \in K(t)$,

$$(w_t, \psi - w) + a(w, \psi - w) \geq (f, \psi - w). \quad (9.5.12)$$

In addition, if $w(x, 0)$ is given, then we say it satisfies the *initial value* PVI problem.

As in a parabolic partial differential equation, we may use an implicit time discretization of the PVI and obtain a sequence of EVIs. In (9.5.12) we make the substitutions

$$w \to w^{k+1} = w(x, (k+1)\Delta t) \in K = K((k+1)\Delta t),$$

$$w_t \to (w^{k+1} - w^k)/\Delta t,$$

$$\psi(x, t) \to \psi \in K.$$

Then (9.5.12) gives the sequence of EVIs

$$a^{\sim}\left(w^{k+1}, \psi - w^{k+1}\right) \geq \left(f^{\sim}, \psi - w^{k+1}\right), \quad (9.5.13)$$

where

$$a^{\sim}(u, v) \equiv (u, v) + \Delta t a(u, v),$$

$$(f^{\sim}, v) \equiv \left(\Delta t f^{k+1} + w^k, v\right).$$

$a^{\sim}$ and $f^{\sim}$ satisfy the assumptions of Theorem 9.2.1 for EVIs. Thus there are unique solutions to the sequence of EVIs in (9.5.13). Furthermore, each solution may be approximated by the finite element method and the algorithm given by (9.3.4) and (9.3.5). We summarize this method.

Implicit Method for PVIs

Step I: Form the EVIs as given by (9.5.13).

Step II: Use the finite element method to form a discrete minimization problem and then use the algorithm (9.3.4), (9.3.5) to approximate its solution.

9.6 OBSERVATIONS AND REFERENCES

A very good general reference on variational inequalities is the book by Elliott and Ockendon [9]. There are additional applications, and other numerical methods are described. A more theoretical text is that by Céa [7]. The report by Cryer and Fetter [8] is also very readable, and as previously noted, Appendix A.4 is essentially taken from this report.

EXERCISES

9-1 Show that the following satisfy the assumptions of Theorem 9.2.1:

$$H = \{u \in H^1(0,1) | u(0) = 0\},$$

$$K = \{u \in H | u(1) = u_1, u \geq \psi_0 \in C[0,1]\},$$

$$a(u,v) = \int_0^1 (m(x) u_x v_x + k(x) uv),$$

where

$$0 < \delta \leq m(x) \leq M_m < \infty \quad \text{and} \quad 0 \leq k(x) \leq M_k < \infty,$$

$$(f,u) = \int_0^1 gu$$

where $g \in L_2(0,1)$.

9-2 Prove Theorem 9.2.1, including all details.

9-3 Consider the discrete problem (9.3.2) in the context of Theorem 9.2.1 where

$$H = \mathbb{R}^{N-1},$$

$$K = \{v \in H | v_j \geq \psi_{0j}\},$$

$$a(u, v) = u^{\mathrm{T}} A v,$$

where A is symmetric and positive definite, and

$$(f, u) = u^{\mathrm{T}} d.$$

When $(u, v)_H = u^{\mathrm{T}} v$, show that the assumptions of Theorem 9.2.1 hold.

9-4 Consider the example problem in Section 9.3 given by (9.1.4), (9.1.5a), (9.1.6a), and (9.1.7)–(9.1.9). Show that its classical solution is also a variational inequality solution. Find the classical solution for $B = 0.5$.

9-5 Consider the problem given in exercise 9-1 where m, k, and g are functions of x. Write a computer program to approximate the solution of the variational inequality and use the algorithm in lines (9.3.4) and (9.3.5). Use the example in exercise 9–4 to debug your program.

9-6 Fill in the details of the proofs of parts (c)–(h) in Proposition 9.4.1.

9-7 Consider the program given in Appendix A.4. Show that all the formulas for the components of the system matrix are correct.

9-8 Consider the one-phase Stefan problem given by (9.5.1)–(9.5.5) with $\rho = c = 1$, $K = 0.001$, variable $L > 0$, and $g(t) = 2$. For the following two methods, use different L and compare the methods.

(a) Enthalpy formulation as given in Section 8.6.

(b) PVI formulation as given in Section 9.5.

APPENDIXES

SOME NONLINEAR PROBLEMS AND THEIR COMPUTER PROGRAMS

A.1 FEMI WRITTEN IN PASCAL: STEADY-STATE HEAT CONDUCTION

This appendix contains a version of FEMI that is written in Pascal. For the most part, we use the same notation as is used in FEMI. However, the input data is obtained by the three procedures (subroutines):

GENXY	generates the (x, y) components
GENBC	generates the boundary conditions
GENNOD	generates the array NOD.

Three sets of computations are done for the steady-state heat-conduction problem

$$-\Delta u = 0$$

with boundary conditions as indicated in Figure A.1.1. The three sets

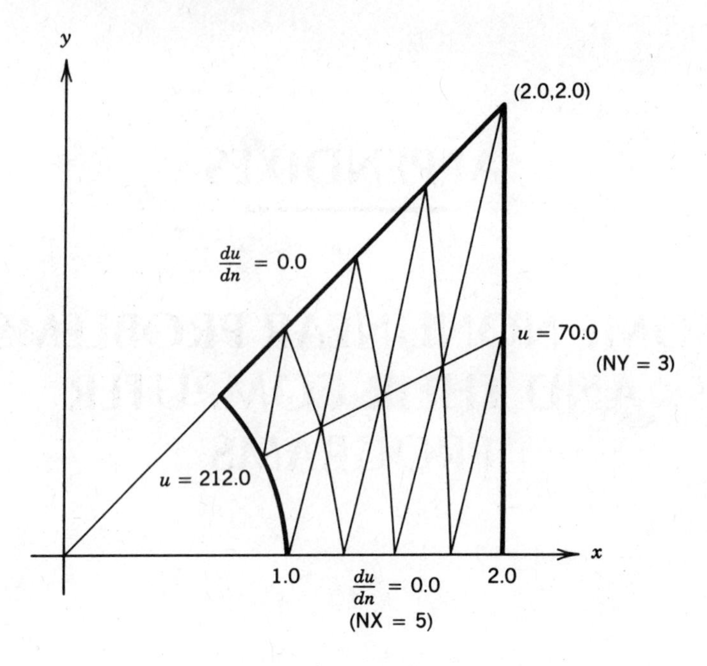

Figure A.1.1

of computations are for (NX = 5, NY = 3), (NX = 5, NY = 5), and (NX = 9, NY = 5). The reader should compare the solution values at the same (x, y) points.

Program FEMI in Pascal

```
PROGRAM FEMI;
(*    THIS PROGRAM IS ALMOST IDENTICAL TO FEMI IN CHAPTER TWO.   *)
(*    HOWEVER, IT IS WRITTEN IN PASCAL, AND THE INPUT DATA IS    *)
(*    GENERATED BY THE SUBROUTINES GENXY, GENBC AND GENNOD.      *)
LABEL
  41,42;
CONST
  N=25;
  NP=26;
  NE=32;
  NPRES=10;
  NX=5;
  NY=5;
TYPE
  ARN=ARRAY[1..N] OF REAL;
  ARSM=ARRAY[1..N,1..NP] OF REAL;
  AR3=ARRAY[1..3] OF REAL;
  ARNOD=ARRAY[1..NE,1..3] OF INTEGER;
  ARBCR=ARRAY[1..NPRES] OF REAL;
  ARBCI=ARRAY[1..NPRES] OF INTEGER;
VAR
  SM:ARSM;
  X,Y,SOL:ARN;
  XX,YY,A,B,C:AR3;
  NOD:ARNOD;
  G1:ARBCR;
  NPT:ARBCI;
  J,ELT,I,LK,LL,ROW,COL,II,JJ,NODE,K:INTEGER;
  AK,DELTA:REAL;
  PR:INTERACTIVE;
(**)
PROCEDURE GENXY(N,NX,NY:INTEGER;VAR X,Y:ARN);
CONST
  A=1.0;
  B=2.0;
  PI=3.1415926;
VAR
  I,J:INTEGER;
  DTHETA,DR,THETA,R:REAL;
BEGIN
  DTHETA:=PI/(4.0*NY-4.0);
  II:=0;
  FOR J:=1 TO NY DO
    BEGIN
      THETA:=(J-1)*DTHETA;
      DR:=(B/COS(THETA)-A)/(NX-1);
      FOR I:=1 TO NX DO
        BEGIN
          R:=A+(I-1)*DR;
          II:=II+1;
          X[II]:=R*COS(THETA);
          Y[II]:=R*SIN(THETA)
        END;
    END;
```

```
END;
(**)
PROCEDURE GENBC(NPRES,NX,NY:INTEGER;VAR NPT:ARBCI;VAR G1:ARBCR);
CONST
  G1INSIDE=212.0;
  G1OUTSIDE=70.0;
VAR
  I:INTEGER;
BEGIN
  FOR I:=1 TO NPRES DO
    BEGIN
      IF I<=NY THEN
        BEGIN
          NPT[I]:=1+NX*(I-1);
          G1[I]:=G1INSIDE
        END;
      IF I>NY THEN
        BEGIN
          NPT[I]:=NX*(I-NY);
          G1[I]:=G1OUTSIDE
        END;
    END;
END;
(**)
PROCEDURE GENNOD(NE,NX,NY:INTEGER;VAR NOD:ARNOD);
VAR
  L,JJ,JJP,K,NYM,NXM:INTEGER;
BEGIN
  L:=-1;
  JJ:=-1;
  NYM:=NY-1;
  NXM:=NX-1;
  K:=NXM;
  FOR I:=1 TO NYM DO
    BEGIN
      K:=K+1;
      L:=L+1;
      FOR J:=1 TO NXM DO
        BEGIN
          K:=K+1;
          L:=L+1;
          JJ:=JJ+2;
          NOD[JJ,1]:=K;
          NOD[JJ,2]:=L;
          NOD[JJ,3]:=K+1;
          JJP:=JJ+1;
          NOD[JJP,1]:=L+1;
          NOD[JJP,2]:=K+1;
          NOD[JJP,3]:=L
        END;
    END;
END;
(**)
PROCEDURE IMAT(N,NP:INTEGER;VAR SOL:ARN;VAR A:ARSM);
```

```
VAR
  I,J,M,L,JJ,K,KP1,NN,IP1:INTEGER;
  BIG,AB,TEMP,QUOT,SUM:REAL;
BEGIN
  M:=N+1;
  L:=N-1;
  FOR K:=1 TO L DO
    BEGIN
      JJ:=K;
      BIG:=ABS(A[K,K]);
      KP1:=K+1;
      FOR I:=KP1 TO N DO
        BEGIN
          AB:=ABS(A[I,K]);
          IF BIG<AB THEN
            BEGIN
              BIG:=AB;
              JJ:=I
            END;
        END;
      IF JJ<>K THEN
        FOR J:=K TO M DO
          BEGIN
            TEMP:=A[JJ,J];
            A[JJ,J]:=A[K,J];
            A[K,J]:=TEMP
          END;
      FOR I:=KP1 TO N DO
        BEGIN
          QUOT:=A[I,K]/A[K,K];
          FOR J:=KP1 TO M DO
            A[I,J]:=A[I,J]-QUOT*A[K,J]
        END;
      FOR I:=KP1 TO N DO
        A[I,K]:=0.0
    END;
  SOL[N]:=A[N,M]/A[N,N];
  FOR NN:=1 TO L DO
    BEGIN
      SUM:=0.0;
      I:=N-NN;
      IP1:=I+1;
      FOR J:=IP1 TO N DO
        SUM:=SUM+A[I,J]*SOL[J];
      SOL[I]:=(A[I,M]-SUM)/A[I,I]
    END;
END;
(**)
BEGIN
  REWRITE(PR,'#5:TEMPO.TEXT');
  FOR I:=1 TO N DO
    BEGIN
      SOL[I]:=0.0;
      FOR J:= 1 TO NP DO
```

```
          SM[I,J]:=0.0
      END;
    (**)
    (*  STEP I:  INPUT DATA  *)
    (**)
    GENNOD(NE,NX,NY,NOD);
    FOR ELT:=1 TO NE DO
      WRITELN(PR,'ELT= ',ELT,'  NOD= ',NOD[ELT,1]:4,NOD[ELT,2]:4,NOD[ELT,3]:4);
    GENXY(N,NX,NY,X,Y);
    FOR I:=1 TO N DO
      WRITELN(PR,'NODE= ',I,'  X= ',X[I]:8:4,'  Y= ',Y[I]:8:4);
    GENBC(NPRES,NX,NY,NPT,G1);
    FOR I:=1 TO NPRES DO
      WRITELN(PR,'PRES NODE= ',I,'  SYSTEM NODE= ',NPT[I]:4,'  G1= ',G1[I]:8:3);
    (**)
    (*  STEP II:  USE THE ENERGY FORMULATION OF THE FEM  *)
    (**)
    (*  STEP III: USE ASSEMBLY BY ELEMENTS  *)
    (**)
     FOR ELT:=1 TO NE DO
       BEGIN
         FOR J:=1 TO 3 DO
           BEGIN
             LK:=NOD[ELT,J];
             XX[J]:=X[LK];
             YY[J]:=Y[LK]
           END;
         FOR J:=1 TO 3 DO
           BEGIN
             LK:=J+1;
             LL:=J+2;
             IF LK<3 THEN
               GOTO 42;
             IF LK=3 THEN
               GOTO 41;
             LK:=1;
             LL:=2;
             GOTO 42;
             41: LL:=1;
             42: A[J]:=XX[LK]*YY[LL]-XX[LL]*YY[LK];
                 B[J]:=YY[LK]-YY[LL];
                 C[J]:=XX[LL]-XX[LK]
           END;
         DELTA:=(C[3]*B[2]-C[2]*B[3])/2.0;
         (*  IN THE NEXT TWO LOOPS THE ASSEMBLY IS DONE  *)
         FOR ROW:=1 TO 3 DO
           FOR COL:=1 TO 3 DO
             BEGIN
               AK:=(B[ROW]*B[COL]+C[ROW]*C[COL])/(4*DELTA);
               II:=NOD[ELT,ROW];
               JJ:=NOD[ELT,COL];
               SM[II,JJ]:=SM[II,JJ]+AK
             END;
       END;
```

```
  (**)
  (*  STEP IV:  THE BOUNDARY CONDITIONS ARE INSERTED  *)
  (**)
  FOR I:=1 TO NPRES DO
    BEGIN
      NODE:=NPT[I];
      FOR K:=1 TO N DO
        SM[NODE,K]:=0.0;
      SM[NODE,NODE]:=1.0;
      SM[NODE,NP]:=G1[I]
    END;
  (**)
  (*  STEP V:  SOLVE THE ALGEBRAIC PROBLEM  *)
  (**)
  IMAT(N,NP,SOL,SM);
  (**)
  (*  STEP VI:  OUTPUT THE SOLUTION IN TABLE FORM  *)
  (**)
  FOR I:=1 TO N DO
    WRITELN(PR,'NODE= ',I,'  SOLUTION= ',SOL[I]:8:3);
CLOSE(PR,LOCK);
END.
(**)
(*    OUTPUT FOR NX = 5 AND NY = 3    *)
(**)
ELT= 1  NOD=    6   1   7
ELT= 2  NOD=    2   7   1
ELT= 3  NOD=    7   2   8
ELT= 4  NOD=    3   8   2
ELT= 5  NOD=    8   3   9
ELT= 6  NOD=    4   9   3
ELT= 7  NOD=    9   4  10
ELT= 8  NOD=    5  10   4
ELT= 9  NOD=   11   6  12
ELT= 10  NOD=    7  12   6
ELT= 11  NOD=   12   7  13
ELT= 12  NOD=    8  13   7
ELT= 13  NOD=   13   8  14
ELT= 14  NOD=    9  14   8
ELT= 15  NOD=   14   9  15
ELT= 16  NOD=   10  15   9
NODE= 1  X=   1.0000  Y=   0.0000
NODE= 2  X=   1.2500  Y=   0.0000
NODE= 3  X=   1.5000  Y=   0.0000
NODE= 4  X=   1.7500  Y=   0.0000
NODE= 5  X=   2.0000  Y=   0.0000
NODE= 6  X=   0.9239  Y=   0.3827
NODE= 7  X=   1.1929  Y=   0.4941
NODE= 8  X=   1.4619  Y=   0.6056
NODE= 9  X=   1.7310  Y=   0.7170
NODE= 10  X=   2.0000  Y=   0.8284
NODE= 11  X=   0.7071  Y=   0.7071
NODE= 12  X=   1.0303  Y=   1.0303
NODE= 13  X=   1.3536  Y=   1.3536
```

```
NODE= 14  X=   1.6768  Y=   1.6768
NODE= 15  X=   2.0000  Y=   2.0000
PRES NODE= 1  SYSTEM NODE=    1  G1=  212.000
PRES NODE= 2  SYSTEM NODE=    6  G1=  212.000
PRES NODE= 3  SYSTEM NODE=   11  G1=  212.000
PRES NODE= 4  SYSTEM NODE=    5  G1=   70.000
PRES NODE= 5  SYSTEM NODE=   10  G1=   70.000
PRES NODE= 6  SYSTEM NODE=   15  G1=   70.000
NODE= 1  SOLUTION=  212.000
NODE= 2  SOLUTION=  168.120
NODE= 3  SOLUTION=  131.582
NODE= 4  SOLUTION=   99.492
NODE= 5  SOLUTION=   70.000
NODE= 6  SOLUTION=  212.000
NODE= 7  SOLUTION=  163.731
NODE= 8  SOLUTION=  125.929
NODE= 9  SOLUTION=   95.258
NODE= 10  SOLUTION=   70.000
NODE= 11  SOLUTION=  212.000
NODE= 12  SOLUTION=  149.463
NODE= 13  SOLUTION=  108.966
NODE= 14  SOLUTION=   83.606
NODE= 15  SOLUTION=   70.000
(**)
(*    OUTPUT FOR NX = 5 AND NY = 5    *)
(**)
NODE= 1  X=   1.0000  Y=   0.0000
NODE= 2  X=   1.2500  Y=   0.0000
NODE= 3  X=   1.5000  Y=   0.0000
NODE= 4  X=   1.7500  Y=   0.0000
NODE= 5  X=   2.0000  Y=   0.0000
NODE= 6  X=   0.9808  Y=   0.1951
NODE= 7  X=   1.2356  Y=   0.2458
NODE= 8  X=   1.4904  Y=   0.2965
NODE= 9  X=   1.7452  Y=   0.3471
NODE= 10  X=   2.0000  Y=   0.3978
NODE= 11  X=   0.9239  Y=   0.3827
NODE= 12  X=   1.1929  Y=   0.4941
NODE= 13  X=   1.4619  Y=   0.6056
NODE= 14  X=   1.7310  Y=   0.7170
NODE= 15  X=   2.0000  Y=   0.8284
NODE= 16  X=   0.8315  Y=   0.5556
NODE= 17  X=   1.1236  Y=   0.7508
NODE= 18  X=   1.4157  Y=   0.9460
NODE= 19  X=   1.7079  Y=   1.1412
NODE= 20  X=   2.0000  Y=   1.3364
NODE= 21  X=   0.7071  Y=   0.7071
NODE= 22  X=   1.0303  Y=   1.0303
NODE= 23  X=   1.3536  Y=   1.3536
NODE= 24  X=   1.6768  Y=   1.6768
NODE= 25  X=   2.0000  Y=   2.0000
NODE= 1  SOLUTION=  212.000
NODE= 2  SOLUTION=  168.710
NODE= 3  SOLUTION=  132.422
```

```
NODE= 4   SOLUTION=  100.189
NODE= 5   SOLUTION=   70.000
NODE= 6   SOLUTION=  212.000
NODE= 7   SOLUTION=  167.659
NODE= 8   SOLUTION=  131.008
NODE= 9   SOLUTION=   99.068
NODE= 10   SOLUTION=   70.000
NODE= 11   SOLUTION=  212.000
NODE= 12   SOLUTION=  164.321
NODE= 13   SOLUTION=  126.623
NODE= 14   SOLUTION=   95.715
NODE= 15   SOLUTION=   70.000
NODE= 16   SOLUTION=  212.000
NODE= 17   SOLUTION=  157.948
NODE= 18   SOLUTION=  118.570
NODE= 19   SOLUTION=   89.837
NODE= 20   SOLUTION=   70.000
NODE= 21   SOLUTION=  212.000
NODE= 22   SOLUTION=  147.118
NODE= 23   SOLUTION=  105.535
NODE= 24   SOLUTION=   80.778
NODE= 25   SOLUTION=   70.000
(**)
(*     OUTPUT FOR NX = 9 AND NY = 5     *)
(**)
NODE= 1   X=    1.0000   Y=    0.0000
NODE= 2   X=    1.1250   Y=    0.0000
NODE= 3   X=    1.2500   Y=    0.0000
NODE= 4   X=    1.3750   Y=    0.0000
NODE= 5   X=    1.5000   Y=    0.0000
NODE= 6   X=    1.6250   Y=    0.0000
NODE= 7   X=    1.7500   Y=    0.0000
NODE= 8   X=    1.8750   Y=    0.0000
NODE= 9   X=    2.0000   Y=    0.0000
NODE= 10   X=    0.9808   Y=    0.1951
NODE= 11   X=    1.1082   Y=    0.2204
NODE= 12   X=    1.2356   Y=    0.2458
NODE= 13   X=    1.3630   Y=    0.2711
NODE= 14   X=    1.4904   Y=    0.2965
NODE= 15   X=    1.6178   Y=    0.3218
NODE= 16   X=    1.7452   Y=    0.3471
NODE= 17   X=    1.8726   Y=    0.3725
NODE= 18   X=    2.0000   Y=    0.3978
NODE= 19   X=    0.9239   Y=    0.3827
NODE= 20   X=    1.0584   Y=    0.4384
NODE= 21   X=    1.1929   Y=    0.4941
NODE= 22   X=    1.3274   Y=    0.5498
NODE= 23   X=    1.4619   Y=    0.6056
NODE= 24   X=    1.5965   Y=    0.6613
NODE= 25   X=    1.7310   Y=    0.7170
NODE= 26   X=    1.8655   Y=    0.7727
NODE= 27   X=    2.0000   Y=    0.8284
NODE= 28   X=    0.8315   Y=    0.5556
NODE= 29   X=    0.9775   Y=    0.6532
```

```
NODE= 30   X=    1.1236  Y=    0.7508
NODE= 31   X=    1.2697  Y=    0.8484
NODE= 32   X=    1.4157  Y=    0.9460
NODE= 33   X=    1.5618  Y=    1.0436
NODE= 34   X=    1.7079  Y=    1.1412
NODE= 35   X=    1.8539  Y=    1.2388
NODE= 36   X=    2.0000  Y=    1.3364
NODE= 37   X=    0.7071  Y=    0.7071
NODE= 38   X=    0.8687  Y=    0.8687
NODE= 39   X=    1.0303  Y=    1.0303
NODE= 40   X=    1.1919  Y=    1.1919
NODE= 41   X=    1.3536  Y=    1.3536
NODE= 42   X=    1.5152  Y=    1.5152
NODE= 43   X=    1.6768  Y=    1.6768
NODE= 44   X=    1.8384  Y=    1.8384
NODE= 45   X=    2.0000  Y=    2.0000
NODE= 1   SOLUTION=   212.000
NODE= 2   SOLUTION=   189.255
NODE= 3   SOLUTION=   168.765
NODE= 4   SOLUTION=   149.988
NODE= 5   SOLUTION=   132.518
NODE= 6   SOLUTION=   116.033
NODE= 7   SOLUTION=   100.270
NODE= 8   SOLUTION=    84.995
NODE= 9   SOLUTION=    70.000
NODE= 10   SOLUTION=   212.000
NODE= 11   SOLUTION=   188.654
NODE= 12   SOLUTION=   167.748
NODE= 13   SOLUTION=   148.717
NODE= 14   SOLUTION=   131.147
NODE= 15   SOLUTION=   114.721
NODE= 16   SOLUTION=    99.186
NODE= 17   SOLUTION=    84.337
NODE= 18   SOLUTION=    70.000
NODE= 19   SOLUTION=   212.000
NODE= 20   SOLUTION=   186.697
NODE= 21   SOLUTION=   164.460
NODE= 22   SOLUTION=   144.643
NODE= 23   SOLUTION=   126.799
NODE= 24   SOLUTION=   110.607
NODE= 25   SOLUTION=    95.842
NODE= 26   SOLUTION=    82.343
NODE= 27   SOLUTION=    70.000
NODE= 28   SOLUTION=   212.000
NODE= 29   SOLUTION=   182.820
NODE= 30   SOLUTION=   158.056
NODE= 31   SOLUTION=   136.841
NODE= 32   SOLUTION=   118.617
NODE= 33   SOLUTION=   103.020
NODE= 34   SOLUTION=    89.813
NODE= 35   SOLUTION=    78.839
NODE= 36   SOLUTION=    70.000
NODE= 37   SOLUTION=   212.000
NODE= 38   SOLUTION=   175.910
NODE= 39   SOLUTION=   146.930
NODE= 40   SOLUTION=   123.613
NODE= 41   SOLUTION=   105.080
NODE= 42   SOLUTION=    90.779
NODE= 43   SOLUTION=    80.338
NODE= 44   SOLUTION=    73.486
NODE= 45   SOLUTION=    70.000
```

A.2 NEWTON'S METHOD: HEAT FLOW IN A RESISTANCE TRANSDUCER

In this appendix we examine the steady-state heat flow in a resistance transducer, which has been discussed in Sandborn, Haberstroh, and Sek [22]. The temperature in the wire is denoted by T_w, and the temperature in the adjacent fluid by T_∞. The wire is cooled by convection, radiation, and Thomson heat effects. The wire is heated by the current moving through the wire. The resulting nonlinear ordinary differential equation is

$$-\left(\pi r^2 k T_{wx}\right)_x = I^2/(\pi r^2)\sigma - 2\pi r(T_w - T_i)h$$

$$-2\pi r\sigma_{sb}\epsilon\left(T_w^4 - T_\infty^4\right) - \mu I T_{wx},$$

$$T_w(0) = T_\infty,$$

$$T_{wx}(0.125) = 0.0.$$

When $u \equiv T_w - T_\infty$, then from the preceding equation we obtain

$$-\left(\pi r^2 k u_x\right)_x + \mu I u_x = f(x,u),$$

$$u(0) = 0.0,$$

$$u_x(0.125) = 0.0,$$

where

$$f(x,u) \equiv I^2/(2\pi r^2)\sigma - 2\pi r u h$$

$$-2\pi r\sigma_{sb}\epsilon\left(u^4 + 4u^3 T_\infty + 6u^2 T_\infty^2 + 4u T_\infty^3\right),$$

$$k = k(x,u) = k_0\left(1 + k_1 u + k_2 u^2\right) = \text{thermal conductivity},$$

$$\sigma = \sigma(x,u) = s_0\left(1 + s_1 u + s_2 u^2\right) = \text{electrical resistivity},$$

$$h = h(x,u) = h_0\left(1 + h_1 u + h_2 u^2\right) = \text{thermal convectivity}.$$

This problem is discretized by the Galerkin formulation of the FEM. Linear shape functions are used, and the formation of the algebraic problem is as discussed in Chapter 8. The terms from $\mu I u_x$ are derived from

$$\int \mu I u_x \psi \to \int \mu I \Big(\sum_j u_j \psi_j \Big)_x \psi_i .$$

When $1 \leq i < N$, this is equal to $\mu I(u_{i+1} - u_{i-1})/2$. When $i = N$, this equals $\mu I(u_N - u_{N-1})/2$.

The resulting nonlinear algebraic system is solved by the Newton–continuation as discussed in Chapter 7.

NOMENCLATURE

U	the present Newton iteration
UP	the previous Newton iteration
PI	π
MU	μ = Thomson heat
CUR	I = current
SB	σ_{sb} = Stefan–Boltzmann constant
EM	ϵ = emissivity
TI	T_∞ = fluid temperature
KC	number of continuation steps
N	number of unknowns
RAD(X)	r = radius
TCOND(U)	$k(u)$ = thermal conductivity
CK(X,U)	$r^2 k(u)$
CKP(X,U)	$r^2 k'(u)$

Program RESTRA

```
      PROGRAM RESTRA
C     THIS IS A CODE WHICH GIVES A NUMERICAL SOLUTION OF A NONLINEAR ODE.
C     THE ODE EVOLVES FROM THE HEAT FLOW IN A RESISTANCE TRANDUCER.
C     THE CONTINUATION METHOD IS USED TO OBTIAN A GOOD FIRST ESTIMATE, AND
C     THEN NEWTON'S METHOD IS USED.
      DIMENSION U(21),UP(21),SOL(20)
      DIMENSION A(20),B(20),C(20),D(20)
      OPEN(6,FILE='*:OUTPUT.TEXT')
      KC = 5
      H = 1.0/KC
      ER = 0.00001
      RESER = 0.000001
      N = 20
      DX = 0.125/N
      DO 10 I=1,N+1
        U(I) = 0.0
        UP(I) = 0.0
   10 CONTINUE
      DO 20 I=1,N
        D(I) = FF(I+1,N,U)
   20 CONTINUE
C     BEGIN THE CONTINUATION METHOD.
      DO 100 K=1,KC
        CALL COEF(A,B,C,N,UP)
        CALL TRID(A,B,C,D,SOL,N)
        DO 30 I=1,N
          U(I+1) = UP(I+1) - H*SOL(I)
          UP(I+1) = U(I+1)
   30   CONTINUE
  100 CONTINUE
      WRITE(6,105) KC
  105 FORMAT('END OF',I4,' CONTINUATION STEPS')
      WRITE(6,110) (UP(I),I=1,N+1)
  110 FORMAT( /F8.3)
C     BEGIN NEWTON'S METHOD.
      DO 200 NEWT=1,100
        DO 120 I=1,N
          D(I) = FF(I+1,N,UP)
  120   CONTINUE
        CALL COEF(A,B,C,N,UP)
        CALL TRID(A,B,C,D,SOL,N)
        NUM1 = 0
        NUM2 = 0
        DO 130 I=1,N
          U(I+1) = UP(I+1) - SOL(I)
          ER1 = ABS((U(I+1) - UP(I+1))/(UP(I+1) + 0.00000001))
          ER2 = D(I)
          IF (ER1.LT.ER) NUM1 = NUM1 +1
          IF (ER2.LT.RESER) NUM2 = NUM2 +1
          UP(I+1) = U(I+1)
  130   CONTINUE
C     CHECK TO SEE IF THE RELATIVE ERROR AND THE RESIDUAL ERROR CRITERIA
C     ARE SATISFIED AT EACH NODE.
        IF (NUM1.EQ.N.AND.NUM2.EQ.N) GOTO 210
```

```
200   CONTINUE
210   CONTINUE
      WRITE(6,220) NEWT
220   FORMAT('CONVERGENCE AFTER',I4,' NEWTON ITERATIONS')
      WRITE(6,110) (U(I),I=1,N+1)
      STOP
      END
C
C    ELECTRICAL RESISTIVITY
      FUNCTION ERES(U)
      S0 = 5.5E-06
      S1 = 5.0E-03
      S2 = 7.0E-07
      ERES = S0*(1.0 + S1*U +S2*U*U)
      RETURN
      END
C
C    THERMAL CONDUCTIVITY
      FUNCTION TCOND(U)
      S0 = 1.82
      S1 = -9.41E-04
      S2 = 6.63E-07
      TCOND = S0*(1.0 + S1*U +S2*U*U)
      RETURN
      END
C
C    THERMAL CONVECTIVITY
      FUNCTION TCONV(U)
      S0 = 35.567E-03
      S1 = 56.2E-06
      S2 = 8.0E-08
      TCONV = S0*(1.0 + S1*U + S2*U*U)
      RETURN
      END
C
C    DERIVATIVE OF ELECTRICAL RESISTIVITY
      FUNCTION ERESP(U)
      S0 = 5.5E-06
      S1 = 5.0E-03
      S2 = 7.0E-07
      ERESP = S0*S1 + 2.0*S0*S2*U
      RETURN
      END
C
C    DERIVATIVE OF THERMAL CONVECTIVITY
      FUNCTION TCONVP(U)
      S0 = 35.567E-03
      S1 = 56.2E-06
      S2 = 8.0E-08
      TCONVP = S0*S1 + 2.0*S0*S2*U
      RETURN
      END
C
C    DERIVATIVE OF THERMAL CONDUCTIVITY
```

```
      FUNCTION TCONDP(U)
      S0 = 1.82
      S1 = -9.41E-04
      S2 = 6.63E-07
      TCONDP = S0*S1 + 2.0*S0*S2*U
      RETURN
      END
C
C   THE NONLINEAR FUNCTION WHOSE ROOT IS TO BE FOUND
      FUNCTION FF(I,N,U)
      DIMENSION U(21)
      RAD(X) = 0.02*X*(X-0.25) + 0.001
      CK(X,U) = PI*RAD(X)*RAD(X)*TCOND(U)
      CKP(X,U) = PI*RAD(X)*RAD(X)*TCONDP(U)
      PI = 3.1415926
      DX = 0.125/N
      MU = 9.61E-06
      CUR = 0.15811388
      X = (I-1)*DX
      IF (I.LE.N) THEN
        FF = -(CK(X+DX,U(I+1)) + CK(X,U(I)))*(U(I+1) - U(I))/(2.0*DX)
        FF = FF +(CK(X,U(I)) + CK(X-DX,U(I-1)))*(U(I) - U(I-1))/(2.0*DX)
        FF = FF + MU*CUR*(U(I+1) - U(I-1))/2.0
        FF = FF - DX*(F(X+DX,U(I+1)) + 4.0*F(X,U(I)) + F(X-DX,U(I-1)))/6.0
      ENDIF
      IF (I.GT.N) THEN
        FF = (CK(X,U(I)) + CK(X-DX,U(I-1)))*(U(I) - U(I-1))/(2.0*DX)
        FF = FF + MU*CUR*(U(I) - U(I-1))/2.0
        FF = FF - DX*(2.0*F(X,U(I)) + F(X-DX,U(I-1)))/6.0
      ENDIF
      RETURN
      END
C
C   THE NONLINEAR FORCING FUNCTION
      FUNCTION F(X,U)
      RAD(X) = 0.02*X*(X-0.25) + 0.001
      CUR = 0.15811388
      PI = 3.1415926
      SB = 5.68E-12
      EM = 0.022
      TI = 273.0
      F = CUR*CUR*ERES(U)/(PI*RAD(X)*RAD(X))
      F = F - 2.0*PI*RAD(X)*U*TCONV(U)
      F = F - 2.0*PI*RAD(X)*SB*EM*(6.0*U*U*TI*TI + 4.0*U*TI*TI*TI)
      F = F - 2.0*PI*RAD(X)*SM*EM*(4.0*U*U*U*TI + U*U*U*U)
      RETURN
      END
C
C   THE DERIVATIVE OF THE NONLINEAR FORCING FUNCTION
      FUNCTION FP(X,U)
      RAD(X) = 0.02*X*(X-0.25) + 0.001
      PI = 3.1415926
      CUR = 0.15811388
      SB = 5.68E-12
```

```
      EM = 0.022
      TI = 273.0
      FP = CUR*CUR*ERESP(U)/(PI*RAD(X)*RAD(X))
      FP = FP - 2.0*PI*RAD(X)*(TCONV(U) + U*TCONVP(U))
      FP = FP - 2.0*PI*RAD(X)*SB*EM*(12.0*U*TI*TI + 4.0*TI*TI*TI)
      FP = FP - 2.0*PI*RAD(X)*SB*EM*(12.0*U*U*TI + 4.0*U*U*U)
      RETURN
      END
C
C    THE COMPONENTS OF THE JACOBIAN MATRIX ARE COMPUTED.
      SUBROUTINE COEF(A,B,C,N,U)
      DIMENSION A(20),B(20),C(20),U(21)
      RAD(X) = 0.02*X*(X-0.25) + 0.001
      CK(X,U) = PI*RAD(X)*RAD(X)*TCOND(U)
      CKP(X,U) = PI*RAD(X)*RAD(X)*TCONDP(U)
      PI = 3.1415916
      DX = 0.125/N
      MU = 9.61E-06
      CUR = 0.15811388
      DO 10 I=2,N+1
        II=I-1
        X =II*DX
        IF (II.EQ.1) A(II) = 0.0
        IF (II.LT.N) THEN
          A(II) = -CKP(X-DX,U(I-1))*(U(I)-U(I-1))/(2.0*DX)
          A(II) = A(II) + (CK(X,U(I))+CK(X-DX,U(I-1)))/(2.0*DX)
          A(II) = A(II) + MU*CUR/2.0 + DX*FP(X-DX,U(I-1))/6.0
        ENDIF
        IF (II.EQ.N) THEN
          A(II) = -CKP(X-DX,U(I-1))*(U(I)-U(I-1))/(2.0*DX)
          A(II) = A(II) + (CK(X,U(I))+CK(X-DX,U(I-1)))/(2.0*DX)
          A(II) = A(II) + MU*CUR/2.0 + DX*FP(X-DX,U(I-1))/6.0
        ENDIF
        IF (II.LT.N) THEN
          B(II) = -CKP(X,U(I))*(U(I+1)-U(I))/(2.0*DX)
          B(II) = B(II) + (CK(X+DX,U(I+1))+CK(X,U(I)))/(2.0*DX)
          B(II) = B(II) + CKP(X,U(I))*(U(I)-U(I-1))/(2.0*DX)
          B(II) = B(II) + (CK(X,U(I))+CK(X-DX,U(I-1)))/(2.0*DX)
          B(II) = B(II) - DX*2.0*FP(X,U(I))/3.0
        ENDIF
        IF (II.EQ.N) THEN
          B(II) = CKP(X,U(I))*(U(I)-U(I-1))/(2.0*DX)
          B(II) = B(II) + (CK(X,U(I))+CK(X-DX,U(I-1)))/(2.0*DX)
          B(II) = B(II) + MU*CUR/2.0 - DX*FP(X,U(I))/3.0
        ENDIF
        IF (II.LT.N) THEN
          C(II) = CKP(X+DX,U(I+1))*(U(I+1)-U(I))/(2.0*DX)
          C(II) = C(II) + (CK(X+DX,U(I+1))+CK(X,U(I)))/(2.0*DX)
          C(II) = C(II) - MU*CUR/2.0 + DX*FP(X+DX,U(I+1))/6.0
        ENDIF
        IF (II.EQ.N) C(I) = 0.0
   10 CONTINUE
      RETURN
      END
```

```
C
C   THE TRIDIAGONAL ALGORITHM IS USED TO SOLVE THE ALGEBRAIC PROBLEMS.
      SUBROUTINE TRID(A,B,C,D,SOL,N)
      DIMENSION A(20),B(20),C(20),D(20),SOL(20),AA(21),DD(21)
      AA(1) = 0.0
      DD(1) = 0.0
      DO 10 I=1,N
        DENOM = B(I) - A(I)*AA(I)
        AA(I+1) = C(I)/DENOM
        DD(I+1) = (D(I) + A(I)*DD(I))/DENOM
   10 CONTINUE
      SOL(N) = DD(N+1)
      DO 20 I=1,N-1
        II = N + 1 - I
        SOL(II-1) = AA(II)*SOL(II) + DD(II)
   20 CONTINUE
      RETURN
      END
C
C         OUTPUT FOR RAD(X) = 0.001
C
END OF    2 CONTINUATION STEPS
    .000   5.516  11.046  16.289  21.244  25.908  30.280
  34.357  38.140  41.624  44.811  47.698  50.283  52.568
  54.549  56.227  57.600  58.669  59.433  59.891  60.044
CONVERGENCE AFTER    2 NEWTON ITERATIONS
    .000   5.539  11.093  16.372  21.373  26.092  30.524
  34.667  38.517  42.070  45.325  48.277  50.926  53.268
  55.302  57.026  58.438  59.538  60.324  60.796  60.953
C
C         OUTPUT FOR RAD(X) = 0.02*X*(X - 0.25) + 0.001
C
END OF    2 CONTINUATION STEPS
    .000  16.263  33.686  51.922  70.932  90.650 110.983
 131.801 152.937 174.185 195.296 215.981 235.913 254.738
 272.082 287.568 300.832 311.540 319.408 324.220 325.840
CONVERGENCE AFTER    4 NEWTON ITERATIONS
    .000  18.052  37.437  57.975  79.662 102.465 126.316
 151.100 176.649 202.730 229.043 255.215 280.800 305.292
 328.137 348.758 366.583 381.081 391.794 398.370 400.587
C
C         OUTPUT FOR RAD(X) = 0.02*X*(X - 0.25) + 0.001
C
END OF    5 CONTINUATION STEPS
    .000  17.108  35.535  54.953  75.340  96.647 118.790
 141.647 165.048 188.770 212.539 236.020 258.828 280.531
 300.667 318.757 334.335 346.966 356.278 361.986 363.909
CONVERGENCE AFTER    3 NEWTON ITERATIONS
    .000  17.972  37.349  57.877  79.554 102.346 126.186
 150.957 176.493 202.560 228.859 255.015 280.586 305.064
 327.896 348.504 366.319 380.808 391.514 398.087 400.303
```

A.3 NONLINEAR GAUSS–SEIDEL METHOD: SOLIDIFICATION OF WATER IN A CHANNEL — THE STEFAN PROBLEM

In this appendix we illustrate the nonlinear Gauss–Seidel method as it is applied to the Stefan problem. Consider a channel whose cross section is a square (see Figure A.3.1). The sides are uniformly lowered below the solidification temperature. We assume there is no supercooling or convection of the liquid. Physical experiments have been done by Saitoh [21], and the position of the solid–liquid interface recorded. The numerical results of this appendix agree with the data in [21].

We use the enthalpy formulation of the Stefan problem. If the walls of channel are uniformly cooled, then we may use the symmetry of the solution and restrict our domain to the triangular region in Figure A.3.1. We use the same functions $\beta(E)$, $F(u) = v$, $F^{-1}(v) = u$, and $H(u)$ as defined in Chapter 8:

$$E_t - \Delta\beta(E) = f,$$

$$d/dn(\beta(E)) = 0 \quad \text{on } x = y \text{ and } y = 0,$$

$$\beta(E) = F(273.1 - 0.05t/60) \quad \text{on } x = 3.95,$$

$$E(x,0) = H(273.1).$$

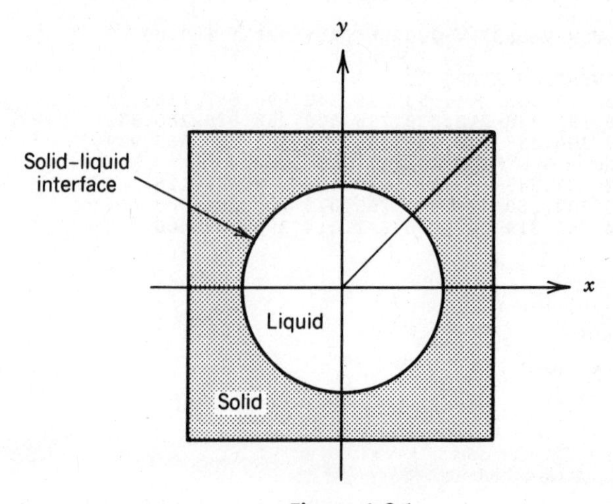

Figure A.3.1

TABLE A.3.1

time	$x = y$ (on the diagonal)
1500	$11.0h$ where $h = 3.95/12$
3000	$9.5h$
4500	$8.0h$
6000	$7.0h$
7500	$5.5h$
9000	$4.0h$
10500	$3.0h$
12000	$1.0h$
12300	$0.5h$
12600	$0.0h$

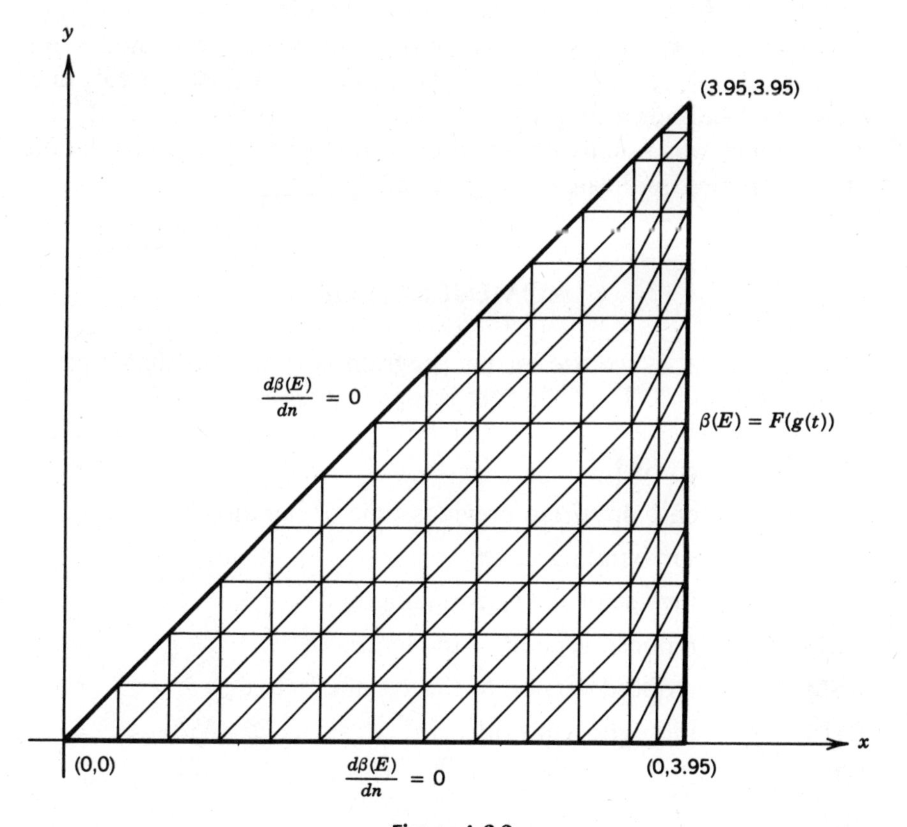

Figure A.3.2

An implicit time discretization and the Galerkin formulation of the FEM with linear shape functions yield at each time step the following nonlinear problem:

$$BE + A\beta(E) = \eta,$$

where

$$B = \left(\iint \psi_i \psi_j\right),$$

$$A = \left(\Delta t \iint (\psi_{ix}\psi_{jx} + \psi_{iy}\psi_{jy})\right),$$

$$\eta = B(\Delta t E^{\sim} + f),$$

and $E^{\sim}$ is the computed enthalpy from the previous time.

In the calculations given in Table A.3.1 for the diagonal nodes, we used $\Delta t = 300$, relative error = 0.0001, number of nodes = 105, and number of unknown nodes = 91. No SOR parameter was used, and convergence was usually obtained in 34 iterations. The triangulation of the domain is indicated in Figure A.3.2.

NOMENCLATURE

The notation that we use in this program is similar to that used in Chapters 2 and 3. The following is a partial list of symbols.

E	enthalpy
D,F	enthalpy from previous time or iteration
NS	surrounding nodes
INS	inverse of the surrounding nodes
RBM	reduced version of the matrix B
RSM	reduced version of the system matrix
RHS	the right-hand side, η
HI	β
HH	H

RI	inverse of R
FI	inverse of F
INVNS	generates INS
GENNOD	generates NOD
GENXC	generates horizontal nodes XC
GS	nonlinear Gauss–Seidel algorithm
GENXY	generates all nodes
GENNX	generates NX
GENNS	generates NS
GENNPT	generates NPT

Program FEMS

```
      PROGRAM FEMS
C   THIS PROGRAM APPROXIMATES THE SOLUTION OF THE STEFAN PROBLEM
C   FOR A FREEZING CHANNEL.  THE ENTHAPY FORMULATION IS USED,
C   AND THE RESULTING NONLINEAR ALGEBRAIC EQUATIONS ARE SOLVED BY
C   THE NONLINEAR GAUSS-SEIDEL METHOD.
      DIMENSION E(105),D(105),F(105),NS(105,8),NX(105),INS(105)
      DIMENSION RBM(105,7),RHS(105),RSM(105,7),NOD(169,3),X(105),Y(105)
      DIMENSION XC(20),A(3),B(3),C(3),XX(3),YY(3),NPT(20),TEMP(105)
      COMMON H1,H2,AS,AL,UF,HF,BR,AR,W,R1,R2,AKS,AKL
      COMMON INS,NS,NPT,XC,NX,X,Y,NOD,N,NPRES,NUMI,NN,DT,ER
      COMMON RBM,RSM,E,D,RHS
C
C  STEP I:  INPUT DATA
C
      FCT(X,Y,T)=0.0
      G(X,Y)=273.1
      G1(X,Y,T)=273.0 - 0.050*T/60
      AKS=0.00530
      AKL=0.00144
      AS=86.9565
      AL=714.2856
      HF=73.6
      UF=273.0
      ER=0.0001
      DT=300.0
      NT=48
      NPRES=14
      NE=(NPRES-1)**2
      N=INT(NPRES*(NPRES+1)/2+.1)
      W=1.0
      OPEN(6,FILE='PRINTER:')
      UF=AKS*UF
      H1=UF*AS
      H2=H1+HF
      W1=W
      NN=N-NPRES
      NSOL=0
C  GENERATE THE ARRAY NPT
      CALL GENNPT(NPRES)
C  GENERATE THE X COMPONENTS ON THE X AXIS
      CALL GENXC(NPRES)
C  GENERATE ALL THE (X,Y) COMPONENTS
      CALL GENXY(NPRES)
C  GENERATE THE ARRAY NS
      CALL GENNS(NPRES)
C  GENERATE THE ARRAY NOD
      CALL GENNOD(NPRES)
C  GENERATE THE ARRAY NX
      CALL GENNX
C  THE REDUCED MARTICES ARE INITIALIZED TO ZERO
      DO 20 I=1,N
        DO 10 J=1,7
        RSM(I,J)=0.0
        RBM(I,J)=0.0
```

```
  10     CONTINUE
  20   CONTINUE
C
C  STEP II:  USE THE GALERKIN FORMULATION OF FEM
C
C  STEP III:  ASSEMBLE THE REDUCED MATRICES BY THE METHOD OF ELEMENTS
C
       DO 105 NELT=1,NE
         DO 30 J=1,3
           LK=NOD(NELT,J)
           XX(J)=X(LK)
           YY(J)=Y(LK)
  30     CONTINUE
         DO 40 J=1,3
           LK=MOD(J+1,3)
           IF (LK.EQ.0) LK=3
           LL=MOD(J+2,3)
           IF (LL.EQ.0) LL=3
           A(J)=XX(LK)*YY(LL) - XX(LL)*YY(LK)
           B(J)=YY(LK)-YY(LL)
           C(J)=XX(LL)-XX(LK)
  40     CONTINUE
         DELTA=(C(3)*B(2)-C(2)*B(3))/2.0
         DO 60 IR=1,3
           II=NOD(NELT,IR)
           CALL INVNS(II)
           DO 50 IC=1,3
             JJ=INS(NOD(NELT,IC))
             GL=(B(IR)*B(IC)+C(IR)*C(IC))/(4.0*DELTA)
             RSM(II,JJ)=RSM(II,JJ)+GL
             RBM(II,JJ)=RBM(II,JJ)+DELTA/12.0
             IF (IR.EQ.IC) RBM(II,JJ)=RBM(II,JJ)+DELTA/12.0
  50       CONTINUE
  60     CONTINUE
  105 CONTINUE
C  THE INITIAL CONDITION IS DEFINED
       T=0.0
       DO 110 I=1,NN
         II=NX(I)
         E(II)=HH(G(X(II),Y(II)))
         D(II)=E(II)
         F(II)=E(II)
  110 CONTINUE
       DO 120 I=1,NPRES
         NODE=NPT(I)
         F(NODE)=HH(G1(X(NODE),Y(NODE),0.0))
  120 CONTINUE
C  THE TIME LOOP BEGINS
       DO 500 KT=1,NT
         T=T+DT
         W=(W1-1.0)*NSOL/NN+1.0
C
C  STEP V:  THE NEW BOUNDARY CONDITIONS ARE INSERTED
C
```

```
      DO 210 I=1,NPRES
         NODE=NPT(I)
         E(NODE)=HH(G1(X(NODE),Y(NODE),T))
         D(NODE)=E(NODE)
  210   CONTINUE
C  THE NEW RIGHT SIDE IS DEFINED
        DO 220 I=1,N
          RHS(I)=0.0
  220   CONTINUE
        DO 240 I=1,NN
          FF=0.0
          II=NX(I)
          JA=NS(II,8)+1
          DO 230 J=1,JA
            IA=NS(II,J)
            FF=FF+RBM(II,J)*(F(IA)+DT*FCT(X(IA),Y(IA),T))
  230     CONTINUE
          RHS(II)=RHS(II)+FF
  240   CONTINUE
C
C  STEP V:  AT EACH TIME STEP APPROXIMATE THE SOLUTION OF THE
C             NONLINEAR PROBLEM BY A GAUSS-SEIDEL ALGORITHM
C
        CALL GS
C
C  STEP VI:  AT EACH TIME STEP THE DIAGONAL VALUES ARE PRINTED
C
        IF (NUMI.EQ.150) GOTO 600
        WRITE(6,400) T,NUMI
  400   FORMAT(1X,'T=',F12.4,4X,'NUMI=',I4)
        DO 410 I=1,N
          F(I)=E(I)
          D(I)=E(I)
          TEMP(I)=FI(HI(E(I)))
  410   CONTINUE
        NSOL=0
        DO 420 I=1,NN
          II=NX(I)
          IF (E(II).LT.H1) NSOL=NSOL+1
  420   CONTINUE
        II=1
        DO 440 I=1,NPRES
          WRITE(6,430) E(II),TEMP(II)
  430     FORMAT(1X,2F10.4)
          II=II+(NPRES-I+1)
  440   CONTINUE
  500 CONTINUE
  600 CONTINUE
      WRITE(6,610) NUMI
  610 FORMAT(1X,'NUMI=',I4)
      STOP
      END
C
      FUNCTION HI(X)
```

```
      COMMON H1,H2,AS,AL,UF,HF,BR,AR,W,R1,R2,AKS,AKL
      IF (X.LT.H1) HI=X/AS
      IF (X.GE.H1.AND.X.LE.H2) HI=UF
      IF (X.GT.H2) HI=(X-HF+UF*(AL-AS))/AL
      RETURN
      END
C
      FUNCTION HH(X)
      COMMON H1,H2,AS,AL,UF,HF,BR,AR,W,R1,R2,AKS,AKL
      VF=UF/AKS
      IF (X.LT.VF) HH=X*AS*AKS
      IF (X.GE.VF) HH=AS*AKS*VF+AL*AKL*(X-VF)+HF
      RETURN
      END
C
      FUNCTION RI(X)
      COMMON H1,H2,AS,AL,UF,HF,BR,AR,W,R1,R2,AKS,AKL
      IF (X.LT.R1) RI=X/(BR+AR/AS)
      IF (X.GE.R1.AND.X.LE.R2) RI=(X-AR*UF)/BR
      IF (X.GT.R2) RI=(X+AR*(HF-UF*(AL-AS))/AL)/(BR+AR/AL)
      RETURN
      END
C
      FUNCTION WW(X)
      COMMON H1,H2,AS,AL,UF,HF,BR,AR,W,R1,R2,AKS,AKL
      IF (H1.EQ.H2) GOTO 10
      IF (X.LT.H1) WW=W
      IF (X.GE.H1.AND.X.LE.H2) WW=(1.0-W)*(X-H1)/(H2-H1)+W
      IF (X.GT.H2) WW=W
   10 CONTINUE
      RETURN
      END
C
      FUNCTION FI(X)
      COMMON H1,H2,AS,AL,UF,HF,BR,AR,W,R1,R2,AKS,AKL
      IF (X.LE.UF) FI=X/AKS
      IF (X.GT.UF) FI=(X-UF)/AKL+UF/AKS
      RETURN
      END
C
      SUBROUTINE INVNS(II)
      DIMENSION INS(105),NS(105,8)
      COMMON H1,H2,AS,AL,UF,HF,BR,AR,W,R1,R2,AKS,AKL
      COMMON INS,NS
      JA=NS(II,8)+1
      DO 10 J=1,JA
        INS(NS(II,J))=J
   10 CONTINUE
      RETURN
      END
C
      SUBROUTINE GENNPT(NPRES)
      DIMENSION INS(105),NS(105,8),NPT(20)
      COMMON H1,H2,AS,AL,UF,HF,BR,AR,W,R1,R2,AKS,AKL
```

```
      COMMON INS,NS,NPT
      II=NPRES
      NPT(1)=NPRES
      DO 10 I=2,NPRES
        II=II-1
        NPT(I)=II+NPT(I-1)
   10 CONTINUE
      RETURN
      END
C
      SUBROUTINE GENXC(NPRES)
      DIMENSION INS(105),NS(105,8),NPT(20),XC(20)
      COMMON H1,H2,AS,AL,UF,HF,BR,AR,W,R1,R2,AKS,AKL
      COMMON INS,NS,NPT,XC
      XC(1)=0.0
      DX=3.95/(NPRES-2)
      DO 10 I=2,NPRES-2
        XC(I)=XC(I-1)+DX
   10 CONTINUE
      XC(NPRES-1)=3.95-DX/2.0
      XC(NPRES)=3.95
      RETURN
      END
C
      SUBROUTINE GS
      DIMENSION E(105),D(105),F(105),NS(105,8),NX(105),INS(105)
      DIMENSION RBM(105,7),RHS(105),RSM(105,7),NOD(169,3),X(105),Y(105)
      DIMENSION XC(20),A(3),B(3),C(3),XX(3),YY(3),NPT(20),TEMP(105)
      COMMON H1,H2,AS,AL,UF,HF,BR,AR,W,R1,R2,AKS,AKL
      COMMON INS,NS,NPT,XC,NX,X,Y,NOD,N,NPRES,NUMI,NN,DT,ER
      COMMON RBM,RSM,E,D,RHS
      NUMI=0
   10 CONTINUE
      ME=0
      NUMI=NUMI+1
      DO 105 I=1,NN
        II=NX(I)
        CALL INVNS(II)
        BR=RBM(II,INS(II))
        AR=RSM(II,INS(II))*DT
        R1=BR*H1+AR*UF
        R2=BR*H2+AR*UF
        GL=0.0
        JA=INS(II)-1
        IF (JA.EQ.0) GOTO 40
        DO 30 J=1,JA
          GL=GL+RSM(II,J)*HI(E(NS(II,J)))*DT
          GL=GL+RBM(II,J)*E(NS(II,J))
   30   CONTINUE
   40   CONTINUE
        U=0.0
        JA=JA+2
        JB=NS(II,8)+1
        IF (JA.GT.JB) GOTO 60
```

```
        DO 50 J=JA,JB
          U=U+RSM(II,J)*HI(D(NS(II,J)))*DT
          U=U+RBM(II,J)*D(NS(II,J))
 50     CONTINUE
 60     CONTINUE
        E(II)=RI(RHS(II)-GL-U)
        E(II)=WW(E(II))*E(II)+(1.0-WW(E(II)))*D(II)
        AE=ABS((E(II)-D(II))/D(II))
        IF (AE.LT.ER) ME=ME+1
105   CONTINUE
      DO 110 I=1,NN
        II=NX(I)
        D(II)=E(II)
110   CONTINUE
      IF (ME.EQ.NN.OR.NUMI.EQ.150) GOTO 200
      GOTO 10
200   CONTINUE
      RETURN
      END
C
      SUBROUTINE GENNOD(NH)
      DIMENSION NOD(169,3),X(105),Y(105),INS(105),NS(105,8)
      DIMENSION XC(20),NX(105),NPT(20)
      COMMON H1,H2,AS,AL,UF,HF,BR,AR,W,R1,R2,AKS,AKL
      COMMON INS,NS,NPT,XC,NX,X,Y,NOD
      NODE=0
      NELT=0
      K=NH-1
      DO 30 NER=1,K
        NELT=NELT+1
        NODE=NODE+2
        NOD(NELT,1)=NODE
        NOD(NELT,2)=NODE+NH-NER
        NOD(NELT,3)=NODE-1
        IF (NH-NER-1.EQ.0) GOTO 20
        KK=NH-NER-1
        DO 10 J=1,KK
          NELT=NELT+1
          NOD(NELT,1)=NODE+NH-NER
          NOD(NELT,2)=NODE
          NOD(NELT,3)=NODE+NH-NER+1
          NELT=NELT+1
          NOD(NELT,1)=NODE+1
          NOD(NELT,2)=NODE+NH-NER+1
          NOD(NELT,3)=NODE
          NODE=NODE+1
 10     CONTINUE
 20   CONTINUE
 30   CONTINUE
      RETURN
      END
C
      SUBROUTINE GENXY(NH)
      DIMENSION NOD(169,3),X(105),Y(105),INS(105),NS(105,8),XC(20)
```

```
      DIMENSION NX(105),NPT(20)
      COMMON H1,H2,AS,AL,UF,HF,BR,AR,W,R1,R2,AKS,AKL
      COMMON INS,NS,NPT,XC,NX,X,Y,NOD
      NODE=0
      DO 20 NR=1,NH
        DO 10 I= NR,NH
          NODE=NODE+1
          Y(NODE)=XC(NR)
          X(NODE)=XC(I)
   10   CONTINUE
   20 CONTINUE
      RETURN
      END
C
      SUBROUTINE GENNX
      DIMENSION NOD(169,3),X(105),Y(105),INS(105),NS(105,8),XC(20)
      DIMENSION NX(105),NPT(20)
      COMMON H1,H2,AS,AL,UF,HF,BR,AR,W,R1,R2,AKS,AKL
      COMMON INS,NS,NPT,XC,NX,X,Y,NOD,N,NPRES
      II=0
      DO 20 I=1,N
        DO 10 J=1,NPRES
          IF (NPT(J).EQ.I) GO TO 20
   10   CONTINUE
        II=II+1
        NX(II)=I
   20 CONTINUE
      RETURN
      END
C
      SUBROUTINE GENNS(NH)
      DIMENSION NOD(169,3),X(105),Y(105),INS(105),NS(105,8),XC(20)
      DIMENSION NX(105),NPT(20)
      COMMON H1,H2,AS,AL,UF,HF,BR,AR,W,R1,R2,AKS,AKL
      COMMON INS,NS,NPT,XC,NX,X,Y,NOD,N,NPRES
      I=0
      DO 20 NR=1,NH
        DO 10 J=NR,NH
          I=I+1
          IF (NR.EQ.1.AND.J.EQ.1) GOTO 1
          IF (NR.EQ.1.AND.J.LT.NH) GOTO 2
          IF (J.EQ.NH.AND.NR.EQ.1) GOTO 3
          IF (J.EQ.NR.AND.NR.LT.NH) GOTO 4
          IF (NR.LT.NH.AND.J.LT.NH) GOTO 5
          IF (NR.LT.NH.AND.J.EQ.NH) GOTO 6
          IF (NR.EQ.NH.AND.J.EQ.NH) GOTO 7
    1     NS(I,8)=2
          NS(I,1)=I
          NS(I,2)=I+1
          NS(I,3)=I+NH-NR+1
          GOTO 10
    2     NS(I,8)=4
          NS(I,1)=I-1
          NS(I,2)=I
```

```
          NS(I,3)=I+1
          NS(I,4)=I+NH-NR
          NS(I,5)=I+NH-NR+1
          GOTO 10
 3        NS(I,8)=2
          NS(I,1)=I-1
          NS(I,2)=I
          NS(I,3)=I+NH-NR
          GOTO 10
 4        NS(I,8)=4
          NS(I,1)=I-NH+NR-2
          NS(I,2)=I-NH+NR-1
          NS(I,3)=I
          NS(I,4)=I+1
          NS(I,5)=I+NH-NR+1
          GOTO 10
 5        NS(I,8)=6
          NS(I,1)=I-NH+NR-2
          NS(I,2)=I-NH+NR-1
          NS(I,3)=I-1
          NS(I,4)=I
          NS(I,5)=I+1
          NS(I,6)=I+NH-NR
          NS(I,7)=I+NH-NR+1
          GOTO 10
 6        NS(I,8)=4
          NS(I,1)=I-NH+NR-2
          NS(I,2)=I-NH+NR-1
          NS(I,3)=I-1
          NS(I,4)=I
          NS(I,5)=I+NH-NR
          GOTO 10
 7        NS(I,8)=2
          NS(I,1)=I-NH+NR-2
          NS(I,2)=I-NH+NR-1
          NS(I,3)=I
10     CONTINUE
20   CONTINUE
     RETURN
     END
```

A.4 VARIATIONAL INEQUALITIES: STEADY-STATE FLUID FLOW IN A POROUS MEDIUM — AN AXISYMMETRIC WATER FILTER

The following program models a water filter, which is described in Chapter 9 and in Cryer and Fetter [8]. We have used the following:

$R = r = 4.8,$ $\qquad RR = \mathrm{R} = 76.6$

$HW = h_w = 12.0,$ $\qquad H = \mathrm{H} = 48.0$

$g(y) = 0.0$,
ER = relative error = 0.01 or 0.0001,
$X(I) = x$ = radial nodes,
WW = SOR parameter = 1.6 or 1.8,
$Y(I) = y$ = vertical nodes,
N = number of rectangles in the x direction,
M = number of rectangles in the y direction.

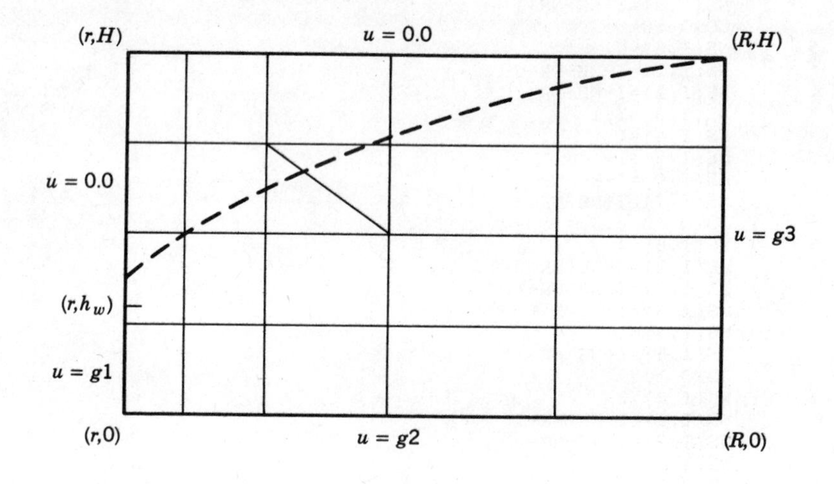

Figure A.4.1

Consequently, the boundary conditions for the Baiocchi function are as indicted in Figure A.4.1 and

$$g1 = (\mathrm{HW} - Y(I)) * (\mathrm{HW} - Y(J))/2.0$$

$$g2 = (HW * HW * \mathrm{ALOG}(RR/X(I))$$

$$+H * H * \mathrm{ALOG}(X(I)/R))/(2 * ALOG(RR/R))$$

$$g3 = (\mathrm{H} - \mathrm{Y(J)}) * (\mathrm{H} - \mathrm{Y(J)})/2.0.$$

The initial estimate of the unknown is the linear interpolation in the x direction. The system matrix may be written in a reduced form

$A(N+1, M+1, 5)$. In Chapter 9 we wrote the components of this matrix as $a(\psi_i, \psi_j)$, where the i corresponds to the pair (I, J). As long as the pairs of elements form rectangles, the only nonzero components of a given row will be given by the columns which correspond to (I, J), $(I+1, J)$, $(I-1, J)$, $(I, J+1)$, and $(I, J+1)$. In order to see how this is established, consider the lower element in Figure A.4.2. The circled numbers correspond to the reduced system nodes in $A(N+1, M+1, 5)$.

$$\text{Part of } A(I, J, 2) \text{ from element } L = \int_L x(\psi_{ix}\psi_{jx} + \psi_{iy}\psi_{jy})$$

$$= \int_L (X(I)(N_1 + N_3) + (X(I) + DX)N_2)(N_{1X}N_{2X} + N_{1Y}N_{2Y})$$

$$= \int_L (X(I)(N_1 + N_3) + (X(I) + DX)N_2)(-1/(DX\,DX) + 0)$$

$$= X(I)(-1/(DX\,DX))\int_L (N_1 + N_2 + N_3) + (-1/DX)\int_L N_2$$

$$= X(I)(-1/(DX\,DX))(DX\,DY/2) + (-1/DX)(DX\,DY/6)$$

$$= XL(-DY/(DX\,2)) = -AXL.$$

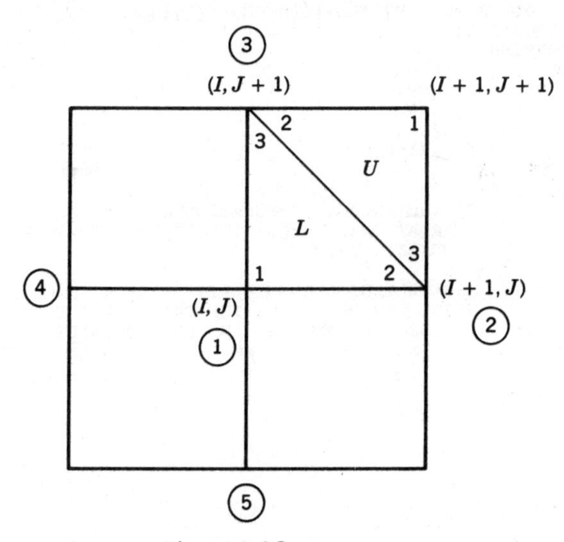

Figure A.4.2

Program POROUS

```
      PROGRAM POROUS
C   THE SOLUTION OF THE STEADY STATE POROUS MEDIUM PROBLEM IS FOUND.
C   THE PROBLEM MAY BE VIEWED AS A VARIATIONAL INEQUALITY OR AS A
C   DISCONTINUOUS NONLINEAR PROBLEM:
C               -(X*Ux)x-(X*Uy)y= EITHER 0.0 FOR (U.LT.0.0),
C                                 OR      -X FOR (U.GE.0.0)
C               PRESCRIBED BOUNDARY CONDITIONS ARE GIVEN FOR THE
C               ENTIRE BOUNDARY.
      DIMENSION A(25,17,5),U(25,17),UP(25,17),B(25,17),X(25),Y(17)
C
C   STEP I:  INPUT DATA. THIS INCLUDES THE PRESCRIBED BOUNDARY DATA,
C            THE X,Y COORDINATES, AND THE INITIAL ESTIMATE OF THE SOLUTION.
C
      DATA M/16/,N/12/,ER/.010/,WW/1.6/
      DATA R/4.8/,RR/76.8/,HW/12./,H/48./
      OPEN(6,FILE='*:TEMP.TEXT')
      NP=N+1
      MP=M+1
      DO 10 I=1,NP
        X(I)=R*EXP((I-1.)/N*ALOG(RR/R))
        U(I,1)=(HW*HW*ALOG(RR/X(I))+H*H*ALOG(X(I)/R))/(2*ALOG(RR/R))
        UP(I,1)=U(I,1)
        U(1,MP)=0.0
        UP(1,MP)=0.0
  10  CONTINUE
      DO 20 J=1,MP
        Y(J)=(J-1)*H/M
        U(1,J)=0.0
        IF (Y(J).LE.12.00001) U(1,J)=(HW-Y(J))*(HW-Y(J))/2.0
        UP(1,J)=U(1,J)
        U(NP,J)=(H-Y(J))*(H-Y(J))/2.0
        UP(NP,J)=U(NP,J)
  20  CONTINUE
      DO 30 I=2,N
       DO 30 J=2,M
         RL=Y(J)/Y(MP)
         U(I,J)=U(I,MP)*RL+(1.0-RL)*U(I,1)
         UP(I,J)=U(I,J)
  30  CONTINUE
      DO 40 I=1,NP
        DO 40 J=1,MP
          B(I,J)=0.0
          DO 40 K=1,5
            A(I,J,K)=0.0
  40  CONTINUE
C
C   STEP II:  USE THE ENERGY FORMULATION OF THE FEM.  IN THIS CASE
C             A VARIATIONAL INEQUALITY IS USED TO FORM THE ALGEBRAIC
C             INEQUALITY PROBLEM.
C
C   STEP III:  ASSEMBLY BY ELEMENTS WITH LINEAR SHAPE FUNCTIONS IS
C              USED.  IN THIS CASE THE TRIANGULAR ELEMENTS ARE PAIRED
C              TO FORM RECTANGLES, AND CONSEQUENTLY, THE SYSTEM MATRIX
C              HAS IN EACH ROW AT MOST FIVE NONZERO COMPONENTS.
```

```
C
      DO 50 I=1,N
        DO 50 J=1,M
          DX=X(I+1)-X(I)
          DY=Y(J+1)-Y(J)
          XL=X(I)+DX/3.
          XU=X(I)+DX*2./3.
          AXL=DY/DX*XL/2.
          AXU=DY/DX*XU/2.
          AYL=DX/DY*XL/2.
          AYU=DX/DY*XU/2.
          IP=I+1
          JP=J+1
          A(I,J,1)=A(I,J,1)+AXL+AYL
          A(IP,JP,1)=A(IP,JP,1)+AXU+AYU
          A(IP,J,1)=A(IP,J,1)+AXL+AYU
          A(I,JP,1)=A(I,JP,1)+AYL+AXU
          A(I,J,2)=A(I,J,2)-AXL
          A(I,J,3)=A(I,J,3)-AYL
          A(I,JP,2)=A(I,JP,2)-AXU
          A(IP,J,3)=A(IP,J,3)-AYU
          A(IP,J,4)=A(IP,J,4)-AXL
          A(I,JP,5)=A(I,JP,5)-AYL
          A(IP,JP,4)=A(IP,JP,4)-AXU
          A(IP,JP,5)=A(IP,JP,5)-AYU
          B(I,J)=B(I,J)+DX*DY*(XL/3.-DX/36.)/2.
          B(IP,J)=B(IP,J)+DX*DY*((XL+XU)/3.+DX/12.)/2.
          B(I,JP)=B(I,JP)+DX*DY*((XL+XU)/3.-DX/12.)/2.
          B(IP,JP)=B(IP,JP)+DX*DY*(XU/3.+DX/36.)/2.
   50 CONTINUE
C
C  STEP IV:  THE BOUNDARY CONDITIONS HAVE BEEN DEFINED IN THE VECTOR U.
C            AS AN ITERATIVE METHOD IS TO BE USED, THE BOUNDARY CONDITIONS
C            DO NOT NEED TO BE INSERTED INTO THE SYSTEM MATRIX, A.
C
C
C  STEP V:  THE NONLINEAR MINIMIZATION PROBLEM IS SOLVED BY A NONLINEAR
C           GAUSS-SEIDEL-SOR ALGORITHM.
C
      NUMI=0
   55 MER=0
      DO 60 I=2,N
        DO 60 J=2,M
          BM = U(I-1,J)*A(I,J,4)+U(I,J-1)*A(I,J,5)
          BP = U(I+1,J)*A(I,J,2)+U(I,J+1)*A(I,J,3)
          U(I,J)=(-B(I,J)-BM-BP)/A(I,J,1)
          U(I,J)=WW*U(I,J)+(1.-WW)*UP(I,J)
          IF (U(I,J).LT.0.0) U(I,J)=0.0
          ERR=ABS((U(I,J)-UP(I,J))/(UP(I,J)+.0000001))
          IF (ERR.LT.ER) MER=MER+1
          UP(I,J)=U(I,J)
   60 CONTINUE
      NUMI=NUMI+1
      NN=(M-1)*(N-1)
```

```
      IF (MER.LT.NN) GO TO 55
C
C  STEP VI:  THE OUTPUT IS GIVEN BY ROW STARTING AT THE TOP.
C
      WRITE(6,61) NUMI
   61 FORMAT('NUMBER OF ITERATIONS =',I4)
      DO 70 JM=1,MP
        J=MP+1-JM
        WRITE(6,62) J
        WRITE(6,63) (U(I,J),I=1,NP)
   62 FORMAT('ROW NUMBER= ',I4)
   63 FORMAT(7F8.2)
   70 CONTINUE
      STOP
      END
C
C         OUTPUT FOR N=12, M=16, ER=0.01 AND WW=1.6
C
NUMBER OF ITERATIONS =  34
ROW NUMBER=   17
     .00     .00     .00     .00     .00     .00     .00
     .00     .00     .00     .00     .00     .00
ROW NUMBER=   16
     .00     .00     .00     .00     .00     .00     .00
     .00     .00     .00     .00     .00    4.50
ROW NUMBER=   15
     .00     .00     .00     .00     .00     .00     .00
     .00     .00     .00    3.14    8.63   18.00
ROW NUMBER=   14
     .00     .00     .00     .00     .00     .00     .00
     .00    1.93    6.88   15.11   26.08   40.50
ROW NUMBER=   13
     .00     .00     .00     .00     .00     .00    1.43
    5.53   12.33   22.39   35.90   52.38   72.00
ROW NUMBER=   12
     .00     .00     .00     .00    1.52    4.96   10.67
   19.31   31.15   46.52   65.47   87.54  112.50
ROW NUMBER=   11
     .00     .22    1.79    4.87    9.89   17.32   27.60
   41.18   58.34   79.25  103.82  131.57  162.00
ROW NUMBER=   10
     .00    3.37    8.32   15.24   24.54   36.67   52.04
   71.04   93.87  120.57  150.95  184.49  220.50
ROW NUMBER=    9
     .00    8.32   18.28   30.35   45.00   62.70   83.87
  108.82  137.70  170.48  206.86  246.29  288.00
ROW NUMBER=    8
     .00   14.80   31.26   49.83   71.01   95.27  123.01
  154.50  189.85  228.98  271.55  317.00  364.50
ROW NUMBER=    7
     .00   22.78   47.19   73.63  102.56  134.38  169.46
  208.08  250.33  296.10  345.04  396.61  450.00
ROW NUMBER=    6
     .00   32.40   66.30  102.00  139.82  180.15  223.34
```

```
  269.64  319.18  371.86  427.35  485.14  544.50
ROW NUMBER=    5
     .00   44.22   89.30  135.52  183.26  232.90  284.81
  339.29  396.47  456.30  518.49  582.60  648.00
ROW NUMBER=    4
    4.50   61.14  117.98  175.33  233.54  293.02  354.13
  417.17  482.29  549.47  618.50  689.01  760.50
ROW NUMBER=    3
   18.00   86.06  154.20  222.57  291.40  360.96  431.57
  503.45  576.74  651.44  727.39  804.36  882.00
ROW NUMBER=    2
   40.50  119.61  198.74  277.97  357.40  437.17  517.43
  598.33  679.94  762.26  845.22  928.69 1012.50
ROW NUMBER=    1
   72.00  162.00  252.00  342.00  432.00  522.00  612.00
  702.00  792.00  882.00  972.00 1062.00 1152.00
C
C         OUTPUT FOR M=16, N=24, ER=0.0001 AND WW=1.8
C
NUMBER OF ITERATIONS=    69
ROW NUMBER=   17
      .00      .00      .00      .00      .00      .00      .00
      .00      .00      .00      .00      .00      .00      .00
      .00      .00      .00      .00      .00      .00      .00
      .00      .00      .00      .00
ROW NUMBER=   16
      .00      .00      .00      .00      .00      .00      .00
      .00      .00      .00      .00      .00      .00      .00
      .00      .00      .00      .00      .00      .00      .00
      .00      .07     2.10     4.50
ROW NUMBER=   15
      .00      .00      .00      .00      .00      .00      .00
      .00      .00      .00      .00      .00      .00      .00
      .00      .00      .00      .00      .00      .99     3.14
     5.75     8.79    13.15    18.00
ROW NUMBER=   14
      .00      .00      .00      .00      .00      .00      .00
      .00      .00      .00      .00      .00      .00      .00
      .00      .34     1.93     4.18     6.94    10.56    15.15
    20.41    26.27    33.13    40.50
ROW NUMBER=   13
      .00      .00      .00      .00      .00      .00      .00
      .00      .00      .00      .00      .29     1.46     3.25
     5.57     8.53    12.39    17.07    22.49    28.77    35.97
    43.92    52.57    62.04    72.00
ROW NUMBER=   12
      .00      .00      .00      .00      .00      .00      .00
      .55     1.58     3.08     5.05     7.57    10.79    14.73
    19.42    24.90    31.27    38.53    46.65    55.66    65.57
    76.28    87.72    99.88   112.50
ROW NUMBER=   11
      .00      .00      .29      .92     1.89     3.24     5.00
     7.25    10.05    13.45    17.50    22.26    27.79    34.14
    41.35    49.46    58.50    68.49    79.40    91.23   103.94
```

```
  117.48  131.74  146.66  162.00
ROW NUMBER=   10
     .00    1.55    3.45    5.75    8.48   11.69   15.44
   19.79   24.78   30.47   36.92   44.17   52.29   61.31
   71.26   82.17   94.06  106.92  120.73  135.47  151.08
  167.50  184.63  202.38  220.50
ROW NUMBER=    9
     .00    4.03    8.41   13.21   18.46   24.23   30.59
   37.58   45.28   53.73   62.99   73.12   84.15   96.13
  109.07  123.00  137.91  153.81  170.64  188.39  206.99
  226.37  246.42  267.04  288.00
ROW NUMBER=    8
     .00    7.27   14.90   22.94   31.44   40.47   50.08
   60.34   71.30   83.03   95.57  108.98  123.30  138.55
  154.75  171.93  190.06  209.15  229.14  250.01  271.68
  294.08  317.11  340.65  364.50
ROW NUMBER=    7
     .00   11.26   22.87   34.89   47.36   60.33   73.87
   88.01  102.83  118.36  134.66  151.78  169.74  188.58
  208.33  228.98  250.53  272.97  296.25  320.34  345.16
  370.65  396.71  423.22  450.00
ROW NUMBER=    6
     .00   16.07   32.48   49.26   66.45   84.08  102.19
  120.84  140.06  159.90  180.41  201.62  223.58  246.32
  269.86  294.21  319.36  345.30  371.99  399.39  427.45
  456.09  485.22  514.74  544.50
ROW NUMBER=    5
     .00   22.04   44.27   66.71   89.40  112.37  135.67
  159.35  183.44  208.01  233.10  258.75  285.01  311.91
  339.47  367.71  396.62  426.20  456.41  487.22  518.57
  550.41  582.66  615.24  648.00
ROW NUMBER=    4
    4.50   32.82   61.16   89.56  118.04  146.65  175.43
  204.42  233.67  263.24  293.16  323.50  354.28  385.54
  417.30  449.59  482.40  515.73  549.56  583.84  618.56
  653.64  689.05  724.70  760.50
ROW NUMBER=    3
   18.00   52.03   86.07  120.13  154.23  188.38  222.62
  256.97  291.47  326.15  361.06  396.22  431.67  467.43
  503.54  540.00  576.82  613.98  651.49  689.32  727.43
  765.80  804.39  843.15  882.00
ROW NUMBER=    2
   40.50   80.05  119.61  159.17  198.75  238.35  277.99
  317.68  357.43  397.27  437.21  477.27  517.48  557.84
  598.37  639.08  679.97  721.04  762.28  803.68  845.23
  886.91  928.70  970.58 1012.50
ROW NUMBER=    1
   72.00  117.00  162.00  207.00  252.00  297.00  342.00
  387.00  432.00  477.00  522.00  567.00  612.00  657.00
  702.00  747.00  792.00  837.00  882.00  927.00  972.00
 1017.00 1062.00 1107.00 1152.00
```

REFERENCES

1. D. R. Atthey, "A finite difference scheme for melting problems," *J. Inst. Math. Appl.* **13**, 353–366 (1974).
2. K. J. Bathe, *Finite Element Procedures in Engineering Analysis*, Prentice-Hall, Englewood Cliffs, N.J., 1982.
3. E. B. Becker, G. F. Carey, and J. T. Oden, *Finite Elements: An Introduction*, Vol. I, Prentice-Hall, Englewood Cliffs, N.J., 1981.
4. A. Berman and R. J. Plemmons, *Nonnegative Matrices in the Mathematical Sciences*, Academic Press, Orlando, Fla., 1979.
5. G. F. Carey and J. T. Oden, *Finite Elements: A Second Course*, Vol. II, Prentice-Hall, Englewood Cliffs, N.J., 1983.
6. G. F. Carey and J. T. Oden, *Finite Elements: Computational Aspects*, Vol. III, Prentice-Hall, Englewood Cliffs, N.J., 1984.
7. J. Céa, *Lectures on Optimization: Theory and Algorithms*, Tata Institute of Fundamental Research, Bombay (Springer-Verlag), 1978.
8. C. W. Cryer and H. Fetter, "The numerical solution of axisymmetric free boundary porous flow well problems using variational inequalities," *MRC Technical Summary Report #1761*, 1977.
9. C. M. Elliott and J. R. Ockendon, *Weak and Variational Methods for Moving Boundary Problems*, Pitman, Boston, 1982.
10. M. S. Engelman, G. Strang, and K. J. Bathe, "The application of quasi-Newton methods in fluid mechanics," *Int. J. Num. Methods Eng.* **17**, 707–718 (1981).
11. A. Friedman, *Partial Differential Equations*, Holt, Rinehart and Winston, New York, 1969.